Hamilton Smith

Hydraulics

The flow of water through orifices, over weirs, and through open conduits and

pipes

Hamilton Smith

Hydraulics
The flow of water through orifices, over weirs, and through open conduits and pipes

ISBN/EAN: 9783337212889

Printed in Europe, USA, Canada, Australia, Japan

Cover: Foto ©berggeist007 / pixelio.de

More available books at **www.hansebooks.com**

HYDRAULICS.

THE FLOW OF WATER

THROUGH

ORIFICES,

OVER

WEIRS,

AND THROUGH

OPEN CONDUITS AND PIPES.

BY

HAMILTON SMITH, Jr.,

Member Am. Soc. of C.E., and Am. Inst. of M.E.

RIGHTS OF TRANSLATION RESERVED.

NEW YORK : JOHN WILEY AND SONS, 15, ASTOR PLACE.
LONDON : TRUBNER AND CO., 57 & 59, LUDGATE HILL, E.C.

1886.

HYDRAULICS.

THE FLOW OF WATER THROUGH ORIFICES, OVER WEIRS, AND THROUGH OPEN CONDUITS AND PIPES.

BY

HAMILTON SMITH, Jr.

1886.

CONTENTS.

CONTENTS.

iii

ATLAS.

INTRODUCTION.

IN this volume will be discussed the flow of water through orifices, over weirs, and through open conduits and circular pipes.

The authorities chiefly relied upon in establishing the values of the co-efficients of discharge or velocity are as follows :

FOR ORIFICES.—Poncelet and Lesbros, "*Expériences hydrauliques sur les lois de l'écoulement de l'eau.*" *Tome III., Savants étrangers, l'Académie des Sciences*, 1832. Lesbros, "*Expériences Hydrauliques.*" *Tome XIII., Savants étrangers, l'Académie des Sciences*, 1852. T. G. Ellis, "*Hydraulic Experiments with Large Apertures at Holyoke, Mass., 1874 ;*" *Transactions Am. Soc. of C.E., February*, 1876. A large number of original experiments by the author, many of which were made with the co-operation of Mr. Clemens Herschel.

FOR WEIRS.—Poncelet and Lesbros, and Lesbros, as above. J. B. Francis, "*Lowell Hydraulic Experiments,*" *New York*, 1868. Fteley and Stearns, "*Experiments on the Flow of Water, made during the Construction of Water Works for Boston;*" *Transactions Am. Soc. of C.E., January, February, and March*, 1883. A number of experiments by the author.

FOR PIPES.—Couplet, "*Recherches sur le mouvement des eaux;*" *Mémoires de l'Académie des Sciences*, 1732. Bossut, "*Traité théorique et expérimental d'Hydro-dynamique, par M. l'Abbé Bossut, Paris*, 1786." "*Principes d'Hydraulique, Paris, 1786 ;*" and later edition of same work printed in 1816. Lampe, "*Untersuch-ungen über die Bewegung des Wassers in Röhren;*" *Der Civilingenieur, Vol. XIX.*, 1873. Stearns, "*Experiments on the Flow of Water in a 48-inch Pipe;*" *Transactions Am.*

B

Soc. of C.E., January, 1885.   Hamilton Smith, Jun., "*Flow of Water through Pipes;*" *Transactions Am. Soc. of C.E., April,* 1883.*

FOR OPEN CONDUITS.—Darcy and Bazin, "*Recherches Hydrauliques, entreprises par M. H. Darcy, continuées par M. H. Bazin. Paris,* 1865" (also published in Tome XIX., Savants étrangers, l'Académie des Sciences, 1865). Fteley and Stearns; very valuable unpublished experiments with a large conduit at Boston, Mass., the results of which have been very kindly communicated to the author by those gentlemen.

The effort has been made to critically examine all the recorded experiments which have been made with weirs and pipes by German, French, English, and American authorities. It is only by careful analysis that it is possible to separate the most reliable experimental data from the great mass thus far contributed by savants and engineers, and in which, it is almost needless to observe, there is a great amount of chaff.

In discussing the laws governing the movement of water under various conditions, many of the views expressed by us have been stated before, either in the same form or in cognate forms. It would needlessly encumber this volume to attempt to refer in detail to the theories suggested in the many works upon Hydraulics, which are in part or in whole repeated here. Our final conclusions, however, possess a fair claim to originality, as many of them considerably differ from the views which have heretofore obtained.

Especial care has been taken to accurately give, with sufficient detail, the results of the experiments from which these final conclusions have been drawn. We have endeavoured to state these results in a perfectly candid spirit, and have in no instance rejected data which seemed to us to possess reasonable claims to accuracy, because they appeared to be antagonistic to our final conclusions.

Original authorities have almost exclusively been consulted, and not unfrequently errors, either in typography or in the reductions of the author, have been corrected. The danger of quoting secondhand is well shown by the fifty-one experiments with pipes by Couplet, Bossut, and Dubuat, given by Prony in his "*Recueil de cinq Tables, Paris,* 1825." Prony makes two blunders in Nos. 10 and 11, giving in both cases the erroneous length of 138.5 inches, instead of 737 inches, the correct values; his final reductions, however, are based upon the correct lengths.[†] These errors, with many additional ones, are generally repeated by English authorities; in no work thus far published in the English language are these experiments given correctly. In spite of the care exercised in the preparation of the material for this volume, it is hardly possible that the results as here

---

* Some of these experiments have been re-calculated by the author, so that the results, as given in this volume, differ slightly from those given in the paper published by the Am. Soc. of C.E..

† These errors were made by Prony in his original table, in his "*Recherches Physico-Mathématiques sur la théorie des eaux-courantes,*" Paris, 1804.

given can be altogether free from error. Should the reader discover any such inaccuracies, he will confer a favor by communicating them to us. The reader, however, who compares our data with the original authorities, must keep in mind that we have often corrected original errors, and also that seeming small errors in the determination of the final co-efficients arise from the omission of unnecessary decimals in the tabulated results.

Perhaps some statement should be made of the reasons which induced us to use the English foot as our unit of measure, instead of adopting the metric system, which would have involved no more labor, as such a large proportion of our material has been derived from French authorities.

In the first place, for the quantities we chiefly have to deal with, the metre is much too long for a convenient unit; volumes in cubic metres being especially objectionable on account of the necessary long decimals required to express most values of Q.

In the next place, this work is designed to be of value particularly to English-speaking scientists and hydraulicians. In order to thoroughly comprehend a given experiment, the values should at once make a direct impression upon the mind of the reader. Few persons have had more experience in examining metrical data, and in transforming them into English expressions, than the author; but we must confess, that in order to obtain a quick and clear idea of the given sizes, especially as to linear measurements, it is always necessary to first, either mentally or upon paper, translate the metre into the foot. We fancy that this is true of almost every one who has not been educated from childhood to *think* in French measures.

The formula, $Q = c \frac{2}{3} (2 g h)^{1/2} l h$, for rectangular vertical weirs, has in all cases been used. As $(2 g)^{1/2}$ is nearly constant, the form adopted by Mr. Francis and others of $Q = c' l h^{3/2}$, and $c' = \frac{2}{3} c (2 g)^{1/2}$, is somewhat more convenient, but the longer expression is the correct one, and the values of c deduced by it are of much advantage in facilitating the comparison of the respective co-efficients of discharge for orifices and weirs.

We have not attempted to discuss the laws governing the distribution of velocities in open conduits and pipes, as we have not experimentally examined this question with sufficient care to enable us to speak with authority upon it. In order to be a competent critic as to the value of experimental data, one should possess a thorough knowledge of all the many details necessary in order to secure trustworthy results. We have hence almost exclusively confined our discussion to those branches of the science of Hydraulics, where we know by practical experience the proper methods of investigation which should be followed by the experimenter.

No attempt will be made to discuss the retarding effect of angles or curves in conduits and pipes. The experimental data at hand are entirely insufficient to permit a satisfactory analysis of this quite complicated subject; in fact, about the only experiments of value are those made by Bossut and Dubuat with small pipes. To the constructing engineer the effect of bends should have but little practical importance, as in building a costly conduit or pipe he would be grossly at fault should he construct curves sharp enough to notably retard the flow.

The selected experiments for orifices, weirs, open conduits, and pipes are numbered consecutively for each order.

In the concluding chapters of this volume will be found a detailed description of the experiments made by the author with orifices, weirs, and pipes.

We are under many obligations to Mr. Ross E. Browne, Mr. A. Fteley, and Mr. F. P. Stearns for valuable suggestions and criticisms ; we also have to thank Professor W. C. Unwin, M. H. Bazin, Herr Iben, and Professor Dr. Lampe for material and suggestions.

NOMENCLATURE.

Where not otherwise expressly stated :

All linear distances will be given in English feet ;

All measures of area in English square feet ;

All measures of capacity in English cubic feet ;

All temperatures by the Fahrenheit scale.

French measures have been reduced to English as follows :

1 metre = 3.2809 English feet.[*]

1 French inch (système ancien) = .027 0699 metre[†] = .088 814 English foot.

The unit of time is 1 second.

The word *orifice* always signifies an opening, the upper side of which is covered by the liquid in the feeding reservoir ; when not qualified, it means an opening pierced in a " thin wall," the escaping jet only touching the regular line formed by the inner edge or corners, with perfect interior contraction.

The term *suppression of contraction* always means for an orifice or a weir, placing the side or sides of the feeding canal or reservoir in line at right angles to the side or sides of the opening ; thus, *contraction suppressed on two sides of a rectangular weir* indicates that the axial line of the feeding canal is normal to the plane of the weir, and that the width of the feeding canal is the same as the length of the weir.

Partial suppression signifies that one side or more of the feeding canal is so near the corresponding side or sides of the opening as not to allow full or normal contraction as the water enters the plane of the opening.

These terms must not be confounded with *total suppression of contraction* which can be only accomplished for an orifice by having the form of approach similar or nearly similar to that of the contracted vein. It is manifest that for a weir there never can be total suppression of contraction.

---

[*] Prof. W. A. Rogers, of Cambridge, U.S.A., has determined this ratio to be 3.28080, with the possibility that it may be more nearly 3.28085.

[†] Ratio adopted by Prony.

The expression *open conduit* signifies a conduit in which the upper surface of the water is exposed to the air; a conduit enclosed with walls on all sides becomes a *pipe* when the conduit is full.

The expression *hydraulic-grade line* signifies: for open conduits or streams, the surface axial line; for pipes, in general the straight line uniting the inlet and outlet ends of the pipe, the ordinate of the end point being the difference in elevation of the two ends, or H, and the abscissa the total length, or l. Strictly speaking, the ordinate for this point should be the "frictional" head, or h.

SYMBOLS.

Unless otherwise expressly stated, the following characters will always have these significations:

$a = $ area.

      For rectangular orifices, $a = l\,w$.

          ,,   ,,   weirs, $a = l\,h$.

          ,,   ,,   conduits, $a = d\,w$.

      ,,   circular orifices, $a = D^2\,\dfrac{\pi}{4}$.

      ,,   ,,   pipes (being mean area of the pipe for its entire length), $a = D^2\,\dfrac{\pi}{4}$.

$a_v = $ area of water section in feeding canal for a weir at the measuring point for H, the section being vertical and normal to the axial line of the canal. $a_v = d\,F$.

$b = $ co-efficient for h_a, in correction for velocity of approach for orifices and weirs, in the expression $h = H + b\,\dfrac{v_a^2}{2\,g} = H + b\,h_a$.

$C = $ approximate co-efficient of discharge for orifices and weirs.

      For rectangular or circular orifices; with the head measured from centre of the orifice to the surface of the still water in the feeding canal or reservoir, $Q = C\,(2\,g\,H)^{1/2}\,a$.

      ,,   weirs; with no allowance for increased head due to velocity of approach, $Q = C\,\tfrac{2}{3}\,(2\,g\,H)^{1/2}\,l\,H$.

$c = $ correct co-efficient of discharge for orifices and weirs.

      For rectangular vertical orifices; $Q = c\,l\,\tfrac{2}{3}\,(2\,g)^{1/2}\,(H_b^{3/2} - H_t^{3/2})$.[*]

      ,,   circular, triangular, and irregularly-shaped vertical or inclined orifices; with formulæ based upon the proposition that each successive horizontal layer of water passing through the orifice has a velocity due to its respective head.

---

[*] Where for an orifice the velocity of approach is notable, H_b and H_t must be corrected for the additional head due to this velocity.

For rectangular vertical weirs; $Q = c \frac{2}{3} (2 g h)^{1/2} l h$.

c_c = co-efficient for rectangular weirs, with full contraction on the three sides (bottom and two ends).

c_d = co-efficient for weirs, with contraction suppressed on both ends.

c' = co-efficient for weirs, with contraction suppressed on one end.

c_l = co-efficient for a weir of infinite length, and being nearly the same as the co-efficient c'.

c_p = co-efficient for weirs, when there is partial suppression of contraction on any one side.

d = vertical depth of water. For open conduits; mean vertical depth in axis of conduit.

D = diameter.

F = width of a rectangular feeding canal, of uniform section, for an orifice or a weir.

g = acceleration of gravity; being the velocity acquired by a body falling freely in vacuo at the expiration of the first second of its fall; g is slightly variable in different latitudes, and at different elevations above sea level.

G = inner depth of a rectangular vertical weir, measured from the crest of the weir to the bottom of the feeding canal; also the inner depth below the lower edge of a vertical orifice.

H = measured head.

    For rectangular or circular, vertical, orifices; the vertical elevation above the centre of the orifice, of the surface of the comparatively still water, determined at a point several feet up-stream from the orifice.

    „ rectangular vertical weirs; the vertical elevation above the crest of the weir, of the surface of the comparatively still water, determined at a point several feet up-stream from the weir.

    „ pipes; the vertical difference in elevation between the surface of the water at the inlet and at the outlet. Where pipes discharge into the air; the vertical difference in elevation between the surface of the water at the inlet, and the centre of the lower or discharge end of the pipe.

H_b = head measured from bottom of rectangular or circular, vertical orifices to surface of still water.

H_t = head measured from top of rectangular or circular, vertical orifices to surface of still water.

H_u = head measured from crest of a submerged weir, or from the centre of a submerged orifice, to surface of still water up-stream.

H_l = head measured from crest of a submerged weir, or from the centre of a submerged orifice, to surface of water down-stream.

H_w = vertical height from crest of a rectangular vertical weir, to mean surface of the water in the plane of the weir.

$h =$ effective head.

For orifices and weirs; H corrected for velocity of approach. $h = H + b\frac{v_a^2}{2\,g}$

  „   submerged orifices; difference in elevation between surface up-stream and surface down-stream. $h = H_a - H_b$

  „   submerged weirs; difference in elevation between surface up-stream and surface down-stream, corrected for velocity of approach. $h = H_a + b\frac{v_a^2}{2\,g} - H_b$

  „   open conduits; difference in elevation of surface for a certain length, after the regimen of flow has been established.

  „   pipes; the "frictional" head, being either H corrected for losses due to primary contraction and to imparting velocity $\left(h = H - \frac{v^2}{2\,g\,\sigma^2}\right)$, or, the difference in elevation of surface in piezometric tubes after the regimen of flow has been established.

$h' =$ for pipes the loss* of head due to contraction at the entrance, and to imparting velocity. $h' = \frac{v^2}{2\,g\,\sigma^2}$

$h_a =$ head due to velocity of approach for orifices and weirs. $h_a = \frac{v_a^2}{2\,g}$

$h'_a = effective$ additional head for orifices and weirs, for proper corrections for velocity of approach, $h'_a = b\frac{v_a^2}{2\,g}$, and $h = H + h'_a$.

$L =$ horizontal distance from the side of a vertical rectangular orifice or weir, to the corresponding side of the rectangular feeding canal. ($L =$ distance on one side; $L' =$ distance on other side; hence $L + l + L' = F$)

$l =$ length.

For rectangular orifices and weirs; horizontal distance between vertical sides or ends.

  „   open conduits; a certain length along their course after the regimen of flow has been established.

  „   pipes; either the total length measured along the line of the pipe, whether it be horizontal, inclined, straight, or curved, from the inlet to the outlet;† or where piezometers are used, the length between the terminal piezometers. (The upper piezometer being placed at a point below where the regimen of flow has been established.)

---

* Looking at the flow through pipes from a dynamic point of view, the *effective* head is that which produces the velocity of the escaping jet from the lower end of the pipe; this is expressed approximately by $\frac{v^2}{2\,g}$. It is, however, more convenient, in discussing the flow through pipes, to consider the "frictional" head as the effective head.

† Where an inlet funnel-shaped mouth-piece of the form of the contracted vein is attached to the pipe, this mouth-piece will not be included in l; where the form of a mouth-piece is considerably dissimilar from the contracted vein, then its length will be included in l.

$l' =$ total length of a pipe when h is determined by piezometers.

$m =$ co-efficient for mean velocity in circular pipes. $v = m \left(\dfrac{D h}{l} \right)^{\frac{1}{2}}$, and $2\,m = n$.

$n =$ co-efficient for mean velocity in open conduits and pipes in Chezy formula of $v = n \, (r \, s)^{\frac{1}{2}}$.

$o =$ co-efficient of contraction for orifices, or for very short pipes.

$p =$ wetted perimeter.

　For rectangular orifices; $p = 2\,l + 2\,w$.

　,, circular 　　,, 　$p = D\,\pi$.

　,, rectangular weirs; 　$p = l + 2\,h$.

　,, 　　,, 　　open conduits; $p = w + 2\,d$.

　,, circular pipes; 　$p = D\,\pi$.

$q =$ absolute quantity of water held by any measuring vessel.

$Q =$ quantity of water discharged in one second of time, through an orifice, over a weir, or through a conduit or pipe. 　$Q = \dfrac{q}{t}$.

$R =$ radius. $R = \dfrac{D}{2}$.

$r =$ hydraulic mean radius of open conduits or pipes. $r = \dfrac{a}{p}$; for circular full pipes,

$$r = \dfrac{D^2 \frac{\pi}{4}}{D \pi} = \dfrac{D}{4}.$$

$s =$ sin of hydraulic inclination for open conduits or pipes. $s = \dfrac{h}{l}$.

$T =$ temperature in degrees of Fahrenheit scale.

$T_c =$ 　　,, 　　,, 　　,, 　　Centigrade scale.

$t =$ time in seconds, generally indicating length of time in measuring q for the particular experiment.

$u =$ maximum velocity per second in an open conduit or pipe.

$v =$ mean velocity per second.

　For orifices at smallest section; $v = \dfrac{Q}{a}$.

　,, rectangular weirs; $v = \dfrac{Q}{l\,h}$.

　,, open rectangular conduits; $v = \dfrac{Q}{d\,w}$.

　,, full circular pipes; $v = \dfrac{Q}{D^2 \frac{\pi}{4}}$.

v_a = mean velocity of approach in the feeding canal for an orifice or a weir, as the water passes the measuring point for H.

For rectangular canals of uniform section ; $v_a = \dfrac{Q}{d\,F} = \dfrac{Q}{a_c}$.

v = width.

For rectangular vertical orifices ; the vertical distance between the horizontal sides.

„ rivers or streams of irregular section ; in general the surface width.

Δ = condition of wetted surface of an open conduit or pipe, so far as relative roughness or smoothness is concerned.

Δ^0 = very smooth surface, such as glass.

Δ^{10} = maximum degree of roughness for conduits.

Δ^1 et cet. = degrees of roughness, varying from Δ^0 to Δ^{10}.   An increased value of Δ hence signifies an increased degree of roughness.

β = angle of convergence or divergence of the side of an interior or exterior adjutage, with the axial line of an orifice, or short pipe.

π = ratio of circumference of a circle to its diameter ; $\pi = 3.141592$, and $\dfrac{\pi}{4} = .785398$.

CHAPTER I.

PROPERTIES OF WATER.

PRESSURE.

WATER, when subjected to great pressures, is appreciably compressed; when the pressure is removed it resumes its original form, thus being perfectly elastic. The most reliable experiments indicate that its compression is in direct proportion to the pressure. Canton, Sturm, Regnault, Oersted, and Grassi have determined that the compressibility of water is from .000 040 to .000 051 for one atmosphere.

M. Grassi obtained the following results* with distilled water;

T.	Maximum Pressure in Atmospheres.	Mean Compressibility for 1 Atmosphere.
32°	7.4	.000 0502
35°	10.0	.000 0515
51°	5.1	.000 0480
56°	8.4	.000 0476
79°	7.2	.000 0455
128	6.3	.000 0440

These experiments showed that compression was in direct proportion to the pressure, and, strange to say, that compressibility diminished with increasing temperatures.

Hence, for each foot of pressure, distilled water will be diminished in *volume* from .000 0015 to .000 0013. This is so minute a change that it can be neglected in the consideration of our future experimental data.

Dubuat,† by observing the oscillations of water in different siphons, believed he had demonstrated that the amount of hydrostatic pressure upon the interior walls of a pipe had no noteworthy effect upon the flow through the pipe.

Dr. Robinson,‡ in a more satisfactory manner, demonstrated the same truth. He used a bent tube, having its axis throughout in the same plane, swinging on hollow trunnions to an inlet and an outlet tank. The outlet end was plugged, and the inlet tank filled to a certain height; the siphon was placed in a horizontal position, the plug

---

* Ann. de Chim. et de Phys., III., 31, 1851.
† Principes d'Hydraulique, 1816, Vol. II., p. 42.
‡ Ency. Brit. Article Rivers. 8th Edition.

withdrawn, and the time noted when the height of the surface of the water in the two tanks became identical. The experiment was then repeated, with the siphon in a vertical position. There was no appreciable difference in the two times; if any difference, the time was shorter for the siphon vertical. This was possibly due to a slight increase in α, caused by the greater pressure.

Darcy, who considered Dubuat's experiments as not entirely conclusive, found that the flow through a pipe with similar hydraulic heads, was practically the same with small as with considerable hydrostatic heads.* These particular determinations are fully confirmed by all of M. Darcy's experiments with pipes; the pressure was always greater upon the longitudinal half of the pipe adjoining the inlet, than upon the other half, and the indicated piezometric heads for the two halves show no evidence of any disturbing effect, caused by this difference in pressure.

Our pipe experiment, No. 356, was with an "inverted siphon," the deepest point in its longitudinal section being 760 feet below the hydraulic-grade line; the discharge through this pipe appears to be normal, that is to say, fairly agreeing with other experiments, where the pipes sustained but little pressure.

Hence, we can assume, *that the loss of head due to friction, cross currents, et cet., as the water passes through a pipe, is not appreciably affected by the amount of pressure to which the interior walls of the pipe are subjected.*

The question of *pressure* will, therefore, have no bearing upon our future discussions.

Impurities.

The water of springs, rivers, and lakes is always slightly heavier than pure (distilled) water, owing to inorganic matter carried either in solution or in suspension. The following statement gives the specific gravity of the water of a number of springs, rivers, and lakes;

		Sp. Gr.	
Rivers.	Garonne	1.000 149	Boisgaraud (D'Aubuisson).
	Thames at Twickenham (1847)	1.000 3	Watt's Dictionary of Chemistry.
	Mississippi (filtered)	1.000 25	Riddell.
Springs.	Holywell, Malvern	1.001 2	Watt's Dictionary of Chemistry.
	Bradford Moor coal-pits	1.000 78	" " "
	Artesian well, Trafalgar-square	1.000 93	" " "
	Carlsbad	1.004 97	" " "
	Cheltenham	1.006 4	" " "
Lakes.	Dead Sea	1.172	" " "

Thick oil, passing through an orifice, has a much larger co-efficient of discharge than water; hence it is probable that water carrying in suspension a very large

* Recherches expérimentales relatives au mouvement de l'eau dans les tuyaux. 1857, pp. 84—86, and p. 12.

quantity of clayey sediment will have a slightly larger co-efficient of discharge than pure water, either through an orifice or over a weir.

For conduits and pipes, it is most probable that for very small values of c, or very low velocities, muddy water will flow with a slower velocity than clear water; the increased viscosity of the water due to the sediment in suspension will, in all probability, with such small values of c or v appreciably retard the flow.

Whether or not with considerable values of v and c, the impurity of water has any *notable* effect *per se* upon the flow is uncertain. A very impure stream, like that generally flowing through a sewer, will with ordinary velocities soon make a slimy deposit upon the walls of the conduit or pipe, thus increasing the value of Δ, and the flow may consequently be indirectly greatly retarded by the impurities.

We are rather inclined to the opinion that very minute and unknown changes in the water sometimes affect notably the flow through small orifices with low heads, and perhaps by analogy notably affect the flow over weirs with such low heads as .2 and less. Possibly this may be due to varying quantities of impurities, such as greasy particles, in the water.

HEAT.

Water is greatly affected by changes in temperature. The following table, deduced from a table compiled by Rossetti,[*] gives the specific gravity and the absolute weight in English avoirdupois pounds of a cubic foot of distilled water for each degree of the Fahrenheit scale from $14°$ to $212°$. The given weights are, of course, in air.

The table of Rossetti is based upon experiments made by himself, Kopp, and others, and embodies the most accurate determinations thus far made upon the density of water with various temperatures.

According to Kupffer,[†] the weight of a cubic centimetre of distilled water at $39.2°$ is not exactly one gramme. If this be so, our given weights of a cubic foot are too low by about $\frac{1}{50000}$th part.

Rossetti considers that the most probable temperature of maximum density is about $4.07°$ Cent., or $39.33°$ Fahr. .

The density of water below freezing point ($32°$) was determined, by taking advantage of the remarkable property of water of remaining unfrozen when kept perfectly quiet, while the temperature is being reduced from above the freezing point to $-10°$ Cent. .

---

[*] Annales de Chimie et de Physique. IV. Series, Vol. 17, 1869.
[†] *Vide* Units and Physical Constants. J. D. Everett.

TABLE I.

Relative Densities, and Weights of a Cubic Foot, of Distilled Water. Fahrenheit Scale. Computed from Rossetti's Deductions from his own and other Experiments.

Temperature	Relative Density	Weight of a Cubic Foot	Temperature	Relative Density	Weight of a Cubic Foot	Temperature	Relative Density	Weight of a Cubic Foot
14°	.99814		47°	.99987	62.416	80°	.99669	62.217
15°	.99831		48°	.99983	62.413	81°	.99654	62.208
16°	.99846		49°	.99979	62.411	82°	.99639	62.199
17°	.99860		50°	.99975	62.408	83°	.99624	62.189
18°	.99873		51°	.99970	62.405	84°	.99608	62.179
19°	.99886		52°	.99964	62.402	85°	.99592	62.169
20°	.99898		53°	.99958	62.398	86°	.99576	62.159
21°	.99910		54°	.99952	62.394	87°	.99560	62.149
22°	.99920		55°	.99946	62.390	88°	.99544	62.139
23°	.99930		56°	.99939	62.386	89°	.99527	62.129
24°	.99938		57°	.99931	62.381	90°	.99510	62.118
25°	.99947		58°	.99924	62.377	91°	.99492	62.107
26°	.99954		59°	.99916	62.372	92°	.99474	62.096
27°	.99961		60°	.99907	62.366	93°	.99456	62.084
28°	.99968		61°	.99898	62.360	94°	.99437	62.073
29°	.99973		62°	.99889	62.355	95°	.99418	62.061
30°	.99979		63°	.99880	62.349	96°	.99399	62.049
31°	.99983		64°	.99869	62.342	97°	.99379	62.036
32°	.99987	62.416	65°	.99859	62.336	98°	.99359	62.024
33°	.99990	62.418	66°	.99848	62.329	99°	.99339	62.011
34°	.99993	62.420	67°	.99837	62.322	100°	.99318	61.998
35°	.99996	62.421	68°	.99826	62.315	101°	.99298	61.986
36°	.99997	62.422	69°	.99814	62.308	102°	.99277	61.973
37°	.99999	62.423	70°	.99802	62.300	103°	.99256	61.960
38°	.99999	62.423	71°	.99790	62.293	104°	.99235	61.947
39.3°	1.	62.424	72°	.99778	62.285	105°	.99214	61.933
40°	.99999	62.423	73°	.99765	62.277	106°	.99193	61.920
41°	.99999	62.423	74°	.99752	62.269	107°	.99171	61.907
42°	.99998	62.423	75°	.99739	62.261	108°	.99149	61.893
43°	.99997	62.422	76°	.99726	62.253	109°	.99127	61.879
44°	.99994	62.420	77°	.99712	62.244	110°	.99105	61.865
45°	.99992	62.419	78°	.99698	62.235	111°	.99082	61.851
46°	.99990	62.418	79°	.99684	62.227	112°	.99060	61.837

TABLE I.—*continued.*

Temperature.	Relative Density.	Weight of a Cubic Foot.	Temperature.	Relative Density.	Weight of a Cubic Foot.	Temperature.	Relative Density.	Weight of a Cubic Foot.
113°	.99037	61.823	147°	.98134	61.259	181°	.97021	60.565
114°	.99014	61.809	148°	.98104	61.241	182°	.96986	60.543
115°	.98991	61.794	149°	.98074	61.222	183°	.96950	60.520
116°	.98968	61.780	150°	.98043	61.203	184°	.96915	60.498
117°	.98944	61.765	151°	.98013	61.184	185°	.96879	60.476
118°	.98920	61.750	152°	.97982	61.165	186°	.96843	60.453
119°	.98895	61.734	153°	.97952	61.146	187°	.96808	60.431
120°	.98870	61.719	154°	.97921	61.126	188°	.96772	60.409
121°	.98845	61.703	155°	.97889	61.106	189°	.96737	60.387
122°	.98820	61.687	156°	.97857	61.086	190°	.96701	60.365
123°	.98794	61.671	157°	.97826	61.067	191°	.96665	60.342
124°	.98768	61.655	158°	.97794	61.047	192°	.96629	60.320
125°	.98741	61.638	159°	.97762	61.027	193°	.96593	60.297
126°	.98714	61.621	160°	.97729	61.006	194°	.96556	60.274
127°	.98688	61.605	161°	.97697	60.986	195°	.96519	60.251
128°	.98661	61.588	163°	.97664	60.966	196°	.96482	60.228
129°	.98634	61.571	163°	.97631	60.945	197°	.96445	60.205
130°	.98608	61.555	164°	.97598	60.925	198°	.96408	60.182
131°	.98582	61.539	165°	.97565	60.904	199°	.96371	60.159
132°	.98556	61.523	166°	.97531	60.883	200°	.96333	60.135
133°	.98530	61.506	167°	.97498	60.862	201°	.96295	60.111
134°	.98503	61.490	168°	.97465	60.842	202°	.96257	60.088
135°	.98476	61.473	169°	.97431	60.821	203°	.96219	60.064
136°	.98449	61.456	170°	.97397	60.799	204°	.96180	60.040
137°	.98421	61.439	171°	.97363	60.778	205°	.96141	60.015
138°	.98394	61.422	172°	.97330	60.757	206°	.96102	59.991
139°	.98366	61.404	173°	.97296	60.736	207°	.96063	59.966
140°	.98338	61.386	174°	.97262	60.715	208°	.96024	59.942
141°	.98309	61.368	175°	.97228	60.694	209°	.95984	59.917
142°	.98280	61.350	176°	.97194	60.672	210°	.95945	59.893
143°	.98251	61.332	177°	.97160	60.651	211°	.95905	59.868
144°	.98222	61.314	178°	.97125	60.629	212°	.95865	59.843
145°	.98193	61.296	179°	.97091	60.608			
146°	.98164	61.278	180°	.97056	60.586			

It will hereafter be shown that the flow of water through a small orifice was quite appreciably diminished by an increase in the temperature from 48° to 132°. Hence it is probable that for both orifices and weirs, an increase in T will somewhat diminish the flow. In the ordinary ranges of T met with in practice, however, this effect will be so very slight, that it is not worth while to make T a factor in the formula, which expresses the discharge. Changes by variation in T will probably only be appreciable with small orifices, or with very low heads for orifices or weirs.

With glass tubes of very small diameter, Poiseuille and Hagen have shown that T is a most important factor, the discharge being increased threefold by an increase in T from 0° Cent. to 45° Cent. . Experiments with pipes of large diameter are not precise enough to determine whether or not T has any notable effect on the flow through such pipes. With $r = .25$ and over, it is likely, with ordinary temperatures and velocities, that T need not be considered. We conjecture though, that with very low velocities in either conduits or pipes of considerable size, where the resistance to the flow in a notable degree is caused by the adhesion of the water to the surrounding walls, that changes in T may then appreciably affect the flow. This supposition is to some extent confirmed by the experiments lately made by Professor Reynolds. Experiments made by Professor Unwin with discs rotating rapidly in water of various degrees of temperature, show that the friction rapidly diminishes with an increase in temperature; this perhaps may indicate that with considerable velocities in pipes or conduits changes in temperature affect the flow.

CHAPTER II.

THEORY OF HYDRAULICS.

THE science of Hydraulics dates its origin from the great discovery of Torricelli, enunciated in his " *De Motu Gravium Naturaliter Accelerato*, 1643," that *the velocity of a fluid passing through an orifice in the side of a reservoir is the same as that which would be acquired by a body falling in vacuo*[*] *from the vertical hight, measured from the surface of the fluid in the reservoir to the centre of the orifice.* This proposition, known as the theorem of Torricelli, is expressed by

$$v = (2\,g\,H)^{\frac{1}{2}}.$$

Upon it rests the whole theory of water actuated by the force of gravity.

Mariotte made many experiments illustrating the truth of this theorem, the results of which were published after his death, in 1686. From this date for nearly a century Hydraulics engaged the attention of the greatest mathematicians of the age, being discussed by Newton, Daniel and John Bernoulli, Euler, Maclaurin, and d'Alembert. But these great men investigated the science only as geometers, and their labors resulted in but little advantage to its development. In 1738 Daniel Bernoulli published in his " *Hydronamica* " his famous equation,

$$v = \left(\frac{2\,g\,H}{1 - \left(\frac{a'}{a} \right)^2} \right)^{\frac{1}{2}},$$

which gave rise to many bitter controversies, all of which were practically barren of good, as the reasoning of the disputants was chiefly based upon mathematical abstractions, instead of resting upon a firm foundation of experimental facts.

The elder Michelotti made a large number of careful experiments,[†] especially with orifices, in which he showed that the velocity of the escaping jet measured at the smallest section of the *vena-contracta* was substantially the same as that due to $(2\,g\,H)^{\frac{1}{2}}$, and that the velocity in the plane of the orifice was about $\frac{62}{100}$ths of the

---

* " A heavy body, falling freely."
† Sperimenti idraulici, et cet. Turin, 1767 and 1771.

maximum velocity of the escaping vein ; the diameter of the *vena-contracta* hence being about $\frac{8}{10}$ths of that of the orifice. Calling o the co-efficient of contraction, we hence have approximately, $v = o\ (2\ g\ H)^{\frac{1}{2}}$, o having a nearly constant value of .62.

The Abbe Bossut published, in 1771, his first work upon Hydrodynamics, and his final edition in 1786. His reflections were founded upon his own careful experiments, and he may be said to have been the first to place the science of Hydraulics upon a proper footing.

The Chevalier Dubuat, using the experimental data of Bossut, reinforced by many experiments of his own, in 1786 published his complete work, " *Principes d'Hydraulique*," in which he discusses with wonderful clearness and ability the laws governing the flow of water through orifices, over weirs, and in natural and artificial conduits. He had in 1779 announced, in a preliminary edition, his great discovery of the law of uniform motion in conduits, which he thus states : " *Quand l'eau coule uniformément dans un lit quelconque, la force accélératrice qui l'oblige à couler est égale à la somme des résistances qu'elle essuie, soit par sa propre viscosité, soit par le frottement du lit.*" From the experimental data which he had at hand he framed, by a most beautiful course of reasoning, a formula for the flow in uniform channels, as follows, in English feet ;

$$v = \frac{88.51\ (r^{\frac{1}{2}} - .0298)}{\left(\frac{1}{s}\right)^{\frac{1}{2}} - \text{Hyper. Log.} \left(\frac{1}{s} + 1.6\right)^{\frac{1}{2}}} - .0894\ (r^{\frac{1}{2}} - .0298).$$

Dubuat unfortunately came to the very erroneous final conclusion that the character of the wetted surface, Δ, has no appreciable effect upon the discharge. He thus expresses this opinion : " Les molécules d'eau s'introduisent dans les pores de la paroi, et remplissent toutes les petites cavités de sa superficie. Ainsi, elles forment elles-mêmes la surface sur laquelle toute la masse doit couler ; d'où il suit, que les différentes matières dont les parois peuvent être composées, ne changent pas sensiblement l'intensité de la résistance." This false supposition arose from the fact that the range of his experiments was too limited. We shall see hereafter that a considerable increase in r will largely increase the chief co-efficient in Dubuat's expression, and that a considerable increase in the value of Δ will largely decrease the same co-efficient. In the data which Dubuat had at hand the effects produced by changes in r and Δ balanced each other ; that is to say, he found his expression would give satisfactory results when applied to a small smooth pipe, and also to a large rough canal, but in the first case r was small and Δ was low, and in the other case r was large and Δ was high. He had unluckily no data* where one of these factors remained constant and the other had widely varying values, which would have enabled him to distinguish their separate effects.

---

* He had at hand, but did not use them, the experiments of Couplet with old pipes. Vide our notice of these experiments in Chapter VIII.

This erroneous dogma of Dubuat was not disproved until 1854, when Darcy, by experiments with pipes having widely varying values of r and Δ, conclusively demonstrated that Δ is a most important factor in formulating the discharge through pipes.

We shall assume that there are three general laws to govern us in our future discussion. The truth of these propositions will be incidentally shown hereafter; they can thus be stated:

FIRST.—*The velocity of water escaping through an orifice placed in the side of a feeding reservoir relatively of large size both in area and in height, is nearly in direct proportion to* $(2 g H)^{\frac{1}{2}}$; *hence,* $Q = C a (2 g H)^{\frac{1}{2}}$.

SECOND.—*The discharge of water flowing over a weir, attached to a feeding canal relatively of large dimensions, is expressed by formulæ based upon the hypothesis that each horizontal layer of water in the plane of the weir, whose vertical height is* H, *is actuated by the force of gravity measured from the summit of the line* H—*surface of still water. That is to say, dividing this plane into minute horizontal sections, having the vertical distances— from their centres—of* H_a, H_b, *et cet. below the summit of* H, *and the respective areas of* a_a, a_b, *et cet., the discharge would be expressed by*

$$Q = C (2 g)^{\frac{1}{2}} (a_a H_a^{\frac{1}{2}} + a_b H_b^{\frac{1}{2}} + \ldots).$$

This same proposition should be applied to vertical or inclined orifices, where the head, H, *from the centre is not largely in excess of the width,* w, *of the orifice.*

THIRD.—*After the regimen of flow has been established in a conduit of uniform section, whether open or closed, the velocity is approximately in direct proportion to the square root of the product of the hydraulic mean radius multiplied by the sin of the hydraulic inclination, or,* $v = n \left(\dfrac{a}{p} \dfrac{h}{l} \right)^{\frac{1}{2}} = n (r s)^{\frac{1}{2}}$; *the coefficient* n *varying with* v, r, *and* Δ.

This expression, well known as the Chezy formula, is a rough approximation of Dubuat's equation before given. Prony states that Chezy proposed this formula in 1775; if this be the case, General Chezy to some extent anticipated Dubuat's discovery of the law governing the motion of water in conduits. *Vide* "Recherches Physico-Mathématiques sur la théorie des eaux courantes. Paris, 1804."

VALUE OF $(2 g)^{\frac{1}{2}}$.

The velocity acquired by a body falling in vacuo at the end of one second of time is expressed by g; its value increases with an increase in latitude, and diminishes with an increase in elevation above sea level.

Assistant C. S. Pierce, of the U.S. Coast and Geodetic Survey, has been kind enough to place the following formulæ, expressing the laws of variation in g, at our disposition.

$$g = g_e (1 + \eta \sin^2 \lambda).$$
$$g_h = y - \frac{2 g h}{a}.$$

In these formulæ the various letters have the following significations, values being stated in metrical measures.

g_e = gravity at the equator at sea level.

g = ,, in any latitude ,, ,,

g_h = ,, ,, ,, ,, any elevation h.

h = elevation in metres above mean sea level.

η = .005 2375 a constant.

λ = latitude.

u = radius of the earth, assumed as a sphere with a radius of 6370 kilos..

A "second" pendulum has a length at the equator at sea level of very nearly 991 mm.. Hence, .991 × π^2 = 9.7808 metres = g_e.

Using this value of g_e the following table has been constructed, giving values of $(2g)^{1/2}$ at various latitudes and at various elevations above sea level, in English feet. *

TABLE II.

Value of $(2g)^{1/2}$.

| Lati- tude. | ELEVATION ABOVE MEAN SEA-LEVEL. | | | | | | | | | | |
	0	500	1000	1500	2000	2500	3000	3500	4000	4500	5000
0°	8.0112	8.0110	8.0108	8.0106	8.0104	8.0102	8.0101	8.0099	8.0097	8.0095	8.0093
5°	8.0114	8.0112	8.0110	8.0108	8.0106	8.0104	8.0102	8.0100	8.0098	8.0096	8.0095
10°	8.0118	8.0117	8.0115	8.0113	8.0111	8.0109	8.0107	8.0105	8.0103	8.0101	8.0099
15°	8.0126	8.0124	8.0122	8.0120	8.0118	8.0117	8.0115	8.0113	8.0111	8.0109	8.0107
20°	8.0137	8.0135	8.0133	8.0131	8.0129	8.0127	8.0125	8.0123	8.0121	8.0119	8.0117
25°	8.0150	8.0148	8.0146	8.0144	8.0142	8.0140	8.0138	8.0136	8.0134	8.0132	8.0130
30°	8.0165	8.0163	8.0161	8.0159	8.0157	8.0155	8.0153	8.0151	8.0149	8.0147	8.0145
35°	8.0181	8.0179	8.0177	8.0175	8.0173	8.0171	8.0170	8.0168	8.0166	8.0164	8.0162
40°	8.0199	8.0197	8.0195	8.0193	8.0191	8.0189	8.0187	8.0185	8.0183	8.0181	8.0180
45°	8.0217	8.0215	8.0213	8.0211	8.0209	8.0207	8.0205	8.0203	8.0202	8.0200	8.0198
50°	8.0235	8.0233	8.0231	8.0229	8.0227	8.0226	8.0224	8.0222	8.0220	8.0218	8.0216
55°	8.0253	8.0251	8.0249	8.0247	8.0245	8.0243	8.0241	8.0239	8.0237	8.0236	8.0234
60°	8.0269	8.0267	8.0265	8.0263	8.0262	8.0260	8.0258	8.0256	8.0254	8.0252	8.0250
65°	8.0284	8.0282	8.0280	8.0278	8.0277	8.0275	8.0273	8.0271	8.0269	8.0267	8.0265

The value of $(2g)^{1/2}$ for intermediate latitudes can be quickly obtained from the foregoing table by interpolation, and with sufficient accuracy.

For convenience the values of $(2g)^{1/2}$ assumed by the various authorities hereafter

* Everett gives in English measures :

$g = 32.173 - .082 \cos 2\lambda - .000003\ h$.

The correction for latitude is nearly identical with that above given, but the correction for elevation is only about one-half as much. Assistant Pierce believes that the old formulæ give too small a correction for elevation.

quoted are followed, although they are sometimes very slightly in error. Such errors, however, are so insignificant that they will not appreciably affect the accuracy of our deductions.

FORMULÆ. ORIFICES.

The form $Q = C (2 g H)^{1/2} a$ represents the discharge with sufficient exactness, when the correction for velocity of approach is an insignificant factor of H, and when H with vertical orifices is greater than 8 w.

For rectangular vertical orifices, with no corrections for v_a, the expression $Q = c \frac{2}{3}$ $(2 g)^{1/2} (H_b^{3/2} - H_t^{3/2})$ is the correct one.[*]

Table III. gives the ratio between these co-efficients C and c, with $\frac{H}{w}$ varying from .5 to 10

TABLE III.

Rectangular Vertical Orifices.

Ratio of C and c in $\begin{cases} Q = C (2 g) H^{1/2} l w \\ Q = c l \frac{2}{3} (2 g)^{1/2} (H_b^{3/2} - H_t^{3/2}) \end{cases}$

$c \frac{2}{3} (H_b^{3/2} - H_t^{3/2}) = C w H^{1/2}$

$\frac{H}{w}$	$\frac{C}{c}$	$\frac{H}{w}$	$\frac{C}{c}$	$\frac{H}{w}$	$\frac{C}{c}$	$\frac{H}{w}$	$\frac{C}{c}$
.5	.9428	.60	.9637	.95	.9878	2.25	.9979
.51	.9466	.62	.9684	1.0	.9890	2.5	.9983
.52	.9498	.64	.9707	1.1	.9910	2.75	.9986
.53	.9525	.66	.9727	1.2	.9925	3.	.9988
.54	.9549	.68	.9745	1.3	.9937	3.5	.9991
.55	.9571	.70	.9762	1.4	.9946	4.	.9993
.56	.9592	.75	.9796	1.5	.9953	5.	.9996
.57	.9610	.80	.9823	1.6	.9959	6.	.9997
.58	.9627	.85	.9843	1.8	.9968	8.	.9998
.59	.9643	.90	.9863	2.0	.9974	10.	.9999

For vertical circular orifices, approximate formulæ only can be used to express our

---

[*] This formula is deduced as follows: Assume two rectangular weirs, having each the length, l, of the orifice, and the respective heads of H_b and H_t. The difference between the discharge of these two weirs will represent the discharge for the orifice having the width $H_b - H_t = w$.

$$Q = c \frac{2}{3} l (2 g)^{1/2} H_b^{3/2}.$$
$$Q' = c \frac{2}{3} l (2 g)^{1/2} H_t^{3/2}.$$
$$\overline{Q = c \frac{2}{3} l (2 g)^{1/2} (H_b^{3/2} - H_t^{3/2})}$$

This supposition conforms with our second general proposition.

second general proposition. In Table IV. will be found the ratio $\frac{C}{c}$, which will answer all our purposes in obtaining the value of c.

TABLE IV.

Vertical Circular Orifices. Ratio of $\frac{C}{c}$.

$\frac{H}{D}$	$\frac{C}{c}$	$\frac{H}{D}$	$\frac{C}{c}$	$\frac{H}{D}$	$\frac{C}{c}$
.5	.9604*	1.25	.9948*	2.2	.9983
.6	.9753	1.3	.9953	2.3	.9984
.625	.9774*	1.4	.9960	2.4	.9986
.7	.9823	1.5	.9965+	2.5	.9987*
.75	.9849*	1.6	.9969	3.	.9991
.8	.9867	1.7	.9973	3.5	.9994*
.875	.9892*	1.8	.9976	4.	.9995
.9	.9897	1.9	.9978	4.5	.9996
1.	.9918*	2.	.9980*	5.	.9997*
1.1	.9933	2.1	.9982	10.	1.*
1.2	.9944				

The values of $\frac{C}{c}$ which have no asterisk attached, have been obtained by interpolation, and some of them may be fully .0001 in error.

In using Tables III. and IV., it must be kept in mind that, when reducing the value of C as determined by experiment to the correct co-efficient c, C must be multiplied by $\frac{1}{\frac{C}{c}}$; on the other hand, when Q is to be computed from a table of values of c, c must be multiplied by $\frac{C}{c}$.

Weirs.

The only weirs which we propose to consider are vertical rectangular ones, the sills or crests being horizontal. The formula which will be used, when no corrections for velocity of approach are necessary, is, $Q = C \frac{2}{3} (2\,g\,H)^{\frac{1}{2}} l\,H.$

This expression is in accordance with our second general proposition, for : dividing the line H into an indefinitely great number, n, of equal parts, distant $H\frac{1}{n}, H\frac{2}{n}$, et cet. from the summit of H, the last number of the series being $H\frac{n}{n} = H$, the discharge will be represented by $Q = C l H (2 g)^{\frac{1}{2}} \times H^{\frac{1}{2}} \dfrac{\left(\frac{1}{n}\right)^{\frac{1}{2}} + \left(\frac{2}{n}\right)^{\frac{1}{2}} + \ldots + 1^{\frac{1}{2}}}{n}$. The sum of the members of this series divided by their number is $\frac{2}{3}$.

French hydraulicians, commencing with Dubuat, have, as a rule, used the height, H_w, of the water in the plane of the weir, in formulating expressions for the discharge over weirs. We believe, however, that this is a vicious method; we prefer to consider the surface curve of the escaping sheet, from the measuring point in still water to the plane of the weir, as the upper portion of the contracted vein; this assumption will be fully discussed hereafter, in tracing the analogies between the discharge over weirs and through orifices.

French authors, when only considering the head H, generally have used the co-efficient $C = \frac{2}{3} C$. American authorities have generally used $C' = \frac{2}{3} (2 g)^{\frac{1}{2}} C$; assuming $(2 g)^{\frac{1}{2}} = 8.020$ as a constant, $C' = 5.347 C$. We prefer the longer, and theoretically correct expression, especially for facilitating comparisons of the values of C or c, for weirs and orifices.

OPEN CONDUITS AND PIPES.

Using the Chezy formula for circular full pipes;

$$v = n (r s)^{\frac{1}{2}} = n \left(\frac{D}{4} \frac{h}{l}\right)^{\frac{1}{2}}.$$

$$h = \frac{4 l v^2}{D n^2} = \frac{l v^2}{r n^2}.$$

$$l = \frac{n^2 D h}{4 v^2} = \frac{n^2 h r}{v^2}.$$

$$D = \frac{4 l v^2}{n^2 h} = \left(\frac{l Q^2}{.1542 n^2 h}\right)^{\frac{1}{5}}.$$

$$Q = D^2 \frac{\pi}{4} n (r s)^{\frac{1}{2}} = .3927 n \left(D^5 \frac{h}{l}\right)^{\frac{1}{2}}.$$

If we consider that $(2 g)^{\frac{1}{2}}$ being a variable, should be a function in the equation, making $f = \dfrac{n}{(2 g)^{\frac{1}{2}}}$, we have, $v = f (2 g r s)^{\frac{1}{2}}$

In the following table are given the properties of a circular conduit partly full. The table is based upon a radius of unity; the given arc is that of the wetted surface μ; d is the axial depth; w is the surface width.

TABLE V.

Properties of a Circle having a Radius of Unity; partly filled with Water.

Arc	p	d	w	a	r	a r^N	Arc	p	d	w	a	r	a r^N
360°	6.283	2.	0	3.1416	.5	2.2214	180°	3.142	1.	2.	1.5708	.5	1.1107
350°	6.109	1.9962	.1743	3.1411	.5142	2.2525	170°	2.967	.9128	1.9924	1.3967	.4707	.9583
340°	5.934	1.9848	.3473	3.1381	.6288	2.2820	160°	2.793	.8263	1.9696	1.2253	.4388	.8116
330°	5.760	1.9659	.5176	3.1298	.7434	2.3072	150°	2.618	.7412	1.9319	1.0590	.4045	.6735
320°	5.585	1.9397	.6840	3.1139	.8575	2.3251	140°	2.443	.6580	1.8794	.9003	.3685	.5465
310°	5.411	1.9063	.8452	3.0883	.5708	2.3332	130°	2.269	.5774	1.8126	.7514	.3312	.4324
30S°	5.376	1.8988	.8767	3.0818	.5733	2.3334	120°	2.094	.5	1.7321	.6142	.2933	.3326
300°	5.236	1.8660	1.	3.0510	.5827	2.3290	110°	1.920	.4264	1.6383	.4901	.2553	.2476
290°	5.061	1.8191	1.1472	3.0006	.5928	2.3103	100°	1.745	.3572	1.5321	.3803	.2179	.1775
280°	4.887	1.7660	1.2856	2.9359	.6008	2.2755	90°	1.571	.2929	1.4142	.2854	.1817	.1217
270°	4.712	1.7071	1.4142	2.8562	.6961	2.2236	80°	1.396	.2340	1.2856	.2057	.1473	.0790
260°	4.538	1.6428	1.5321	2.7613	.6085	2.1540	70°	1.222	.1809	1.1472	.1410	.1154	.0479
257°	4.485	1.6225	1.5632	2.7299	.6086	2.1297	60°	1.047	.1340	1.	.0906	.0865	.0260
250°	4.363	1.5736	1.6383	2.6513	.6077	2.0670	50°	.873	.0937	.8452	.0533	.0611	.0132
240°	4.189	1.5	1.7321	2.5274	.6034	1.9632	40°	.698	.0603	.6840	.0277	.0396	.0055
230°	4.014	1.4226	1.8126	2.3901	.5954	1.8443	30°	.524	.0341	.5176	.0118	.0225	.0018
220°	3.840	1.3420	1.8794	2.2413	.5837	1.7123	20°	.349	.0152	.3473	.00352	.0101	.0003
210°	3.665	1.2588	1.9319	2.0826	.5682	1.5699	10°	.175	.0038	.1743	.00044	.0025	.00002
200°	3.491	1.1736	1.9696	1.9163	.5490	1.4199	0°	0	0	0	0	0	0
190°	3.316	1.0872	1.9924	1.7449	.5262	1.2657							

From the preceding table it will be observed that with a circular section r is greatest with a wetted arc of 257°, when it is .6086 with D = 2; it has the value of $\frac{D}{4}$ = .5 when the arc is either 360° or 180°. Assuming that the discharge is approximately in proportion to $a\,r^{½}$, it will be greatest when the wetted arc is about 308°; roughly speaking, a pipe will discharge 5 per cent. more water when filled to $\frac{18}{20}$ths of its diameter than it will when completely filled; also, the discharge from a circle will be double that from a semi-circle of the same diameter.

CHAPTER III.

FLOW THROUGH ORIFICES.

Mariotte, Bossut, Michelotti, and other physicists have investigated the flow of water through orifices of different shapes; the results of their experiments are given in brief in the many text books treating upon Hydraulics.

These investigators, with heads varying from a few inches to 25 feet, proved conclusively the truth of the fundamental principle, that *the velocity of a jet varies substantially as* $(2 g h)^{\frac{1}{2}}$.

With square-edged orifices, with the jet escaping freely into the air, through a thin side—so that the only contact was with the inner sharp corners of the orifice—it was found that the discharge was represented by $Q = C a \, (2 g H)^{\frac{1}{2}}$, the co-efficient C having a nearly constant value of .62. This discharge of only $\frac{62}{100}$ths of that due to the effect of gravity, was shown to be almost entirely caused by the contraction or convergence of the fillets or veins of water as they formed into place; the jet after its escape from the orifice assuming a much smaller section, termed the *rena-contracta*, whose velocity was nearly that due to $(2 g H)^{\frac{1}{2}}$.

Michelotti found that the diameter of circular jets, measured at the smallest section of the *rena-contracta*, was about $\frac{79}{100}$ths of the diameter of the orifice; hence the area of the smallest section was about $\frac{62}{100}$ths of that of the orifice. He determined that the co-efficient of discharge, C, of these orifices was from .60 to .62; thus showing that the mean velocity of the *rena-contracta* at its smallest section was almost exactly $(2 g H)^{\frac{1}{2}}$. Hence some authors have assumed that the co-efficients of contraction and discharge (efflux) are identical.

It was also shown that an increased discharge could be obtained by rounding the inner corners of an orifice, and that by adding a trumpet-shaped mouth-piece, converging towards the orifice, the value of C could be increased to .95 or slightly more; *i.e.*, $Q =$ very nearly $a \, (2 g h)^{\frac{1}{2}}$, the area a being taken at the outer or smallest section of the orifice.

These early experiments were sufficiently precise to demonstrate these general principles, but were not made with enough care to warrant the deduction of the minor laws of variation in C, due to changes of form in the orifices, differences of H, variation

E

in T, et cet.. For instance, comparing the discharge through square and circular orifices with D = side of the square, with full contraction and equal heads, some experimenters have found C larger for the square section than for the circular, while others have found just the reverse.[*] We hope, by the aid of the experiments made by MM. Poncelet and Lesbros, those made by M. Lesbros, and the very careful experiments lately made by ourselves, to be able to draw final trustworthy conclusions as to the general effect upon C or c, of variation in section, head, and temperature.

The ratio of the area of the contracted section to the area of the orifice is the co-efficient of contraction. The ratio of the mean velocity at the contracted section to the theoretic velocity is the co-efficient of velocity. The co-efficient of discharge (often called co-efficient of efflux) is the product of these ratios.

Weisbach has found that with a head of 3.4 metres the co-efficient of velocity was .978 for a circular orifice with full contraction, and we can assume that this co-efficient is always less than unity.

The co-efficients of contraction and velocity can be obtained by direct measurements of the contracted vein, and the parabolic curve of the jet; from these quantities the co-efficient of discharge, C or c, can be readily deduced. This is an objectionable method, and should never be employed where accuracy is desired. Even with a circular jet it is difficult to make a fairly accurate measurement of the contracted section, owing to the slight vibrations in the jet; the cruciform section of the jet from a rectangular orifice can never be measured with any reasonable degree of accuracy.

As we do not intend to discuss the dynamic effect of jets from orifices, we will entirely neglect the co-efficients of contraction and velocity, and only obtain the co-efficient of discharge, which, with properly conducted experiments, may be determined with great accuracy.

In all of the experiments about to be given the velocity of approach was so inconsiderable that no correction for u, need be applied. Hence H is always practically identical with h.

The value of the correct co-efficient of discharge, c, will generally be given.

EXPERIMENTAL DATA.

Lesbros.— Vertical Rectangular Orifices.

In the year 1827 MM. Poncelet and Lesbros commenced at Metz an elaborate series of experiments upon the discharge through rectangular vertical orifices of various sizes, the largest being .656 square. These experiments appear to have been executed with a care before unknown in experimental Hydraulics. The results were published in the Mémoires of the Academy of Sciences, Savants étrangers, Paris, 1832.

M. Lesbros, during the years 1828-1835, continued these investigations, repeating some of the experiments, and making very many additional ones. His results are published in the same Mémoires, Paris, 1852.

The experiences gained in conducting the first series of experiments doubtless

---

[*] In the elaborate treatise upon Hydraulics, lately published by M. Graeff, the author assumes that both theory and experiment demonstrate the flow will be greater through a circular orifice than through a rectangular one of the same area. M. Graeff roughly places the value of C at .00 or .61 for rectangular openings, and at .64 for circular openings. It will be seen hereafter that our experimental results lead to very different conclusions.

Vide "Traité d'Hydraulique, par M. A. Graeff, Paris, 1883." Tome deuxième, p. 24 ; and " Essai sur la théorie des eaux courantes," by M. Boussinesq.

enabled M. Lesbros to continue his labors with still greater accuracy, and hence these later experiments are probably somewhat more reliable than the first. A careful analysis of the results obtained by him impresses one with the belief that his investigations were conducted with care, and that his stated facts are honestly given.

Lesbros used various forms of approach for the canal feeding the orifices experimented upon ; the forms of approach which were used in the experiments hereafter selected are shown by Figs. 1 to 12 on Plate I. Reference will be made in the following tables and diagrams to the particular figure which represents the form of the feeding canal for the several experiments quoted.

In all of the experiments selected from Lesbros there was a perfectly free discharge into the air ; he also made a large number of experiments with the discharge into an uncovered canal or flume, but which will not be discussed by us, owing to the uncertainty attending such submerged discharges, due to the irregular or curved form of the axial surface line in the canal.

The quantity discharged was directly measured in vessels of ample size.

The height of the surface of the still water in the feeding reservoir was measured by a vertical rod, sharply pointed at its lower end, placed at a distance of 11.48 feet above the orifice, where the water was practically stagnant, or unaffected by the discharge. Hence in these experiments the additional head or force due to the velocity of the water, as it passed the measuring point for H, need not be considered.

In some of the forms of approach, particularly those shown by Figs. 6 and 7, Plate I., a portion of the head, H or h, was absorbed by " friction" and adhesion with the sides of the feeding canal, and also by primary contraction as the water entered the mouth of the canal. The height of the surface of the water at a distance of only .0656 foot above the orifice, was also determined, by which these losses of head were indicated with more or less accuracy.

The largest orifice, .6562 square (2 decimetres), was cut in a copper plate, having a thickness of .013 ft. (4 mm.) with beveled outer sides, firmly fixed to a wooden frame. In order to obtain orifices with the same length and smaller width,* a copper gate of the same thickness (.013) was employed, sliding on the inner side of the fixed plate, and moved vertically by means of an iron rod. The lower edge of this gate, which formed the upper side of the orifices of less width than .6562, was sharply beveled on its *inner* face. The widths of the opening were determined by a scale attached to the iron rod, and also by templates.

The smaller fixed orifice was cut in a copper plate, .013 thick, the opening being 1.9685 × .0656 (.6 m. × .02 m.), and the plate being firmly fastened to a wooden frame. It was first placed with its long side horizontal, $l = 1.9685$ and $w = .0656$; it was then

---

* It must be be kept in mind that for vertical rectangular orifices, the length l is the horizontal side, and the width is the vertical side or height.

reversed, with $l = .0656$ and $w = 1.9685$. In this last position, in order to obtain smaller widths, a copper gate was employed, sliding on the inner face of the fixed plate.

In regard to the foregoing methods, we may remark : that the two fixed copper plates were altogether too thin ; any swelling of the wooden frames on which they were mounted would very likely have resulted in distorting the form of the orifice. The experiments made by the aid of the gates, especially those where w was small, have the following dangers or chances of experimental error : warping of the copper gate under considerable pressures ; this was in part guarded against by stiffening the outer side of the gate by a wooden block, supported by a screw ; the inner bevel of the lower edge of the gate was objectionable, as it doubtless somewhat added to the flow, compared with the normal flow through an opening with a plane inner face, such as was the case with the two fixed plates ; when the gates were employed, the methods of measuring w cannot be considered as exact. It is much to be regretted that MM. Poncelet and Lesbros did not employ fixed brass plates, at least .02 thick, for each dimension of orifice experimented upon. The several dimensions of these orifices could have been accurately determined by means of a delicate measuring or comparing apparatus.

We hence regard the experiments with the two fixed plates—orifices .6562 × .6562 and 1.9685 × .0656—as much the most reliable of the series. Errors of measurement in the larger orifice would be less apt to notably affect the values of the co-efficient c, than in the smaller orifice.

When the gates were employed, errors in w are probably nearly constant for a given width. It is, however, almost impossible to immovably hold in place, even for a few hours, such a gate ; changes in the temperature of the comparatively long iron moving rod, as M. Lesbros has pointed out, would appreciably change the width of the opening ; there is always more or less " lost motion " in such apparatus, and the lower edge of the gate doubtless was sometimes appreciably out of parallel with the lower edge of the fixed plate.

In the following table are given the distances from the edges of the two fixed orifices to the bottom and sides of the respective feeding canals employed, as shown by Plate I.

These distances were the same not only for the two fixed orifices, but also for the orifices formed by the two sliding gates.

TABLE VI.

Distances from Sides of Orifices to respective Sides of Feeding Canals in some of Lesbros' Experiments.

Length (Horizontal) of Orifice. l	Lower Edge of Orifice to Bottom of Canal. G	Vertical Sides of Orifice to respective Sides of Canal. L	L'	Width of Approach on Inner Face of Orifice. $L + l + L' = F$	Number of Figure, *vide* Plate I.
	1.772	5.709	5.709	12.074	Fig. 1
	1.772	1.772	5.709	8.137	,, 2
	1.772	1.772	1.772	4.200	,, 3
	0	5.709	5.709	12.074	,, 4
	0	.066	5.709	6.431	,, 5
.6562	0	.066	.066	.788	,, 6
	0	0	0	.656	,, 7
	1.772	.066	5.709	6.131	,, 8
	1.772	.066	.066	.788	,, 9
	1.772	0	0	.656	,, 10
	0	.066	.066	.788	,, 11
	1.772	.066	.066	.788	,, 12
.0656	1.772	6.004	6.004	12.074	,, 1
1.9685	1.772	5.053	5.053	12.074	,, 1

For Figs. 11 and 12 the side approaches were inclined at an angle of 45°, *vide* Plate I.

Owing to the swelling of the wooden sides of the canals, the above distances slightly varied from time to time. These sides were of planed plank, tongued and grooved.

In the following table are given the results of Lesbros' experiments, with the forms of approach shown by Figs. 1, 2, and 3, Plate I. Lesbros assumes $(2\,g)^{\frac12}$ as 8.0227, which will be taken as its value in all the following reductions, both from Lesbros, and Poncelet and Lesbros. In the table, H is the height from the centre of the orifice to the surface of the still water in the reservoir, taken at a point 11.48 feet above the plane of the orifice; H_u is the height from the centre of the orifice to the surface of the water, measured at a point .0656 foot above the plane of the orifice, or up-stream. The values of the co-efficient C are deduced from H, and are generally the means of several determinations with constant values of H. The co-efficient c has been deduced from C, by aid of Table III.

The experiments made by the author with orifices are numbered from 1 to 155 inclusive, and will be given after our résumé of Lesbros, and Poncelet and Lesbros.   Hence we begin with No. 156 in the following table.

Lesbros measured his heads to .0001 metre.   In the transfer to feet we have given them to .0001 foot, for the purpose of accurately comparing the differences between H and H_v, which are often minute.

TABLE VII.

Lesbros.—Flow through Vertical Orifices, with Free Discharge into Air, and Full, or very nearly Full Contraction. H, or effective Head, measured 11.48 Upstream from Orifice. $C = \dfrac{Q}{l \, w \, (2\,g\,H)^{\frac{1}{2}}}$.

No.	Lesbros' Nos.	Figure Plate I.	l	w	H	H_	Q	C	Means.	
									h	c
156 {	1	2	.6562	.6562	3.5413	5.5096	4.9008	.6043 }	5.50	.6032
	2	"	"	"	5.4883	5.4866	4.8711	.6019 }		
157	3	"	"	"	2.9905	2.9886	3.6135	.6049	2.99	.6052
158	4-5	"	"	"	1.3370	1.3399	2.4024	.6015	1.34	.6030
159	6-8	..	"	"	.8317	.8281	1.8782	.5962	.83	.6002
160	9-11	"	"	"	.3970	.3842	1.2507	.5746	.40	.5946
161 {	12	3	.6562	.6562	5.8272	5.8239	5.0147	.6014 }		
	13	"	"	"	5.7682	5.7649	5.0132	.6043 }		
	14	"	"	"	5.4706	5.4673	4.8832	.6044 }	5.60	.6032
	15	"	"	"	5.4640	5.4607	4.8620	.6021 }		
	16	"	"	"	5.4663	5.4630	4.8738	.6035 }		
162	17-18	"	"	"	3.0384	3.0348	3.6518	.6065	3.04	.6068
163	19-20	"	"	"	1.3980	1.4009	2.4617	.6027	1.40	.6042
164	21-22	"	"	"	.8238	.8278	1.8685	.5960	.82	.6001
165	23-36	"	"	"	.4039	.3904	1.2611	.5715	.40	.5935
166	311-313	2	.6562	.16404	5.6471	5.6487	1.2661	.6170	5.65	.6170
167	314-315	"	"	"	1.3666	1.3794	.6814	.6304	1.37	.6305
168	316-317	"	"	"	1.0876	1.0860	.5692	.6320	1.09	.6321
169	318-319	"	"	"	.7366	.7379	.4696	.6336	.74	.6339
170	320-321	"	"	"	.1152	.0997	.1788	.6152	.113	.6308
171	322-324	3	.6562	.16404	5.7531	5.7527	1.2774	.6167	5.75	.6167
172	325-326	"	"	"	3.2300	3.2297	.9724	.6265	3.23	.6265
173	327-329	"	"	"	1.6290	1.6385	.6965	.6319	1.63	.6320
174	330-334	"	"	"	.6447	.6359	.4400	.6346	.64	.6350
175	661-662	2	.6562	.03281	5.8711	5.8754	.2571	.6142	5.87	.6142
176	663-664	"	"	"	3.2891	3.2904	.1982	.6328	3.29	.6328
177	665-666	"	"	"	1.6355	1.6401	.1427	.6461	1.64	.6461

TABLE VII.—*continued.*

No.	Lesbros' Nos.	Figure. Plate I.	l	w	H	H_m	q	C	Means h	Means c
178	667-669	2	.656 2	.032 81	.7989	.8019	.101 5	.6572	.80	.6572
179	670-671	"	"	"	.0804	.0801	.034 2	.6992	.080	.7004
180	672-673	3	.656 2	.032 81	6.1553	6.1573	.263 2	.6142	6.16	.6142
181	674-676	"	"	"	3.2530	3.2560	.196 4	.6305	3.25	.6305
182	677-679	"	"	"	1.5404	1.5335	.139 6	.6511	1.54	.6511
183	680-681	"	"	"	.8219	.8265	.103 2	.6588	.82	.6588
184	682-685	"	"	"	.3724	.3793	.070 5	.6685	.372	.6685
185 {	948	1	1.968 5	.065 62	5.6169		1.523 2	.6202	5.61	.6214
	949-950	"	"	"	5.6054		1.527 4	.6225		
186 {	951	"	"	"	3.3514		1.191 5	.6281	3.35	.6268
	952-953	"	"	"	3.3416		1.181 7	.6254		
187	954-956	"	"	"	.8022		.588 7	.6343	.80	.6343
188	957-959	1	.065 62	1.968 5	4.8410		1.426 3	.6256	4.84	.6267
189	960-961	"	"	"	2.4000		1.009 6	.6289	2.40	.6335
190	962-963	"	"	"	1.0778		.656 5	.6103	1.08	.6380
191	964-966	1	.065 62	.656 2	5.4971		.501 6	.6193	5.50	.6194
192	967-968	"	"	"	2.9118		.373 5	.6336	2.91	.6340
193	969-970	"	"	"	1.6388		.283 0	.6399	1.64	.6410
194	971-973	"	"	"	.6480		.179 4	.6451	.65	.6524
195	974-976	1	.065 62	.164 04	5.7432		.127 7	.6172	5.74	.6172
196	977-978	"	"	"	3.1579		.097 12	.6329	3.16	.6329
197	979-980	"	"	"	1.8849		.075 65	.6380	1.88	.6380
198	981-982	"	"	"	.8957		.052 73	.6451	.90	.6453
199	983-985	1	.065 62	.065 62	5.7934		.051 24	.6164	5.79	.6164
200	986-988	"	"	"	3.2071		.039 13	.6326	3.21	.6326
201	989-991	"	"	"	1.9341		.030 58	.6366	1.93	.6366
202	993-995	"	"	"	.9449		.021 63	.6447	.94	.6447
203	996-999	"	"	"	.4331		.014 83	.6525	.433	.6527

Comparing the heights H and H_w in the preceding table, it will be observed that their differences follow no general order. We hence feel inclined to attribute these differences chiefly to experimental error, except for the lowest heads, where the surface in the reservoir was not very far above the top of the orifice. In this latter case the surface immediately above the orifice must have been appreciably depressed. In the second series (orifice .6562 × .6562, Fig. 3) the velocity of approach in the feeding canal was much greater than for any other of the series ; the effective head due to this velocity could not have amounted to more than .0006, or .0002 metre, a quantity barely appreciable with the methods of measurement employed.

In the following table are given the results of the experiments made by Poncelet and Lesbros in 1828, the form of approach being the same as Fig. 1, Plate I. There was hence for these experiments almost perfect contraction. For the orifices, where w was less than .6562, the same objections apply, so far as great accuracy was concerned, which we have stated in regard to the Lesbros experiments where the sliding gate was used.

The second column in this table, headed " P. and L.," gives the number of determinations by Poncelet and Lesbros, the head being constant and Q the variable ; C has been deduced from the average value of Q, as was also the case with the Lesbros experiments.

The head due to velocity of approach for all these experiments was not appreciable ; *i.e.*, at the measuring point for H_w. As before, we regard the differences between H and H_w given in the first series, as chiefly due to experimental error, except for Nos. 212 and 213 ; for these experiments the given depression at the orifice is much greater than that shown by Lesbros, with the same orifice, and Figs. 2 and 3.

TABLE VIII.

Poncelet and Lesbros.—Flow through Vertical Orifices, with full Contraction. Effective Head
measured 11.48 Up-stream from Orifice.

l in all cases = .6562 $C = \dfrac{Q}{l\,w\,(2\,g\,H)^{\frac{1}{2}}}$

No.	P and L	w	H	H_a	Q	C	Means h	Means c	No.	L	w	H	Q	C	Means h	Means c
204	1	.6562	4.8295	4.8288	4.5744	.6026	4.83	.6028	224	3	.164 04	.1529	.2083	.6168	.153	.6248
205	2	"	4.6992	4.6986	4.5140	.6028	4.70	.6030	225	2	"	.1191	.1814	.6088	.119	.6225
206	1½	"	4.3144	4.3137	4.3319	.6037	4.31	.6039								
207	2	"	4.0191	4.0184	4.1701	.6022	4.02	.6024	226	2	.098 43	4.5309	.6869	.6228	4.53	.6228
208	3½	"	3.1251	3.1241	3.6906	.6044	3.13	.6047	227	2	"	3.5893	.6160	.6375	3.59	.6275
209	6	"	1.8668	1.8655	2.8437	.6025	1.87	.6033	228	2	"	1.5633	.4090	.6313	1.5?	.6306
210	1	.	1.3110	1.3120	2.3764	.6001	1.31	.6017		2	"	1.5259	.4031	.6298		
211	3	"	.7940	.7897	1.8324	.5953	.79	.5997	229	2	"	.6808	.2705	.6326	.68	.6327
212	1½	"	.5349	.5141	1.4706	.5876	.52?	.5982	230	4	"	.1660	.1429	.6392	.186	.6411
213	4	"	.4003	.3727	1.2527	.5732	.400	.592?	231	3	"	.0659	.0833	.6263	.066	.6433
										1	.065 62	4.5939	.4603	.6217		
214	2	.3281	5.1031		2.3842	.6111	5.10	.6111	232	1	"	4.5083	.4586	.6232	4.51	.6238
215	2	"	3.3045		1.9315	.6152	3.30	.6153		1	"	4.4374	.4544	.6244		
216	3	"	1.5807		1.3403	.6172	1.58	.6175	233	3	"	3.2323	.3934	.6335	3.23	.6335
217	2	"	.3740		.6422	.6080	.374	.6132	234	2	"	1.2861	.2517	.6425	1.29	.6425
218	2	"	.2067		.4653	.5925	.207	.6111	235	2	"	.3655	.1367	.6545	.365	.6547
									236	2	"	.0666	.0589	.6604	.067	.6675
	1	.164 04	5.5450		1.2571	.6182										
	1	"	5.5447		1.2549	.6171			237	1	.032×1	4.5998	.2298	.9204	4.60	.9204
219	1	"	5.5434		1.2548	.6171	5.54	.6175	238	2	"	4.3514	.2231	.6191	4.35	.6191
	1	"	5.5329		1.2550	.6178			239	2	"	3.2576	.1964	.6301	3.26	.6301
220	2	"	3.5460		1.0164	.6250	3.55	.6250	240	3	"	1.6336	.1419	.6426	1.63	.6426
221	6	"	1.5650		.6786	.6281	1.56	.6282	241	3	"	.6398	.1090 27	.6534	.64	.6535
222	6	"	.6972		.4544	.6302	.70	.6306	242	4	"	.1936	.051 36	.6738	.194	.6760
223	3	"	.3471		.3199	.6288	.347	.6303	243	3	"	.0571	.028 71	.6958	.057	.6982

Comparing the three curves formed by plotting, with *h* and *c* as co-ordinates, the three series of experiments with the respective forms of approach shown by Figs. 1, 2, and 3, it will be seen that they closely agree. Any divergence between these curves is not in excess of probable experimental error. Hence we can draw the conclusion that the narrowing of the feeding canal, from a width of 12.1 as in Fig. 1, to the width of 4.2 as in Fig. 3, had, at the utmost, a very slight effect upon the co-efficient of discharge. Therefore with an orifice .66 square, when the sides of the feeding canal are 1.77 from the respective sides of the orifice, the discharge will be practically the same as though these distances were infinite.

M. Lesbros notes that with the highest heads employed for Fig. 2, Experiments Nos. 156-160, 166-170, and 175 to 179, the vein after its escape from the orifice converged a little towards the prolonged direction of the face of the canal nearest the orifice. This phenomenon indicates that the side of the canal nearest the orifice had some effect upon the discharge. But it may be remarked, this is a test of marvellous delicacy ; we are of the opinion, judging from our own experiences, that such a vein could be perceptibly inclined after its escape, without affecting the co-efficient of discharge more than .000 05, a quantity much below the limits of appreciation.

M. Lesbros draws some general conclusions from the variation of the curves—form of approach Figs. 1, 2, and 3, with size of orifice constant—which we do not consider tenable. He did not, in our judgment, sufficiently recognise the danger, or one may say certainty, of experimental errors.

The escaping jet or vein, for Experiments 188-190, with the high and narrow orifice took a remarkable shape, with flanges at the summit and base. *Vide* Plate VI. of Lesbros.

The contracted section of the jet from the larger fixed orifice (.6562 square) was measured with much care, in order to obtain the co-efficients of contraction and velocity. There are so many mechanical difficulties in the way of an accurate measurement of the jet from a square or rectangular orifice, and so many intricate theoretical considerations to be taken into account, that we do not feel inclined to give much weight to these determinations.

The results of a number of the Lesbros experiments, with various forms of approach as indicated by Figs. 4 to 12 inclusive, Plate I., are given in the following table. The length of the orifice was in all cases .656.

TABLE IX.

Lesbros.—Flow through Vertical Orifices, with the Contraction more or less suppressed by the Sides of the Feeding Canal.

Orifice .656 × .656

Fig. 4.		Fig. 5.		Fig. 6.		Fig. 7.		Fig. 8.		Fig. 9.		Fig. 10.		Fig. 11.		Fig. 12.	
h	c	h	c	h	c	h	c	h	c	h	c	h	c	h	c	h	c
5.9	.621	5.3	.637	5.9	.661	5.8	.670	5.3	.611	5.8	.627	5.3	.637	5.6	.641	5.0	.611
5.6	.623	2.9	.637	5.5	.661	5.5	.671	3.0	.612	3.0	.629	3.0	.639	3.9	.612	3.0	.612
4.7	.625	1.1	.636	4.2	.663	1.6	.675	1.5	.610	1.3	.633	1.5	.641	2.6	.644	.86	.609
4.2	.625	.68	.633	2.6	.666	2.9	.677	.83	.607	1.1	.633	.80	.647	1.5	.646	.40	.610
3.7	.624					1.4	.673	1.2	.692	.41	.607	.42	.649	.54	.658	.49	.649
2.5	.624					.83	.684										
1.3	.624					.73	.695										
.93	.624																
.67	.621																

Orifice .656 × .328   Orifice .656 × .164

Fig. 4.		Fig. 5.		Fig. 6.		Fig. 7.		Fig. 9.		Fig. 4.		Fig. 5.		Fig. 6.		Fig. 9.	
h	c	h	c	h	c	h	c	h	c	h	c	h	c	h	c	h	c
6.1	.642	5.9	.653	6.0	.670	6.0	.679	6.0	.629	6.1	.664	5.8	.670	6.0	.676	6.1	.634
5.0	.644	5.4	.654	4.1	.673			3.1	.630	5.2	.665	4.2	.671	4.2	.679	3.2	.635
4.2	.645	2.6	.656	3.1	.675			1.5	.633	4.3	.666	3.0	.673	2.9	.680	1.6	.637
3.3	.647	.67	.658	2.1	.678			.71	.636	3.3	.666	1.6	.676	1.8	.682	.82	.642
2.8	.648	.33	.652	.87	.681					2.9	.667	.53	.676	.75	.687	.12	.663
1.6	.648			.37	.689					1.6	.668	.20	.672	.45	.689		
.67	.649									.69	.671			.26	.693		
.39	.644									.28	.670						

Orifice .656 × .098   Orifice .656 × .033

Fig. 4.		Fig. 5.		Fig. 6.		Fig. 9.		Fig. 4.		Fig. 5.		Fig. 6.		Fig. 8.		Fig. 9.	
h	c	h	c	h	c	h	c	h	c	h	c	h	c	h	c	h	c
6.0	.675	5.8	.677	6.0	.680	6.2	.634	6.0	.694	5.8	.695	6.0	.698	5.7	.622	6.1	.652
4.3	.675	4.2	.681	4.3	.682	2.9	.641	4.9	.698	3.7	.701	4.3	.701	3.3	.634	3.3	.662
2.8	.676	3.0	.683	2.9	.686	1.6	.648	3.1	.701	1.4	.702	3.1	.701	1.7	.649	1.6	.669
1.7	.680	1.6	.682	1.6	.692	.85	.655	2.0	.703	.78	.708	1.7	.701	.71	.600	.71	.685
.55	.682	.81	.684	.73	.694	.08	.679	1.3	.707	.26	.719	.76	.707	.31	.677	.28	.710
.22	.687	.22	.685	.22	.699			.49	.716	.06	.757	.26	.716	.04	.710	.04	.766

In some of these series of experiments there was a notable loss of head from primary contraction, as the water entered the feeding canal from the reservoir proper. This was especially the case with the forms of approach represented by Figs. 6 and 7. The "frictional" loss of head in the canal could hardly have been appreciable, even for Fig. 7, where contraction was suppressed on three sides of the orifice.

The given co-efficients for these two forms of approach (Figs. 6 and 7) are doubtless appreciably too low. Error of this kind could have been avoided by having the upper entrance of the feeding canal trumpet-shaped; had this been done the effect of velocity of approach could have been readily determined, by a comparison of the height H, and the surface height a foot or so above the orifice.

The difference between H and H_w, as given by Lesbros, approximately represents the head due to velocity of approach, of which only a portion is lost, plus the head absorbed in overcoming the primary contraction, all of which is lost.[*]

Hamilton Smith, Jun.

The experiments with orifices made by the author, are described in detail in Chapters IX. and XI. It will only be necessary here to give a résumé of them, stating the approximate sizes and heads, and the co-efficients c.

[*] With comparatively high velocities of approach, as was the case with Fig. 7, the water "piled up" immediately above the inner side of the orifice, thus taking an abnormal level at the point where H_w was measured. It is possible that in some of these experiments this abnormal elevation of the surface at H_w, may have fully represented the total head H minus the loss by primary contraction.

TABLE X.

Smith.—Flow through Vertical Orifices with Free Discharge into Air.

California Experiments, 1874-1876.

No.	No. of Determinations	h	c	Size.	Material.	Remarks.
1	3	.583	.6161	Rectangular, $w = .167$; $l = 1.17$	Wood.	Full contraction.
2	2	1.00	.5988	,, $w = 1.00$; $l = 1.06$	,,	,,
3	1	1.33	.6087	Circular, $D = .254$	Iron.	,, ,,
4	1	1.20	.6041	,, $D = .419$	,,	,, ,,
5	1	1.09	.5939	,, $D = .66$	,,	,, ,,
6	1	1.05	.5913	,, $D = 1.01$	,,	,, Vortex above orifice.
7		322.3	1.040	Circular nozzle. Least $D = .053$	,,	Converging mouth, *vide* Fig. A, Pl. XV.
8		314.5	1.004	,, ,, $D = .085$		,, ,, ,, ,, B, ,, ,,
9		312.2	.986			
10		316.1	1.006			,, ,, ,, ,, C.
11		332.7	1.007	,, ,, $D = .087$	,	,, ,, ,, ,, C, ,, ,,
12		336.0	1.005			
13		317.9	1.011			
14		315.6	1.014	,, ,, ,, $D = .102$	,,	,, ,, ,, D. ,, ,,
15		316.3	.615	Circular ring. $D = .060$	Steel.	Contraction slightly suppressed, Fig. E.
16		312.6	.665	,, ,, $D = .085$	,,	,, somewhat ,, F.
17		312.2	.662			
18		127.3	.647	,, ,, $D = .182$	,,	,, ,, ,, ,, G.

The co-efficients c for Experiments Nos. 2, 6, and 5 are in all probability within a small fraction of the truth. Nos. 1, 4, and 3 are less reliable, owing to danger of error in the measurement of w and D.

Nos. 7 to 18 inclusive can be only considered as approximations, because for them Q was indirectly determined by the flow over a weir. For Nos. 7 to 14 it is altogether improbable that c exceeded .995 or .998 at the utmost. The chances of error in these experiments are discussed in Chapter IX. They possess, however, considerable value in proving for very great heads, that with converging mouth-pieces c has a value of about 1, and that with full contraction c will be about .60 for small circular orifices. They demonstrate conclusively the truth of our fundamental proposition, the theorem of Torricelli.

In the following table is given a summary of the experiments made in 1885, under normal conditions.

TABLE XI.

Smith.—Flow through Vertical Orifices with Full Contraction and Free Discharge into Air. 1886.

No.	Number of Determinations	h	c	Description.	No.	Number of Determinations	h	c	Description.
124	3	.739	.6495	Circular in steel.	58	3	4.55	.6014	
123	4	2.43	.6298	D=.020	17	3	4.60	.6013	
122	4	3.19	.6264	T=40°-46°	60	4	1.80	.6061	Circular in iron.
19	4	.185	.6525	Circular in brass.	59	3	1.81	.6041	D=.100
20	2	.190	.6611	D=.050	61	3	2.81	.6033	T=60.5°-63°.
21	4	.200	.6475	T=19-54°.	62	3	4.68	.6026	
22	4	.203	.6481		63	2	.313	.6410	Square in brass.
23	5	.240	.6438		64	2	.457	.6354	.050×.050
24	2	.283	.6336		65	2	.651	.6286	T=51°-52°.
25	1	.282	.6453		66	3	.877	.6238	
26	1	.282	.6472		67(1)	1	1.70	.6149	No. 67 somewhat defective.
27	3	.283	.6457		68	3	1.79	.6157	
28	4	.335	.6330		69	1	2.81	.6127	
29	4	.336	.6376		70	3	3.70	.6113	
30	2	.401	.6377		71	3	4.63	.6097	
31	4	.437	.6301		72	3	.181	.6292	Square in brass.
32	5	.536	.6265		73	2	.480	.6184	.100×.100
33	3	.720	.6199		74	2	.677	.6157	T=49°-50°
34	3	.910	.6160		75	3	.939	.6139	
35	6	.929	.6194		76	3	1.71	.6084	
36	3	1.74	.6113		77	3	2.75	.6076	
37	2	2.73	.6070		78	3	3.74	.6060	
38	3	3.57	.6060		79	3	4.59	.6065	
39	3	4.63	.6051		80	3	.261	.6476	Rectangular in brass
40	3	.129	.6337	Circular in brass.	81	2	.442	.6361	l=.300; w=.050
41(1)	1	.264	.6288	D=.100	82	3	.665	.6312	T=19°-52°.
42	2	.457	.6155		83	3	.917	.6280	
43	3	.661	.6120	No. 41 known to be defective.	84	3	1.82	.6203	
44	2	.900	.6096		85	1	2.72	.6180	
45	3	1.73	.6042	T for Nos. 40-47 = 49°-51°.	86	2	2.83	.6184	
56	3	1.87	.6038		87	3	3.75	.6176	
57	2	2.05	.6038	T for Nos. 56-58 = 62°-62.5°.	88	3	4.70	.6168	
46	2	3.18	.6025						

The foregoing experiments were made with much care, and under conditions favorable to great accuracy. Those with the smallest circular orifice are the least trustworthy, owing to greater danger of comparative error in the measurement of *D*.

TABLE XII.

Smith.—Flow through Vertical Submerged Orifices, pierced in Brass Plates, with Full Contraction. Holyoke, 1884.

Depth of submergence from .57 to .73

Circular	$D = .05$		Circular.	$D = .10$		Square.	.05 × .05		Square.	.10 × .10		Rect .	.30 × .05	
No.	*h*	*c*	No.	*h*	*c*	No.	*h*	*c*	No.	*h*	*c*	No.	*h*	*c*
97	4.08	.6016	100	3.97	.5992	109	4.06	.6068	112	3.95	.6048	119	2.77	.6188
98	2.16	.6041	101	3.57	.5987	110	2.21	.6092	113	3.11	.6052	120	1.63	.6207
99	4.37	.6183	102	2.99	.5989	111	.350	.6201	114	2.32	.6040	121	.614	.6219
			103	2.58	.5997				115	1.52	.6055			
			104	2.00	.6006				116	.771	.6053			
			105	1.51	.6006				117	.410	.6091			
			106	.985	.6025				118	.207	.6117			
			107	.648	.6027									
			108	.256	.6048									

The above experiments are considerably less reliable than those in Table XI., for the reasons given in Chapter XI. Nos. 119-121 do not fairly represent submerged discharge, owing to the thickness of the sides of the orifice, which formed a slight divergent adjutage, and thereby increased abnormally the co-efficient *c*.

A number of other experiments, illustrating effect of temperature, et cet., et cet., will be described in Chapter XI. They need not be summarized here.

Ellis.

Mr. T. G. Ellis, in the year 1874, made at Holyoke, Massachusetts, an extensive series of experiments with vertical and horizontal orifices of large sizes, with heads up to 18 feet. His results were published in the Transactions of the Am. Soc. of C.E., February, 1876.

The discharge or *Q*, which in some cases reached the large amount of 48 cubic feet per second, was measured over a sharp crested weir, the flow being computed by the weir formula of Mr. J. B. Francis. This indirect method of obtaining the value of *Q* adds considerably to the chances of experimental error, and these experiments hence cannot be considered as exact as most of those of Lesbros and the author. The Francis

formula, as will be hereafter pointed out, gives results pretty near the truth for the
measuring weirs used by Ellis, and therefore the resulting co-efficients of discharge from
his orifices should give quite smooth curves, when his results are shown graphically.
There are, however, frequent irregularities in these curves, which indicate rather
inaccurate methods of observation.

The final results given by Ellis can therefore only be considered as approximative,
although probably without errors of very serious consequence. They are the best
authority to be had for large orifices with high heads.

The experimenter states that in all cases the heads above the orifices were sufficiently large to prevent the forming of vortices.

In the following tables we will only give the approximate heads, and the correct
co-efficient c, deduced from the average values of C for nearly similar heads given by
Ellis. These represent the means of several hundred determinations.

TABLE XIII.

*Ellis.—Flow through Vertical Rectangular Orifices, nearly full Contraction and free Discharge into Air.
Rigid Iron Plates.*

$l=2.00$, $w=2.00$			$l=2.00$, $w=1.00$			$l=2.00$, $w=.50$			$l=1.00$ $w=1.00$		
No.	h	c	No.	h	c	No.	h	c	No.	h	c
244	2.07	.616	247	1.81	.599	256	1.12	.612	266	1.49	.587
245	3.05	.600	248	3.03	.600	257	2.91	.612	267	3.69	.599
246	3.54	.608	249	4.66	.598	258	4.75	.608	268	4.80	.599
			250	5.67	.597	259	6.36	.608	269	5.49	.596
			251	6.87	.598	260	8.54	.607	270	6.72	.599
			252	7.69	.598	261	9.63	.606	271	9.80	.596
			253	8.48	.599	262	11.56	.604	272	9.90	.602
			254	9.65	.600	263	13.51	.604	273	12.00	.600
			255	11.31	.605	264	15.06	.602	274	13.63	.601
						265	16.97	.600	275	15.13	.601
									276	17.56	.597

In the above experiments there appears to have been a partial suppression of
bottom contraction for the 2 × 2 orifice, the horizontal floor of the discharging basin
being .50 below the bottom edge of the orifice ; for the other orifices the suppression of
bottom contraction was less.

TABLE XIV.

Ellis.— Flow through Vertical Circular Orifices, nearly full Contraction, and free Discharge into Air.   Rigid Iron Plates.

D = 2.00			D = 1.00			D = .50		
No.	h	c	No.	h	c	No.	h	c
277	1.77	.595	284	1.15	.578	294	2.15	.600
278	2.60	.596	285	2.36	.589	295	4.16	.602
279	4.47	.604	286	4.81	.590	296	6.35	.605
280	5.83	.610	287	7.97	.586	297	7.30	.601
281	6.93	.613	288	7.92	.589	298	8.01	.601
282	8.34	.613	289	10.88	.591	299	9.06	.602
283	9.64	.615	290	12.48	.594	300	10.51	.601
			291	14.13	.595	301	11.97	.600
			292	15.66	.596	302	12.98	.601
			293	17.72	.600	303	14.47	.601
						304	15.46	.601
						305	15.85	.603
						306	17.26	.596

Contraction appears to have been somewhat suppressed on the bottom for the largest of these three orifices, and very slightly suppressed for the smallest orifice.

The following experiments were made with a submerged discharge, the two orifices, which for free discharge had been placed in a vertical position, now being placed in a horizontal position.   The converging mouth-piece for the 1. × 1. orifice was formed by an inner facing of wood .5 thick bolted to the iron plate and curved on each of the four sides in the shape of a quarter of an ellipse, whose semi-diameters were .50 and .33, with the larger diameter at right angles to the plane of the plate.

Contraction was probably nearly perfect for the plates without the convergent adjutage.

TABLE XV.

Ellis.—Flow through Horizontal Submerged Orifices.

Circular $D=1.00$ Nearly full Contraction.			Square 1.00×1.00 Nearly full Contraction.			Square 1.00×1.00 Curved Mouth-piece.		
No.	h	c	No.	h	c	No.	h	c
307	2.60	.607	316	2.32	.600	323	3.04	.952
308	4.71	.590	317	3.92	.602	324	5.77	.946
309	6.41	.606	318	7.99	.606	325	10.54	.943
310	8.10	.599	319	11.58	.605	326	13.57	.943
311	8.80	.600	320	14.31	.611	327	18.22	.944
312	12.09	.600	321	16.22	.606			
313	14.25	.601	322	18.45	.606			
314	16.29	.602						
315	18.66	.599						

General Ellis measured the head upon his measuring weir, by means of a tube placed near the inner face of the weir. This is a vicious method, as will be hereafter shown, but in these experiments the velocity of approach was never large, and serious errors should not have resulted.

Note our remarks on some weir determinations by the same authority, Chapter V.

Weisbach.

Herr Julius Weisbach, in his " Lehrbuch der Ingenieur," 4th revised edition, and in " Der Civilingenieur," Vol. X., gives the following values of c for circular orifices " in a thin wall ";

D	H						
	.066	.33	.82	2.0	3.0	45.	340.
.033	.711	.665	.637	.628	.641	.632	.600
.066			.629	.621			
.10			.622	.614			
.13			.614	.607			

For an orifice, with $D=.033$, and a well rounded mouth-piece : $h=.066$, $c=.959$; $h=1.64$, $c=.967$; $h=11.5$, $c=.975$; $h=56$, $c=.994$; $h=338$, $c=.994$.

He states that c is much greater for long narrow orifices than for those which are circular or regular, and that c is about $1\frac{1}{2}$ per cent. smaller for a submerged than for a free discharge.

He found the following values for c, with the three given liquids, with a head dropping from 1.08 to .30 (?), the opening in each case having a diameter of about .02 :

—	Thin Plate.	Conoidal Convergent Mouth-piece.	Cylindrical Pipe, with $l = 3d$.
Water ...	.709	.942	.885
Mercury ...	.670	.989	.900
Rape seed oil	.674	{ .430 with $T = 54°$ { .665 ,, $T = 103°$	{ .363 with $T = 54°$ { .604 " $T = 102°$

Unwin.

Professor W. C. Unwin, in the "Philosophical Magazine" for October, 1878, gives the following results illustrating effect of changes in temperature.

With a head dropping from 1.47 to .98 in a vertical cylindrical reservoir, and with an opening in each case having a diameter of .033 :

	Thin Plate.		Convergent Conoidal Mouth-piece.		
T	t	c	T	t	c
205°	{ 149.6 { 149.4	.593 .594	190°	{ 89. { 88.5	.986 .991
140°	148.8	.596	130°	90	.975
65°	149.2	.595	60°	{ 93 { 92.5	.943 .949
61°	{ 148.0 { 148.0	.600 .600			

Professor Unwin states that the given values of c may possibly be in error 2 or 3 per cent., from errors in measurement of H and D.

Francis.

Mr. J. B. Francis has made a number of experiments with orifices, an account of which will be found in his "Lowell Hydraulic Experiments." The values of Q were unfortunately obtained by weir measurement, and often with minute heads. Such determinations are especially unreliable.

With a converging cycloidal mouth-piece, and a minimum or outer diameter of .1018, with a submerged discharge he found a value of .815 for c, with $h = .034$; this value steadily increased with increasing heads, until it was .944 for $h = 1.52$.

With a circular submerged orifice, $D = .1017$ with full contraction, he found values of c of from .56 to .59, with a maximum head of 1.5. In all probability for these experiments Q was under estimated fully 5 per cent. .

Steckel.

Mr. R. Steckel[*] has lately made a large number of experiments with small orifices under various conditions.

With a horizontal circular orifice in a thin plate, $D = .032$, with H measured from the plane of the orifice, he found, with free discharge into air and full contraction, the following results ; $H = 4.2, c = .621$; $H = 2.4, c = .628$; $H = 1.0, c = .628$.

With a horizontal circular orifice, with $D = .033$, having an iron cylinder of $D = .015$ placed normal to the centre of the orifice, thus forming an annular opening, and with nearly constant heads of .25 above the plane of the orifice, he obtained the following results :

With the cylinder removed...	...	...	...	...	...	$c = .673$
,, ,, base of the cylinder .025 above plane of orifice			...	$c = .673$
,, ,, ,, ,, ,, ,, .0004 ,, ,, ,, ,,			...	$c = .729$
,, ,, ,, ,, ,, in ,, ,, ,,			...	$c = .726$
,, ,, ,, ,, ,, .0083 below ,, ,, ,,			...	$c = .685$
,, ,, ,, ,, ,, .017 ,, ,, ,, ,,			...	$c = .680$

For the first two experiments, the area taken for establishing c is the area of the orifice ; for the last four experiments, a is the area of the ring.

The foregoing results are doubtless chiefly due to suppression of contraction by the introduction of the cylinder.

Bazin.

For the experiments with open conduits, Darcy-Bazin series,[†] Q was determined by the flow through one or more of 12 similar adjoining orifices. Each orifice was .656 × .656, with a constant head of 2.624 above the centre of the orifice. The co-efficients of discharge for these orifices, C, were determined by using a section of the experimental conduit as a measuring vessel. The following values of C were obtained :

One gate open ...	$C = .633$
Two gates ,,	$C = .642$
Three ,, ,,	$C = .646$
Four ,, ,, ...	$C = .649$
Five to twelve gates open ...	$C = .650$

The value of C appears to have increased when two or three adjoining gates were opened, aside from the effect of velocity of approach, which for only three gates open must have been barely appreciable.

---

[*] Essay on the Contracted Liquid Vein. A paper read before the Royal Society of Canada. Ottawa, 1884.

[†] Recherches Hydrauliques. Darcy-Bazin. First part, page 61.

We have ourselves found similar results with several orifices 1. × 1.05 placed a few inches apart.

Castel.

M. Castel made a number of experiments with conical convergent tubes, which are described by M. D'Aubuisson in the "Annales des Mines, 1838." The tubes were of smoothly polished metal ; Q was determined by the measurement of q in vessels of known size.

In the following statement we give the general results obtained ; the heads employed varied, for each given experiment, from .2 m. to 3.0 m., but as there was only a very slight difference* in the value of C with varying values of H for the same adjutage, it will only be necessary to give the mean values of C.

Castel-D'Aubuisson.—Converging Mouth-pieces.

l = length of adjutage. C = co-efficient of discharge at smallest section of adjutage.   Given angle = angle of prolonged sides.   All measures metrical.

l = .040 D = about .0155		l = .035 D = about .0155		l = .050 D = about .020	
$2\,\beta$	C	$2\,\beta$	C	$2\,\beta$	C
0	.829	9° 14′	.929	2° 50′	.914
1° 36′	.866	10° 28′	.945	5° 26′	.930
3° 10′	.895	12° 42′	.951	6° 54′	.938
4° 10′	.912	16° 02′	.940	10° 30′	.943
5° 26′	.924	19° 06′	.926	12° 10′	.930
7° 52′	.930			13° 40′	.936
8° 58′	.934	l = .030 D = about .0150		15° 02′	.949
10° 20′	.938			18° 10′	.939
12° 04′	.942	15° 44′	.941	23° 04′	.930
13° 24′	.946	19° 14′	.931	33° 52′	.920
14° 28′	.941				
16° 36′	.938			l = .100 D = about .020	
19° 28′	.924				
21°	.919			11° 52′	.965
23°	.914			14° 12′	.958
29° 58′	.895			16° 34′	.951
40° 20′	.870				
48° 50′	.847				

* If there was any difference due to variation in H, it was a slight increase in C—say .001 or .002—for the greatest head compared with the smallest head.

Castel also determined the co-efficient of velocity for the jets from these mouth-pieces.

Note our remarks on the experiments of Castel, given in the Section on Short Weirs, Chapter V.

Bornemann.

Herr K. R. Bornemann has investigated the flow under sluice gates placed across a rectangular flume or canal. He has published an account of his elaborate series of experiments in " Der Civilingenieur," Vol. XXVI., 1880.

These experiments were necessarily often complicated by a high velocity of approach, as the length of the orifice was very nearly the width of the canal.

The following sketch shows the points where H_g and H_l were measured.

Adding for velocity of approach $\frac{v_a^2}{2\,g}$, or effective head $h = H_v - H_l + \frac{v_a^2}{2\,g}$, the co-efficients c range in 63 experiments from .668 to 1.045. In general, the high values of c are when $\frac{v_a^2}{2\,g}$ is a large part of h. These abnormal values of c are due in part to an under estimate of the effect of velocity of approach, and still more to the incorrect point taken for the measurement of the lower head H_l. It seems to us clear that this head should be measured in the lower portion of the wave as shown at a in the sketch. In any event it should not be measured on the crest of the lower water, which is piled up by the energy of the water escaping from the orifice.

Such an escape presents many interesting mathematical problems. Unfortunately, so far as experimental proof is concerned, it is not practicable to measure the head at a with any reasonable degree of accuracy, owing to the boiling of the water.

This same kind of discharge has been investigated by Boileau and by Lesbros. We regard the investigation of the flow through such orifices as of no value for the practical gauging of water. The quantity of flow can be determined by almost any other method with greater accuracy.

Bossut and Others.

In the following table are given the results of experiments with orifices " in a thin wall" made by Bossut, Castel, and others. Several of these determinations have been taken from the " Traité d'Hydraulique " of D'Aubuisson.

TABLE XVI.
Flow through Orifices.

	CIRCULAR					RECTANGULAR			
Authority.	D	H	C	Remarks.	Authority.	Size.	H	C	Remarks.
Bossut ...	.044	12.5	.614	Horizontal	Bossut ...	.089 × .022	12.5	.612	Horizontal
"	.089	12.5	.617	"	"	.089 × .089	12.5	.617	"
"	.178	12.5	.618	"	"	.178 × .178	12.5	.618	"
"	.044	9.6	.613	Vertical	Castel ...	.033 × .033	.16	.655	
"	.089	9.6	.617	"	Michelotti	.089 × .089	12.5	.607	
"	.044	4.3	.616	"	"	.089 × .089	22.4	.606	
"	.089	4.3	.619	"	"	.178 × .178	7.3	.603	
"	.089	.052	.649	Surface inclines towards orifice	"	.178 × .178	12.6	.603	
Castel ...	.033	2.1	.673		"	.178 × .178	22.2	.602	
"	.033	1.0	.654		"	.27 × .27	7.4	.616	
"	.049	.45	.632		"	.27 × .27	12.6	.619	
"	.049	.98	.617		"	.27 × .27	22.4	.616	
"	.098	.55	.629		Bidone ...	.030 × { .061	1.1	.620	
Michelotti	.18	7.2	.607		"	.121	1.1	.620	
"	.27	12.5	.612		"	.485	1.1	.625	
"	.33	6.9	.619						
"	.53	12.0	.619						

For the horizontal orifices of Bossut *H* was measured from the plane of the orifice. Bossut, however, was of the opinion that for a vertically descending jet, *H* should be measured from the section of the contracted vein, after its escape from the orifice.

Rennie and other experimenters have made a great number of determinations of the value of *c* for small orifices, but their results are still more discordant than the ones given in Table XVI. Their errors probably chiefly arose from inaccurate measurements of the size of the orifice, and sometimes doubtless from a slight rounding of the inner corners.

The orifice, with *D* = .089 and *H* = .052, of Bossut was the old standard French pouce d'eau, being 1 French inch in diameter, with a constant head of 1 line above the top of the opening. Mariotte and Couplet had before determined the flow through this standard or module; their resulting values of *C* are considerably higher than that found by Bossut.

EFFECT OF TEMPERATURE.

Before attempting to determine what values should be assigned to *c* for orifices, it will be well to first discuss what effects, if any, are produced upon the discharge by changes in the temperature of the water, by slightly altered conditions of the edges of the orifices, by an irregular feeding supply, and by possible variations in the character of the water.

In regard to the effect of temperature we have a number of experiments described in Chapter XI., and those made by Prof. Unwin.

The Greenpoint experiments with a circular orifice, $D = .020$, with full contraction and with a head dropping from 3.2 to .56 in a cylindrical reservoir, gave the following results.—Nos. 125-142. Table CVI.;

With a temperature of about 48° the most probable time of discharge was 253.8"; as the temperature was increased from 48° the times became longer, until at 130° t was about 257.5". As c is in the inverse ratio of t, this shows that for this quite small orifice an increase of T from 48° to 130° diminished c about $1\frac{1}{2}$ per cent.. We do not conceive it possible that experimental error could have resulted in such a marked difference.

Prof. Unwin with a similar but larger orifice, $D = .033$, with a head dropping from 1.5 to 1.0, found that c was diminished about 1 per cent. by an increase in T from 61° to 205°.

For a much larger circular orifice, $D = .10$, c had a value of .6013 with $T = 49°$, and of .6014 for $T = 62.5°$—vide Experiments Nos. 47 and 58, Table CII. This indicates that for such an orifice, with $h = 4.6$, a change in T of 13.5° has no appreciable effect. It was thought, however, that a change in T of 25° or 30° for this same orifice, would change c for the same head, perhaps .0005, or about $\frac{1}{10}$th of 1 per cent.. Also, that for a smaller orifice, $D = .05$, an increase in T of 30', for heads less than one foot, notably diminished the flow. The proof in regard to the foregoing is, however, not absolute. *

We are of the opinion that with orifices "in a thin wall," the effect of T increases as a and h diminish. That is to say, the smaller the head and the smaller the orifice, the greater the effect which will be produced by changes in T; c diminishing as T is increased.

With a large head such as 10 feet, it is not probable that a change in T of 50°— which can be assumed to be the maximum range in general practice—will cause any appreciable variation in c for any orifice larger than $D = .02$, or $.02 \times .02$. For a large orifice, such as $D = 1.$, or 1. × 1., it is doubtful if such a change in T will have any noticeable effect upon c, with any head, no matter how small.

With a short convergent mouth-piece, with $D = .033$, with a head dropping from 1.5 to 1., Prof. Unwin found a definite increase of flow by raising T from 60° to 190°. Weisbach, with a lower head, and a convergent mouth-piece, found with rape-seed oil

---

* Note our remarks on experiments made in 1884, at Holyoke, Chapter XI.

H

that an increase of T from 54° to 102° increased the flow about one half. Our experiments and those of Weisbach prove that the co-efficient c, with such mouth-pieces, will be practically unity with great heads such as 300 feet. In our experiments, Nos. 7 to 14, the temperature of the water was about 60°. Now there is no likelihood that an increase in T to 200° would with such a head have appreciably increased the flow, which, with the lower temperature, was already at a maximum. With such a great head, even with a very viscous liquid such as oil, we fancy that variation in T would have no notable effect upon c.

For the flow of water through converging mouth-pieces, we can conclude that an increase in T will increase c in the inverse ratio of a and h. With a large orifice, as, for instance, $D = 1$., it is doubtful if a change in T of 50°, will, for any head, have an appreciable effect upon c.

Condition of Edges.

In Table CVII., Nos. 143—148, Chapter XI., is given an account of experiments made with the inner face and edges of a circular orifice ($D = .020$) well wet with a mineral oil of fair body. The flow was diminished at first nearly 2 per cent. from the normal flow with the orifice perfectly clean ; this diminution became less and less as the escaping jet washed the oil away.

With a much larger orifice, $D = .10$, with a normal condition of the plate c had a value of .6013, with $h = 4.60$ (No. 47, Chapter XI.). The inner face and edges of the orifice were then well wet with sperm oil, and with the same head c was .5996—No. 54. The experiment was repeated, and c had a value of .6004—No.55 ; the plate was then examined, and the inner edges found to be nearly free from oil.

These experiments show that even for a very small orifice, when reasonable care is taken to keep the inner edges clean, no abnormal flow will occur from slight particles of greasy matter adhering to the edges ; also with a large orifice, such as $D = 1$., even if the inner edges should be wet with oil, it would have no appreciable effect upon c.

We regard the diminished flow in these experiments as being perhaps entirely due to the very thin film of oil, which reduced the section of discharge.

Experiments Nos. 59 to 62, Chapter XI., were made with an orifice in an iron plate having as nearly as was possible the same diameter as the orifice in a brass plate with $D = .10$. The edges of this orifice appeared in August, 1885, when it was compared by a microscope of pretty high power with the orifice in brass, to be somewhat more perfect, the edges being better defined with less rounding. The co-efficient c should hence have presumably been very slightly higher for the orifice in brass than for the one in iron. By comparing the values of c for Nos. 59-62 with the curve on Plate III. for $D = .10$ (brass plate), it will be noticed that they are about $\frac{1}{600}$th higher, instead of being lower, as was expected. This is not a large difference, and may possibly be due to experimental error. These experiments show that for orifices in either iron or brass, there is no notable change in the value of c, due to differences in the two metals.

So far as very slight imperfections or roundings of the inner edges of an orifice are concerned, it is apparent that the resulting effect upon c will be in the inverse ratio of the area.

Irregular Supply.

Careful experiments, with the circular orifice $D = .10$, were made by us at Holyoke, to determine what effect, if any, was produced by an "irregular" feeding supply. For these experiments, the normal arrangement is shown by Fig. 1, Plate XVII., where a screen or rack enclosed the feeding water entering the tank **A** from the lower end of the iron supply pipe. When this screen was in place, for the largest head of 4.6 the escaping jet appeared to be perfectly true always for a distance of .2 from the orifice; after this distance the jet sometimes began to twist, and at other times appeared to be perfect for a distance of 1.5 from the orifice. (Experiments Nos. 46, 47, and 58, Chapter XI.)

With the rack removed, and a vertical partition substituted, 3.92 high, placed across the tank in the same place as the vertical portion of the rack and open on top, the jet twisted more than it did when the enclosing rack was employed, and at times seemed to be perceptibly imperfect near the orifice. (Nos. 53 and 52.)

With a vertical partition 1.86 high, the jet was ragged and twisting. (No. 51.)

With a vertical partition .93 high, the jet was ragged and twisting, but not apparently as much so as with No. 51. (No. 50.)

With no protection whatever for the escaping water from the iron supply pipe, the jet was exceedingly ragged and twisting, being very far from smooth immediately at the orifice, and apparently being more ragged than in any of the other experiments. (Nos. 48 and 49.)

The following statement shows the values of c for the foregoing experiments;

Head. h	Rack.		Partitions, Open on Top.			No Protection.	
			3.92 high.	1.86 high.	.93 high.		
3.2	.6025 .6024	.6024	.6020 .6020 .6023	.6021	.6026	.8039 .6037	.6038	.6025 .6024	.6024
4.6	.6013 .6013 .6012	.6013 .6012 .6014 .6015	.6011	.6017 .6015 .6017 .6008 .6007	.6013			.6015 .6010	.6012

It will be observed that there was no variation in the values of c of consequence, except with the partition .93 high, when it was about $\frac{1}{100}$th higher than for the other experiments with the same head of 3.2. These results were exceedingly surprising to us, as it seemed almost certain that with such a very ragged and distorted jet which escaped when there was no protection whatever over the supplying stream, there must be a notable variation in c from its value when the jet was nearly perfect.

From the foregoing experiments it will be seen what a wonderfully delicate test of the evenness of supply is furnished by the form of a jet escaping from an orifice, carefully pierced "in a thin wall."

It is perhaps a misnomer to call the supply an "irregular" one, in the preceding experiments. The supply was very nearly regular, but the particles of water evidently found their way towards the orifice in an irregular manner.

Variations in Water.

The experiments made at Holyoke in 1885 with a circular orifice, having a diameter of .05, showed, with heads of less than one foot, astonishingly large variations in the deduced values of c for nearly equal heads, considering the great care employed. The variation in temperature for these experiments only amounted to 3°; this may have been a slightly disturbing cause, but there is no likelihood that it had sufficient influence to account for the very irregular curve formed by plotting the results, with h and c as co-ordinates. These irregularities are shown by Tables XI. and CII., and graphically by Plate III. The large number of determinations for this orifice with low heads were made for the purpose of endeavoring to fully trace these discrepancies.

The chances of experimental error in these experiments will be fully discussed in Chapter XI. Taking all these chances of error into account, it seems to us almost impossible that they could have amounted to more than $\frac{1}{100}$th of c for the low heads of .2 to .3, and to more than $\frac{1}{210}$th of c for a head of .8. It will be observed that there are variations in the means of c of over $\frac{1}{50}$th, with $h = .28$ (Nos. 24-27), and with the comparatively large head of .9, a variation of $\frac{1}{100}$th. For Experiments Nos. 34 and 35, consisting of 9 determinations with nearly equal heads, c ranged from .6152 to .6200, showing a variation of $\frac{1}{120}$th. These discrepancies seem to us to be beyond the limits of possible experimental error.

For the circular orifice in a brass plate of .10 diameter, being 4 times larger, the values of c are exceedingly close for the largest head employed. In 13 determinations with this head (4.6), c ranged from .6007 to .6017, as will be noticed by reference to the statement in the preceding section. The chances of experimental error for these last experiments were very little less than for the .05 circular orifice with $h = .9$, and yet for the latter orifice the variation in c is five times greater.

An examination of Table CII. will show that variation in the deduced values of c for nearly equal heads, as a rule, increases as both a and h diminish.

From these results we can conclude that with small orifices, such as $D = .05$ and less, and with heads of less than 1., it is practically impossible to obtain single results with certain accuracy. Also, this uncertainty appears to disappear either with larger orifices or with greater heads.

We conjecture that these excessive variations in c are due to some unknown change in the character of the water. Possibly it may be partly due to the varying quantities of greasy matter carried by the water.

We make the foregoing suggestions in a tentative spirit, for we fully realize how easy and natural it is to underestimate the chances of experimental error. These errors sometimes aggregate in one way in the most surprising manner. We trust that some future experimenter may be able to definitely decide whether or not our conjecture is well founded.

ORIFICES IN A THIN WALL.

Values of c for Orifices in a thin Wall.

Our first task is to endeavor to determine what values shall be given to c, for vertical orifices with full contraction ; with a free discharge into the air ; with the inner face of the plate, in which the orifice is pierced, plane ; and with sharp inner corners, so that the escaping vein only touches these inner edges, which constitute, as nearly as may be possible, a mathematical line. To this form, following the French expression, we give the name of "an orifice in a thin wall."

In considering the weight to be given to the experimental data at hand, we think the experiments made by ourselves, and given in Table XI., should rank first. They were made under very favorable circumstances for a high degree of accuracy, as the orifices were pierced in rigid plates of metal, with very nearly perfect inner edges, and whose dimensions were determined with an unusual degree of care by delicate micrometer measurement. The measuring vessels were also practically free from leakage, and of unchanging form. We trust that our readers who carefully analyze these experiments, as given in detail in Chapter XI., will not regard this preference for our own work as egotistic.

Next in authenticity will come the Lesbros experiments with his largest orifice, $.66 \times .66$, in a fixed copper plate, and after these our experiments Nos. 2, 6, and 5 in Table X.

The experiments with the five orifices in brass plates, and with the small orifice in steel, made under normal conditions in March and May, 1885, have been plotted on Plate III., with c as ordinates, and h as abscissæ. For the five larger orifices (brass plates) T had a range from 48° to 54° ; for the smallest orifice T ranged from 40° to 46°. Hence, for all these experiments the variation in T was so slight, that it could not have notably affected the general results.

From the plotted experimental points, six curves have been drawn on Plate III., representing the most probable curve for c, for each of the orifices. These curves rarely vary more than $\frac{1}{120}$th part from the mean experimental value of c, except for the .05 circular orifice with h less than 1..

Examining these curves, we see that for each of them the value of c decreases as the head increases,[*] and that they become asymptotic in nature with the higher heads, indicating that with sufficiently high heads c will practically become constant. The curves all begin to rise very rapidly for heads less than .3. As before stated, for these minute heads the danger of experimental error is comparatively very large ; hence, we cannot be nearly as certain in regard to the forms of the beginning of the curves, as for their forms with $h = 1$, and upwards.

Comparing the curves for the three circular orifices, having the respective diameters of .02, .05, and .10, it will be observed that the smaller the orifice the higher is the curve, and also as the heads increase the curves constantly approach each other. This is especially noticeable in the curve for the smallest orifice, Experiments Nos. 128 and 122.

From these facts it seems evident that for these three circular orifices, c will have practically the same values when h is sufficiently great. Probably with $h = 50$ or thereabouts, c will not differ more than .002 for the three, having a value of near .595.

Comparing the two curves for the square orifices, we see similar results, the curves constantly approaching each other as h increases. It will also be noticed for heads above .6, where the curves begin to be fairly reliable, that for the same head, the difference in c between the $.1 \times .1$ and $.05 \times .05$ curves, is very nearly the same as between the curves for the circular orifices, $D = .10$ and $D = .05$. For these two square orifices we can assume that c will have a value of about .600, with $h = 50$.

Comparing the curves for the $.1 \times .1$ and $D = .1$, and those for $.05 \times .05$ and $D = .05$, it will be seen that for all heads above .6, c is about .005 higher for the square than for the respective circular orifice. If there be any change in this difference as h increases, it is that the difference increases. Hence with great heads we may fairly assume that c will always be larger for the square than for the circular orifice.

Comparing the rectangular orifice, $l = .3$ and $w = .05$ with the square orifice, $.05 \times .05$, it will be observed that the curves continually diverge as h increases. Hence with great heads, the rectangular orifice will always have larger values of c than the square orifices ; with $h = 50$, c for the rectangular orifice will probably have a value from .610 to .612.

On the same plate are plotted the 6 Columbia Hill experiments, Nos. 1-6. Comparing No. 2 for an orifice nearly square with $l = 1.06$ and $w = 1.00$, with No. 6 for

---

[*] Experiment No. 79, with the orifice $.1 \times .1$, appears to be slightly divergent from this law, but not beyond the limit of possible experimental error.

a circular orifice with $D = 1.01$, we have the respective values of c of .5988 and .5913 with a head in both cases of about 1.0. Both these values seem to accord very fairly with the Holyoke results. Nos. 5, 4 and 3, with circular orifices having the respective diameters of .66, .42 and .25 are not so harmonious; assuming the Holyoke results as correct, c for No. 3 should have been about .603, instead of its given value of .6087, thus showing an apparent error of about 1 per cent.; c for No. 4 should have been about .601 instead of .604, thus showing an apparent error of about $\frac{1}{2}$ per cent. No. 1, with a long rectangular opening ($l = 4.17$ and $w = .17$), is in accord with the Holyoke .3 × .05 orifice. The considerably lower value of c for No. 1 compared with the curve for the .3 × .05 orifice at the same head, being fairly attributable to the much greater width (.17 : .05).

We were unpleasantly surprised after the computation of the Holyoke work, to find these disagreements with the Columbia Hill experiments. Although these defective results were obtained by single determinations we regarded them as quite reliable, and thought the given values of c should be within $\frac{1}{500}$th of the truth. As the Holyoke experiments were conducted with a greater degree of care than those made at Columbia Hill we must regard the latter as defective, when the comparative results are contradictory.

The heavy dotted curve on the same plate, No. III., represents the most probable curve for c for the Lesbros orifice, .66 × .66. (Experiments Nos. 156-160, 161-165, and 204-213.) The lowering of the values of c with the small heads appears to be due to the comparatively small depth of water above the summit of the orifice, or more strictly to the surface inclination towards the orifice. Computing c' from H_a for experiments 210-213, Table VIII., we have the following results;

No. 210	$H = 1.3140$	$H_a = 1.3120$	$c = .6017$	$c' = .6023$
„ 211	.7940	.7897	.5997	.6014
„ 212	.5249	.5141	.5982	.6050
„ 213	.4003	.3727	.5927	.6186

From the foregoing we see that with c deduced from the head immediately above the orifice, its value rises with diminished heads.

The curve for the Lesbros orifice is apparently about .003 too high, in comparison with the Holyoke .10 × .10 orifice, and the Columbia Hill 1.06 × 1.00 orifice.

The Ellis experiments give about the following results;

Square,	2. × 2.	$H = 2.1$ to 3.5	$c = .607$
„	1. × 1.	$H = 3.7$ to 18.	$c = .600$
Rectangular,	2. × 1.	$H = 3.0$ to 11.	$c = .600$
„	2. × .5	$H = 1.4$ to 17.	$c = .606$
Circular,	$D = 2.$	$H = 1.8$ to 10.	$c = .606$
„	$D = 1.$	$H = 2.4$ to 18.	$c = .592$
„	$D = .5$	$H = 2.1$ to 17.	$c = .601$

They approximately indicate that for such large orifices, with heads from 3. to 17., c remains constant; also that probably c is smaller for circular than for square orifices, with D = side of square. These results, considering their chances of experimental error, are fairly in accord with our experiments and those just referred to from Lesbros.

The experiments quoted from Weisbach, for circular orifices, show results for the smallest diameter agreeing pretty well with our Holyoke curves; his values of c for the three other orifices appear to be 1 per cent. or more too high.

The experiments of Prof. Unwin[*] show considerably lower values of c than we found. The experiments of Bossut, Michelotti, Rennie, et cet. et cet., generally show larger values of c than those given by us.

On Plate II. are plotted most of the experiments of Lesbros and Poncelet-Lesbros, given in Tables VII. and VIII., with c and h as co-ordinates. Comparing these results with the curves drawn on Plate III. for the Holyoke experiments, they present wide divergencies; the values of c for all the Lesbros orifices less than .66 × .66, being higher than is indicated by Plate III. for similar orifices and the same heads. For instance, with a square orifice .066 × .066, with h = 1.93, Lesbros finds a value of c of .6366. (Experiment No. 201.) According to our Holyoke curves, for such an orifice and head, c should be about .612. This is a very great difference, amounting to nearly 4 per cent..

We can suggest no reason for this disagreement, unless it be the errors resulting directly and indirectly from the use of the sliding gate employed by Lesbros for decreasing the height of his orifices. This conjecture is in part confirmed by the fact that with his small orifice in a fixed plate, l = 1.97 and w = .066, the values of c appear to differ not more than 1 per cent. from our Holyoke results.[†] Our Chapter on Weirs was written before the Holyoke experiments had been made, and in our determination of the co-efficients of weir discharge, we have made considerable use of Lesbros' experiments with weirs, which seemed to us reliable within about 1 per cent. in the values of c. It was hence especially disagreeable to find our orifice experiments differing as much as 4 per cent. from those made by an authority we had so largely used. The Holyoke experiments were made with such care, the resulting curves are so symmetrical in themselves, and so harmonious one with another, that we cannot conceive it possible that they are wrong, and those of Lesbros are right.

Considering the experimental values shown on Plate III. as authentic, we can now enunciate the following general propositions for orifices in a thin wall, with a not over 1. and any side not less than .01.

For similar forms, no matter how a may differ, with a great head such as 100 feet, the co-efficient of discharge, c, will be practically the same.

---

[*] In these experiments t was the important quantity, for the determination of effect of changes in temperature. The value of c was only determined incidentally, as its exact value had no bearing upon the experiments.

[†] The values of c for this Lesbros orifice should probably be a little higher than those for the Holyoke .3 × .05 orifice, the heads being the same.

Each form with great heads will have a distinctive and constant value of c. With great heads, c will have a value of about .592 for circular forms; about .598 for square forms; and about .610 for rectangular forms with l = 6 w.

It is quite probable that the foregoing propositions will apply to orifices of any size, the feeding reservoir always being large enough to insure perfect contraction.

The Lesbros experiments, as given on Plate II., indicate that with great heads c will be about the same for all his orifices. M. Graëff states that with an iron gate or valve at the Furens dam, c is constant under great pressures, with the valve in different positions. *Vide* "Traité d Hydraulique," 1883.

In the following tables, constructed from the curves on Plate III. as a basis, are given the values of c for square and circular orifices, for areas up to 1., and with heads from .3 to 100.

The heads must be measured some distance from the orifice, whenever the vertical height of the orifice is greater than $\frac{H}{2}$. The given values of c are very nearly identical with C, when $\frac{H}{w}$ or $\frac{H}{D}$ is over 4; the co-efficient C can be readily obtained by the use of Tables III. and IV. To insure a high degree of accuracy the orifice must be carefully pierced through a rigid metallic plate.

TABLE XVII.

Values of Coefficient c for Square Vertical Orifices, with sharp Edges, full Contraction, and free Discharge in Air.

NOTE.— To obtain Coefficient C, in $Q = C a (2 g h)^{1/2}$, use Table III.

Head from Centre of Orifice h.	.02	.03	.04	.05	.07	.10	.12	.15	.20	.40	.60	.80	1.0
.3 (?)				.642	.632	.624	.617	.612					
.4			.643	.637	.628	.621	.616	.611					
.5		.648	.639	.633	.625	.619	.614	.610	.605	.601	.597		
.6	.660	.645	.636	.630	.623	.617	.613	.610	.605	.601	.598	.596	
.7	.656	.642	.633	.628	.621	.616	.612	.609	.605	.602	.599	.598	.596
.8	.652	.639	.631	.625	.620	.615	.611	.608	.605	.602	.600	.598	.597
.9	.650	.637	.629	.623	.619	.614	.610	.608	.605	.603	.601	.599	.598
1.0	.648	.636	.628	.622	.618	.613	.610	.608	.605	.603	.601	.600	.599
1.2	.644	.633	.625	.620	.616	.611	.609	.607	.605	.604	.602	.601	.600
1.4	.642	.630	.623	.618	.614	.610	.608	.606	.605	.604	.602	.601	.601
1.6	.640	.628	.621	.617	.613	.609	.607	.606	.605	.605	.603	.602	.601
1.8	.638	.627	.620	.616	.612	.609	.607	.606	.605	.605	.603	.602	.602
2.0	.637	.626	.619	.615	.612	.608	.606	.606	.605	.605	.604	.603	.602
2.5	.634	.624	.617	.613	.610	.607	.606	.606	.605	.605	.604	.603	.602
3.0	.632	.622	.616	.612	.609	.607	.606	.606	.605	.605	.604	.603	.603
3.5	.630	.621	.615	.611	.609	.607	.606	.605	.605	.605	.604	.603	.602
4.	.628	.619	.614	.610	.608	.606	.606	.605	.605	.605	.603	.603	.602
5.	.626	.617	.613	.610	.607	.606	.605	.605	.604	.604	.603	.602	.602
6.	.623	.616	.612	.609	.607	.605	.605	.605	.604	.604	.603	.602	.602
7.	.621	.615	.611	.608	.607	.605	.605	.604	.604	.604	.603	.602	.602
8.	.619	.613	.610	.608	.606	.605	.604	.604	.604	.603	.603	.602	.602
9.	.618	.612	.609	.607	.606	.604	.604	.604	.603	.603	.602	.602	.601
10.	.616	.611	.608	.606	.605	.604	.604	.603	.603	.603	.602	.602	.601
20.	.606	.605	.604	.603	.602	.602	.602	.602	.602	.601	.601	.601	.600
0. (?)	.602	.601	.601	.601	.601	.600	.600	.600	.600	.600	.599	.599	.599
100. (?)	.599	.598	.598	.598	.598	.598	.598	.598	.598	.598	.598	.598	.598

TABLE XVIII.

Values of Co-efficient c for Circular Vertical Orifices, with sharp Edges, full Contraction, and free Discharge in Air.

Note.—*To obtain Coefficient C, in Q = C a (2 g h)ʰ, use Table IV.*

Head from Centre of Orifice.	.02	.03	.04	.05	.07	.10	.12	.15	.20	.40	.60	.80	1.0
.3 (?)				.637	.628	.621	.613	.608					
.4			.637	.631	.624	.618	.612	.606					
.5		.643	.633	.627	.621	.615	.610	.605	.600	.596	.592		
.6	.655	.640	.630	.624	.618	.613	.609	.605	.601	.596	.593	.590	
.7	.651	.637	.628	.622	.616	.611	.607	.604	.601	.597	.594	.591	.590
.8	.648	.634	.626	.620	.615	.610	.606	.603	.601	.597	.594	.592	.591
.9	.646	.632	.624	.618	.613	.609	.605	.603	.601	.598	.595	.593	.591
1.0	.644	.631	.623	.617	.612	.608	.605	.603	.600	.598	.595	.593	.591
1.2	.641	.628	.620	.615	.610	.606	.604	.602	.600	.598	.596	.594	.592
1.4	.638	.625	.618	.613	.609	.605	.603	.601	.600	.599	.596	.594	.593
1.6	.636	.624	.617	.612	.608	.605	.602	.601	.600	.599	.597	.595	.594
1.8	.634	.622	.615	.611	.607	.604	.602	.601	.599	.599	.597	.595	.595
2.0	.632	.621	.614	.610	.607	.604	.601	.600	.599	.599	.597	.596	.595
2.5	.629	.619	.612	.608	.605	.603	.601	.600	.599	.599	.598	.597	.596
3.0	.627	.617	.611	.606	.604	.603	.601	.600	.599	.599	.598	.597	.597
3.5	.625	.616	.610	.606	.604	.602	.601	.600	.599	.598	.598	.597	.596
4.	.623	.614	.609	.605	.603	.602	.600	.599	.599	.598	.597	.597	.596
5.	.621	.613	.608	.605	.603	.601	.599	.599	.598	.598	.597	.596	.596
6.	.618	.611	.607	.604	.602	.600	.599	.599	.598	.598	.597	.596	.596
7.	.616	.609	.606	.603	.601	.600	.599	.599	.598	.598	.597	.596	.596
8.	.614	.608	.605	.603	.601	.600	.599	.598	.598	.597	.596	.596	.596
9.	.613	.607	.604	.602	.600	.599	.599	.598	.597	.597	.596	.596	.595
10.	.611	.606	.603	.601	.599	.598	.598	.597	.597	.597	.596	.596	.595
20.	.601	.600	.599	.598	.597	.596	.596	.596	.596	.596	.596	.595	.594
50. (?)	.596	.596	.595	.595	.594	.594	.594	.594	.594	.594	.594	.593	.593
100. (?)	.593	.593	.592	.592	.592	.592	.592	.592	.592	.592	.592	.592	.592

We have not sufficient data to warrant the construction of a table, giving the co-efficients for rectangular orifices. We conjecture that an orifice having *w* = 1., and various lengths greater than *w*, will give values of *c* not very much greater than for an orifice 1. × 1.. The experiments of Ellis, Nos. 247-255, seem to verify this conjecture. If this be true, an orifice with *w* = .01 and *l* = 6 *w*, should show a greater increase of *c*, compared with an opening .01 × .01, than does our orifice of .3 × .05 compared with the one .05 × .05. If the orifice .3 × .05 had been increased in length considerably, say to 1., the resulting co-efficients would probably only have been a trifle larger.

The following general principles must be kept in mind, when it is required to assume a co-efficient of discharge for rectangular orifices, whose least dimension is less than 0.8. *The co-efficient c increases as the least dimension decreases; this rate of increase is greatest with very low heads, and but slight with such a considerable head as 10. For a rectangular orifice c will have a larger value than for a square orifice, whose side is equal to the least dimension of the rectangle; as the ratio between the two sides of the rectangle increases, the least dimension being constant, the increase in c becomes less and less rapid.*

By keeping these principles in view, a pretty accurate guess, as to the proper value of c for rectangular orifices, can be made by a study of Plate III.

Measurement of H.

With a horizontal orifice, should H be measured from the plane of the orifice, or from the proper section of the contracted vein? We had hoped to have experimentally demonstrated this problem at Holyoke, but the investigations made there with vertical orifices consumed so much time, that we could not make the desired experiments with horizontal openings. The proper way to determine this interesting question is, to first obtain the curve for c with a given orifice placed vertically, and then to ascertain the discharge when it is placed in a horizontal position. It is most probable that H should be measured from the contracted vein.

Now if this be so, should not H for a vertical orifice be measured also from the section of the contracted vein in order to obtain accurate theoretical results? With small heads the jet begins to drop appreciably immediately after its escape from the plane of the opening. Hence were H measured from the contracted vein for such small heads, its value would be greater than when measured from the centre of the orifice, and the deduced value of c would consequently be smaller.

Shape of Escaping Vein.

A very singular phenomenon was noticed in Experiment No. 40, with the circular orifice (D = .10) with the lowest head. The jet was perceptibly flattened soon after its escape from the orifice, and then enlarged and diminished three times before the jet finally struck the floor, some 7 feet vertical below the orifice. The section of the descending jet, however, became smaller and smaller, as it was accelerated by gravity. That is to say, looking at the jet sideways, the eye being normal to the plane of the jet, the swelling in the jet was at a point where the stricture was seen by the eye in the plane of the jet. The following sketch illustrates this phenomenon; the heavy line indicating the outline of the jet from the side point of view, and the dotted lines its outline from the front point of view.

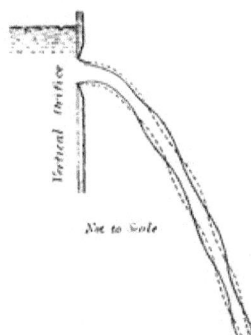

This peculiar form was doubtless caused by the velocity of the particles of water escaping from the bottom of the orifice, being much greater than that of the particles at the top of the orifice. Why the descending jet, falling freely in air, should maintain this recurring elliptical form, is an interesting problem for mathematical enquiry. Somewhat similar phenomena have been discussed by Bidone, Buff, Magnus, Savart, and lately by Lord Rayleigh in the Proceedings of the Royal Society, Vol. XXIX., p. 71, 1879. Lord Rayleigh is of the opinion that the explanation for such recurring forms given by Buff is the correct one, which is that it is caused by capillary action.

In this connection we are reminded of a singular phenomenon in long, uncovered, rectangular, inclined "shoots" for water, where the wave action is rhythmic. In one of these "shoots" at North Bloomfield, California, the length was about 1500 feet with a steep incline; the width of the "shoot" or wooden flume was about 3½ feet; the supply of water was regular, and sometimes amounted to 70 cubic feet a second. The water soon after its entrance at the head of the "shoot" assumed a wave motion, each wave being of great length and separated from the succeeding wave by quite a little interval of time. It would be an interesting philosophical study to investigate such wave movements and determine in what manner they are affected by changes in quantity, inclination, and depth. Such "shoots" are very common in the Sierra Nevada of California, and the phenomenon could there be readily investigated, under widely varying conditions.

SUBMERGED DISCHARGE.

Experiments Nos. 97 to 121, Table XII., were made with the 5 brass plates at Holyoke, placed in a vertical position. The details of these experiments are fully given in Chapter XL; the resulting values of c will be found on Plate V. The solid curved lines represent the most probable values of c, for the orifices with a free discharge; the dotted curved lines represent the most probable value of c for the submerged discharge. The values of c (for submerged discharge) for the .3 x .05 orifice are abnormally high, owing to the comparatively thick divergent sides of this orifice. For the other orifices c is also slightly too high from a similar cause.

Examining Plate V. it will be observed that the dotted curves with increasing heads drop pretty rapidly for the two smaller orifices, and much more slightly for the two larger orifices (.1 × .1 and $D = .10$). The values of c for submerged are in all cases less than those for free discharge, for these four orifices ; this difference becomes less marked as the orifices increase in size, and as the heads increase. The maximum difference between c for submerged and for free discharge is .0189 for the orifice .05 × .05, with $h = .35$; this is about 3 per cent. . The minimum difference is for the orifice .1 × .1, with $h = 3.95$, when it is .0014, or about ⅛th of 1 per cent. . The retardation of discharge caused by submerged discharge is probably measured by the wetted perimeter p, while the quantity is measured by $a\,(b)^{\frac{3}{2}}$.

We can hence generalize as follows, with reasonable safety ;

The co-efficient c for submerged orifices will be always less than for the same orifice with free discharge, h being in both cases the same, but with great heads this difference will become inappreciable. This difference diminishes as the size of the opening increases, so that for orifices as large as 1. × 1., or D = 1., probably it becomes inappreciable, except perhaps for quite small heads.

In speaking of orifices with free discharge, we mean those having a sufficient head above the summit of the opening to insure full contraction. An orifice 1. × 1., with a head of say .2 above the summit of the opening, will have an abnormally low value of c, especially if H be measured several feet back from the orifice. The same orifice submerged one or two feet, and with the same head of .7, or .7 = ½ + .2, will in all probability have a higher value of c, than that deduced from the free discharge as above.

The Ellis experiments with submerged orifices, 1. × 1. and $D = 1$. indicate that c is about 1 per cent higher than for free discharge. Weisbach states that c for submerged will be about 1½ per cent. smaller than for free discharge. Francis thinks that c for submerged is about 5 per cent. less than for free discharge. Steckel, with horizontal orifices, found that c for submerged was 4½ to 5 per cent. less than for free discharge : in his experiments, however, c for the free discharge was deduced from H measured from the plane of the orifice.

Quicksilver and Oil.

On Plate IV. will be found curves for c for the discharge, through the same circular vertical orifice having a diameter of .02, of thick oil, water and quicksilver. These experiments—Nos. 152-155, 122-130, and 149-151—are fully described in Chapter XI.

The experimental data for the water and quicksilver are sufficient to establish the curves for c, with h from .5 to 3. For the oil only one value of c was determined, being that for a head dropping from 3.2 to .58 in a cylindrical reservoir. Nos. 122 to 124, were made with constant heads for water, and are much more reliable than Nos. 125 to 130, which were made with "dropping" heads. It will be observed that Nos. 126 and 129 for the "dropping" heads, give slightly too low values for c, compared with the curve deduced from the constant heads.

The form of the curve for the quicksilver does not differ greatly from that for the water. This establishes the interesting fact that quicksilver in its flow from such an orifice follows the same law as water; i.e., c diminishes as the head increases, and the curve becomes asymptotic with large heads. With great heads, such as 100 and over, it is probable that quicksilver for circular orifices may have a practically constant value for c of about .580 or .590. It is possible, however, that with such great heads quicksilver may have the same co-efficient as water, which for circular orifices we have assumed would be about .592.

Comparing the three liquids, it will be seen that the flow of the most viscous liquid was much the greatest, and that of the least viscous the least. Hence we can say for ordinary heads, such as 4 feet and less, that *the more viscous a liquid the greater will be the value of c.* Sufficient liquidity of course must be maintained, in order that the flow shall be uniform of any very viscous liquid. It would be very interesting to determine the co-efficient of discharge of thick oil from an orifice with a head of 100 feet or over. It is not at all improbable, that with such a head it would be about the same as for water. If this should prove to be the case, the proposition might be warranted that with great heads, and the same form of orifice, all liquids have the same value of c, disregarding the retarding effect of the air, which would be notable with very light liquids.

The very large value of c found by us for the oil, tends to show that, especially with low heads, if water has much oily matter in it, thus increasing its viscosity, the quantity of flow may be somewhat changed from what would be discharged with purer water.

Weisbach found that quicksilver had a lower value of c than water, which agrees with our results. For rape-seed oil he found c slightly higher than for the quicksilver, and considerably lower than for the water.

In our experiments—Nos. 152 to 155—the only uncertainty as to reasonable accuracy, was caused by the jet of oil in places adhering more or less to the divergent sides of the orifice. The jet, however, at no time filled the divergent "tube" formed by the sides of the orifice, and it is not likely that the dripping along the lower side added largely to the flow, and very possibly it did not increase the flow at all.

In concluding this discussion of the flow of liquids through a thin wall it may be remarked, that in our judgment, Weisbach and others have placed the loss of energy in jets escaping from a thin wall at too high a figure. With large heads, such as 100 feet and over, we conceive that the loss of energy is inappreciable. Some experiments made by us, with a head of about 300 feet, showed that a jet escaping from a ring, with nearly full contraction, was slightly more effective than a jet from a converging mouth-piece, where c was practically unity.

Authors, when discussing the flow through orifices in a thin wall, very frequently speak of losses by "friction"—this seems to us an incorrect phrase. The inner edges of a perfect orifice form a geometrical line, on which there can be no friction proper. With low heads there may be losses by cohesion of the liquid particles, and by eddying movements, but these losses can hardly be large. The outgoing water feeds the escaping vein in a uniform manner, so that there can be no considerable losses by eddies. This is in part indicated by our experiments Nos. 48 and 49, where an "irregular" supply, which distorted and greatly twisted the escaping jet, had no appreciable effect upon the co-efficient of discharge.

It seems clear to us that the loss of energy for a square or circular orifice must diminish as h or u increases. For the loss will be approximately measured by ρ, while the quantity is measured by $u\,(h)^8$.

The law of discharge is just the reverse, for c is greatest with the smallest orifices and the least heads. This is perhaps due to the form of contraction being modified by the narrow limits of the sides, and which modification disappears with the great velocity due to high heads.

The diminished discharge of circular orifices, compared with square ones, leads to the conclusion that the increased discharge results from the peculiar form of contraction at the four corners of the square opening. But if this be so, one would suppose that in a rectangular opening, having its least dimension equal to the side of a square, the value of c would be less than that for the square, as the influence at the corners must be proportionately less; we have seen that just the reverse of this is true.

Contraction Modified.

The effect of partial or complete suppression of contraction on the sides of an orifice, by bringing the side or sides of the feeding canal so close to the respective side or sides of the orifice as to interfere with contraction, can be determined pretty fairly by the Lesbros experiments, with the forms of approach shown by Figs. 4-12, Plate I. The results of these experiments are given in Table IX.

In the following table are given the values of c, with $h = 1., 2., 3.,$ and $5.,$ for the forms of approach shown by Figs. 1, and 4 to 12 inclusive; these values have been obtained by interpolation from the given values of c in Table IX., and from Plate II. The orifice was square, in a fixed plate, the side being .656.

TABLE XIX.

Lesbros.— Values of c for various Heads, with a Vertical Orifice .656 × .656. Forms of Approach variously modified, as shown by Figures on Plate I. Free Discharge.

Figure	Distance to Sides of Canal, from Respective Sides of Orifice.			Heads, in Feet.				Remarks.
	G	L	L'	1.0	2.0	3.0	5.0	
1	1.8	5.7	5.7	.601	.604	.605	.603	Full contraction.
4	0	5.7	5.7	.624	.624	.624	.624	Suppressed on bottom.
5	0	.066	5.7	.636	.636	.637	.637	,, ,, ,, and nearly on one side.
6	0	.066	.066	.681	.670	.665	.662	,, ,, ,, ,, ,, both sides.
7	0	0	0	.695	.685	.677	.673	Suppressed on both sides and bottom.
8	1.8	.066	5.7	.608	.610	.612	.611	Nearly suppressed on one side.
9	1.8	.066	.066	.635	.631	.629	.627	,, ,, both sides.
10	1.8	0	0	.645	.640	.639	.637	Suppressed on both sides.
11	0	.066	.066	.647	.645	.644	.641	
12	1.8	.066	.066	.609	.611	.612	.611	

For Figs. 11 and 12 the side approaches were inclined at an angle of 45°; vide Plate I.

As before remarked, in some of the forms of approach—notably Fig. 7—there was a considerable loss of head, due chiefly to primary contraction as the water entered the mouth of the feeding canal, the head H having been measured in the still water above the canal; hence the foregoing results do not sufficiently indicate the increased flow caused by suppression of contraction. The head producing the velocity of approach in the feeding canal is, of course, included in the measured head H.

By reference to Table XIX., it will be seen that as suppression becomes more and more complete, the value of c increases. We can show this clearly by the following statement of the percentage of increase above c for full contraction, with different degrees of suppression for the 4 heads.

Fig.	Amount of Suppression.	$H=1$	$H=2$	$H=3$	$H=5$
8	Contraction nearly suppressed on one side	1.2	1.0	1.2	1.3
4	,, suppressed on one side	3.8	3.3	3.1	3.5
9	,, nearly suppressed on two sides	5.7	4.5	4.0	4.0
5	,, suppressed on one side and nearly suppressed on another side	5.8	5.3	5.3	5.6
10	,, ,, two opposite sides	7.3	6.0	5.6	5.6
6	,, ,, one side and nearly suppressed on two other sides	13.3	10.9	9.9	9.8
7	,, ,, three sides	15.6	13.4	11.9	11.6

These results are fairly uniform, showing a constant increase in c as suppression of the wetted perimeter increases, and also that as the heads increase the effect of suppression diminishes. This latter result, especially for Fig. 7, is in part produced by the loss of head spoken of.

The side approaches, placed at an angle of 45°, show a discharge considerably less than when the sides are normal to the orifice, as will be seen by comparing the results in Table XIX.; those of Fig. 11 with Fig. 6, and those of Fig. 12 with Fig. 9.

For the Lesbros orifices with $l = .656$, and $w = .328$ and less, we can compactly show the effect of four forms of approach in the following statement, which gives the percentage of increased discharge caused by suppression for each orifice.[*] It is based upon the results given in Tables VIII. and IX.

Orifice		Fig. 4.			Fig. 9.			Fig. 5.			Fig. 6.		
		Heads.			Heads.			Heads.			Heads.		
l	w	1.	3.	5.	1.	3.	5.	1.	3.	5.	1.	3.	5.
.656	.656	3.8	3.1	3.5	3.7	4.0	4.0	5.8	5.3	5.6	13.3	9.9	9.8
	.328	5.4	5.2	5.4	3.2	2.4	3.1	7.0	6.7	7.0	10.7	9.8	10.0
	.164	6.3	6.5	7.4	1.6	1.4	2.4	7.3	7.5	8.4	8.9	8.6	9.5
	.098	7.8	7.6	8.5	3.4	2.1	2.4	8.1	8.8	9.2	9.6	9.2	9.5
	.033	8.9	11.1	12.4	4.5	5.1	5.6	8.5	11.1	12.2	8.5	11.1	12.7

The above percentages show fairly reasonable results for the form of approach, Fig. 4; they constantly increase as w diminishes, because the ratio of suppression to the wetted perimeter, p, constantly increases; for $w = .656$, calling the suppressed line S, we have $\frac{S}{p} = .25$, while for $w = .033$, $\frac{S}{p} = .48$. The results for Fig. 9 do not seem to be so satisfactory; in this form of approach there was full bottom contraction, and L and L' were each .066; therefore as w diminishes it is almost certain that the percentages should decrease; especially for the smallest value of w, .033, it seems to us that there must have been nearly perfect contraction, and hence for this width the percentage should have been barely appreciable.

As we have before observed, these Lesbros experiments with w less than .656, where the sliding gate was employed, are probably much less exact than those made with the orifice .656 × .656.

---

[*] Let e be the co-efficient of discharge, with full contraction, for any one of the three heads, and for any one of the five orifices; taken from Table VIII. Let e' be the co-ef. for the same head and orifice, and for any one of the four forms of approach; taken from Table IX. Then $\frac{e'-e}{e}$ will be the percentage of increase as given in the statement.

Bidone's experiments showed[*] that the increased discharge by partial suppression of contraction of rectangular orifices, was nearly in proportion to 15.2 per cent. for complete suppression, or $N = p$. This for square orifices would give ;

Contraction suppressed on one side, 3.8 per cent. increase.
" " " two sides, 7.6 " "
" " " three " 11.4 " "

These determinations for one and two sides agree closely with Lesbros' results, with $h = 1.$; for three sides Bidone's percentage is less than that of Lesbros.

With circular orifices, Bidone found that partial suppression up to $\frac{7}{8}$ths of the perimeter, increased the discharge in the ratio of 12.8 per cent. for complete suppression.

If contraction be suppressed completely, the approach becomes a tube ; this form will be hereafter discussed in the Chapter on Pipes.

With convergent mouth-pieces, the co-efficient c will be about .95 for small heads, when the mouth-piece is formed with care. For very high heads c practically becomes unity, even when the mouth-piece has straight converging sides, with the angle β rather small. It also seems established by our experiments (Nos. 15-18), that with great heads the channel of approach need not be as large in proportion to the orifice, in order to produce complete contraction, as with small heads. The results of Lesbros in general confirm this proposition.

Experiments Nos. 89 to 96, Chapter XI., were made by bisecting an orifice with $l = .30$ and $e = .05$, by brass vertical sheets of various thicknesses. A very thin sheet (Nos. 89-92) had very little effect either upon the co-efficient c, or upon the form of the escaping jet. A sheet .04 thick, which reduced the area from .015 to .013, increased the value of c nearly 1 per cent.. These experiments (Nos. 94-96) illustrate the effect of an incomplete suppression of contraction, as the escaping jets united at a short distance from the plane of the orifice.

* Vide D'Aubuisson's " Traité d'Hydraulique."

CHAPTER IV.

VELOCITY OF APPROACH.

WHEN, for either an orifice or a weir, the feeding stream passes with an appreciable current the point where the surface height of the water is measured, then an additional force due to this velocity of approach presents itself for consideration. This additional force will evidently increase the discharge.

In all the experiments we have selected of the discharge through orifices, the velocity of approach was so inconsiderable that it could be neglected without sensible error, and in general with orifices the feeding canals are made so large in proportion to the quantity discharged, that v_a—the mean velocity in the feeding canal as it passes the measuring point for the head—is an inconsiderable factor. With weirs, however, and especially with those having end contractions suppressed, v_a is often a considerable quantity, and in extreme cases may largely increase the discharge. Hence, before discussing the flow over weirs, it becomes necessary to determine what effect v_a produces.

The head, h_a, required to produce v_a is approximately represented in $v_a = (2 g h_a)^{\frac{1}{2}}$, or $h_a = \dfrac{v_a^2}{2 g}$.

This expression is not strictly accurate, as the threads of water in the cross section of the canal at the point for H, have widely varying velocities. If we could divide this section into a number, n, of minute equal sections, and determine the velocities $v_1, v_2,$ et cet., of the several sections, the sum of the squares of the several velocities, divided by n, would much more nearly give the *true* value of v_a^2. This result would unquestionably somewhat differ from the square of the mean velocity. It is sometimes assumed that the true velocity head is necessarily greater than $\dfrac{v_a^2}{2 g}$.

The question to be solved is, what portion of the kinetic energy, due to the head h_a for the section a_e, produces a useful effect at the opening of discharge?

As to this point there has been a great diversity of opinion among hydraulicians and mathematicians. Some have thought that the entire kinetic energy of the water, as it passes the measuring point for H, should be considered in computing the effective head, h. Others substantially make $h = H + h_a$, thus assuming that the additional force is that due to the velocity of a section of the feeding stream, having the same area as the opening, the velocity of this section being taken as r_a. This subject has engaged the attention of many great minds during the past two centuries, and the discussions and controversies in regard to it, if collected, would fill several good-sized volumes. If the same amount of thought and time, devoted to these almost fruitless discussions, had been employed in ascertaining what the *facts* really were, we would know a good deal more about the effect of velocity of approach than we do at present.

EXPERIMENTAL DATA.

Lesbros.

Lesbros made $h = H + 1.56 \; h_a = H + \dfrac{(1.25 \; v_a)^2}{2 \, g}$, in a number of his experiments, where the crest of the weir was placed at various heights above the bottom of the feeding canal. The quantity $1.25 \; v_a$ was assumed by him to be the central surface current in the canal.

These experiments of Lesbros have, however, but little value, as the problem was greatly complicated by a primary velocity as the feeding water escaped from an orifice with considerable pressure, into a feeding canal of comparatively short length.

Fteley and Stearns.

Messrs. Fteley and Stearns have lately made a large number of experiments for the purpose of determining the effect of velocity of approach. A full description of these investigations has been published in the Transactions of the Am. Soc. of C.E., pp. 1-118, 1883. From their results we will, in a great measure, draw our final conclusions.

The experiments made by these gentlemen bear every evidence of great care, skill, and honesty. The data they have given illustrating the effect of velocity of approach, and their determinations of the quantity of flow over weirs of various kinds, constitute one of the most valuable contributions to experimental Hydraulics, which has been given to the world in late years.

They placed a horizontal sharp-crested weir, 5 ft. long, with vertical sides, at the end of a horizontal rectangular open flume or canal, having the same width as the length of the weir; the end contractions were hence entirely suppressed. The plane of the weir was normal to the axial line of the canal. This canal had a false bottom, moving vertically, but always retaining its horizontal position, extending from the weir a distance of 17 feet up-stream; this false bottom could be placed at will at the respective distances of .5, 1., 1.7, 2.6, and 3.56 feet below the crest of the weir.

A constant volume of water was admitted to the upper end of the canal, and the surface height above the crest was measured, as the section of the feeding stream was changed by lowering or elevating the false bottom of the canal, thus forming one series of experiments. The supply of water was then increased, and another series of surface heights measured. There were 21 series of experiments thus made, Q being constant for each series, with H ranging from .1930 to .9443, when there was the maximum inner depth, G, of 3.56 below the crest.

With the greatest value of Q;

$$H, \text{ with } G = 3.56, \text{ was } .9443$$
$$H, \quad \text{„} \quad \text{„} = .50, \quad \text{„} \quad .8266$$

As the false bottom was raised for each series, the section of the feeding stream was diminished, and consequently c_a increased. Hence by comparing the varying values of H—observed head above crest—with the varying values of h_a, deduced from c_a, it was thought that the effective value of h_a, or h_a', could be determined.

The measuring hook-gauge, reading to .0001, determined the surface elevation of the water above the crest, at a point on the side of the canal 6 feet above the weir.

The discharge was free into the air, except that the sides of the canal were continued beyond the crest; this projection, however, not extending lower than the crest. There were 94 experiments made with end contraction suppressed.

In such experiments it seems apparent to us that there are three causes to be taken into consideration, which produce the changes in H, with varying values of G, Q remaining constant. They are;

FIRST.—That force due to c_a, which increases v,[*] and hence lowers H.

SECOND.—The partial suppression of contraction, caused by bringing the bottom of the canal nearer the crest, which changes the shape of the escaping contracted vein. and increases H.

THIRD.—As the section of the feeding stream diminishes, there results an increased amount of loss of head between the measuring point for H and the crest, due to increased resistance between these points; this decreases v, and hence elevates H. This is represented to some extent by the diminished value of v in the canal, $v = \dfrac{Q}{l'}$, and is further affected by the increase in c_a; an approximate determination of the amount of this increased resistance presents a most complicated problem.

We have seen with orifices, that the effect of suppression by straight approaching sides at right angles to the plane of the orifice, adds as much as 15.6 per cent. to the discharge, when suppression extends to three sides of a square orifice; and with contraction partially suppressed on two opposite sides—orifice .656 square, and channel of

---

[*] It must be kept in mind that v is the velocity in the plane of the weir, having a section with the length, l, of the weir, and the height H or h, depending upon our conception of the effect of h_a; hence $v = \dfrac{Q}{l H}$, or $v = \dfrac{Q}{l h}$. We will in no case regard v as the actual velocity in the plane of the weir, which would be expressed by $\dfrac{Q}{l H_s}$.

approach .787 wide—the discharge was increased $5\frac{1}{2}$ per cent.; *vide* Table XIX. *et seq.*, the normal discharge being with full contraction. It is apparent that with weirs, suppression or partial suppression of contraction will produce similar results; this will be directly shown hereafter, in our analysis of weir experiments.

The third cause is, with the 94 experiments now under consideration, of minor importance compared with the first and second causes, and its separate consideration can be omitted without producing serious errors. In certain cases, however, this increased loss of head, probably chiefly produced by adhesion of the water to the three sides of the feeding canal, may produce a greater effect in elevating H, than the lowering effect upon H of complete suppression on both sides and the bottom of the weir; this will be pointed out in our future discussion of some of Lesbros' weir determinations.

We will now proceed to analyze the experimental results in these 21 series of Fteley and Stearns, with the view to determine as nearly as may be possible, the effect of the first cause, and the combined effects of the last two causes, in increasing the velocity v.

We will first assume that as H increases, the inner depth below the crest remaining constant, the percentage of increased velocity caused by partial suppression will increase; that is to say, with a constant inner depth of .5 below the crest, partial suppression will result in increasing the value of c—co-efficient of discharge—more and more as H is increased. This assumption has already been shown to be true for orifices, and will be hereafter dwelt upon in discussing the effect of partial suppression upon the discharge over weirs.

We will also assume, that with the maximum inner depth of 3.56 below the crest, the effect of partial suppression is barely sensible, even with the maximum value of .944 for H. We have seen with an orifice .656 square, that the discharge was not affected sensibly when the sides of the feeding canal were placed 1.77 from the sides of the orifice; hence it seems safe to suppose, that with a weir having an inner depth 4 times greater than the depth upon the weir, the effect of partial suppression will be very slight. This question will be more elaborately discussed in the following Chapter on Weirs.

Messrs. Fteley and Stearns, in their reduction of these experiments, propose a variable factor b, in $h = H + b\frac{v_o^2}{2g}$; b having a range from 1.33 to 1.87, and representing the *entire* effect of variation in G upon H.

After several trials we found that by giving b a constant value of $1\frac{1}{3}$, we obtained results fairly agreeing with our conception of the problem. That is to say, we assume that $\frac{4}{3}\frac{v_o^2}{2g}$, or $\frac{(1.1547\,v_o)^2}{2g}$, approximately represents the additional head and area of discharge, due to the velocity of approach; hence $Q = c \frac{2}{3}(2g)^{1/2}\left(H + \frac{(1.1547\,v_o)^2}{2g}\right)^{3/2} l$

$$\left(H + \frac{(1.1547 \; v_a)^2}{2 \, g}\right) = c \, \tfrac{2}{3} \, (2 \, g)^{\frac{1}{2}} \, h^{\frac{3}{2}} \, l.$$ For, it is obvious in these experiments, that an increase in v_a diminished both the observed head and the observed area.

The following table has been constructed upon this basis, showing the results from the 94 experiments. The various columns represent;

1st Column.—The number of series and experiment.

2nd Column.—d, or the inner depth from the crest of the weir to the bottom of the canal.

3rd Column.—H, or measured head above crest of weir, at a point six feet up-stream from weir.

4th Column.—$h_a = \frac{v_a^2}{2 \, g}$. This has been taken from Fteley and Stearns, and was determined by them with approximate accuracy, by computing Q for each series by their formula for weir discharge ; for each experiment, $v_a = \frac{Q}{(d + H) \, l}$.

5th Column.—$h_a' = h_a b$, b having a constant value of $1\frac{1}{3}$.

6th Column.—The assumed value of h, or $H + h_a'$, being the effective head, taking only into consideration the effect of velocity of approach. The figures in small type at the beginning of each series, give the assumed value of h for that series, with the effect of partial suppression entirely eliminated.

7th Column.—The percentage of increased discharge, or more strictly the percentage of increased value of the co-efficient of discharge, c, assumed to be due to our *second* and *third* causes combined. This computation is based upon the principle that the co-efficient of discharge will vary inversely as $h^{\frac{1}{2}}$; the percentage is hence obtained by comparing the $\frac{3}{2}$ power of h at the head of each series, with the same power of the succeeding values of h—column 6—for that particular series.

Note our remarks in last section of this chapter, in regard to the inaccurate conception embodied in formula here used—$Q = c \, \tfrac{2}{3} \, (2 \, g)^{\frac{1}{2}} \, h^{\frac{3}{2}} \, l.$

TABLE XX.

Fteley and Stearns.—Weir Experiments. Showing Effect of increasing v_a, also attended with Partial Suppression of Bottom Contraction.

Weir with End Contractions suppressed.

No.	G	H	k_a	k_a'	h	Per cent.	No.	G	H	k_a	k_a'	h	Per cent.
I.					.1931		**VI.**					.4277	
1	3.56	.1930	.0001	.0001	.1931	0	25	3.56	.4263	.0009	.0012	.4275	0.1
2	1.70	.1924	.0004	.0005	.1929	0.2	26	2.60	.4257	.0015	.0020	.4277	0
3	1.00	.1913	.0009	.0012	.1925	0.5	27	1.70	.4230	.0030	.0040	.4270	0.2
4	.50	.1884	.0027	.0036	.1920	0.9	28	1.00	.4154	.0068	.0091	.4245	1.1
II.					.2690		29*	.50	.3997	.0169	.0225	.4222	1.9
5	3.56	.2685	.0002	.0003	.2688	0.1	**VII.**					.4952	
6	2.60	.2685	.0004	.0005	.2690	0	30	3.56	.4933	.0013	.0017	.4950	0.1
7	1.70	.2676	.0009	.0012	.2688	0.1	31	2.60	.4923	.0022	.0029	.4952	0
8	1.00	.2649	.0022	.0029	.2678	0.7	32	1.70	.4886	.0044	.0059	.4945	0.2
9	.50	.2595	.0060	.0080	.2675	0.8	33	1.00	.4784	.0097	.0129	.4913	1.2
III.					.3366		34	.50	.4597	.0230	.0307	.4904	1.5
10	3.56	.3361	.0004	.0005	.3366	0.1	**VIII.**					.5168	
11	2.60	.3358	.0008	.0011	.3369	0	35	3.56	.5148	.0015	.0020	.5168	0
12	1.70	.3341	.0016	.0021	.3362	0.3	36	2.60	.5132	.0025	.0033	.5165	0.1
13	1.00	.3308	.0038	.0051	.3359	0.4	37	1.70	.5088	.0050	.0067	.5155	0.4
14*	.50	.3188	.0100	.0133	.3321	2.1	38	1.00	.4965	.0108	.0144	.5109	1.7
IV.					.3578		39*	.50	.4706	.0256	.0341	.5047	3.5
15	3.56	.3570	.0005	.0007	.3577	0	**IX.**					.5647	
16	2.60	.3561	.0009	.0012	.3573	0.2	40	3.56	.5620	.0018	.0024	.5644	0.1
17	1.70	.3549	.0019	.0025	.3574	0.2	41	2.60	.5596	.0031	.0041	.5637	0.3
18	1.00	.3495	.0044	.0059	.3554	1	42	1.70	.5550	.0062	.0083	.5633	0.4
19	.50	.3404	.0114	.0152	.3556	0.9	43	1.00	.5409	.0132	.0176	.5585	1.6
V.					.4232		44	.50	.5187	.0302	.0403	.5590	1.5
20	3.56	.4219	.0008	.0011	.4230	0.1	**X.**					.6026	
21	2.60	.4205	.0015	.0020	.4225	0.2	45	3.56	.5994	.0022	.0029	.6023	0.1
22	1.70	.4183	.0030	.0040	.4223	0.3	46	.50	.5509	.0345	.0460	.5969	1.4
23	1.00	.4106	.0067	.0089	.4195	1.3							
24	.50	.3976	.0165	.0220	.4196	1.3							

L

TABLE XX.—*continued*

1	2	3	4	5	6	7	1	2	3	4	5	6	7
No.	G	H	h_a	h_a'	h	Per cent.	No.	G	H	h_a	h_a'	h	Per cent.
XI.					.6177		**XVI.**					.7922	
47	3.56	.6143	.0024	.0032	.6175	0	72	3.56	.7859	.0046	.0061	.7920	0
48	2.60	.6120	.0040	.0053	.6173	0.1	73	1.70	.7695	.0142	.0189	.7884	0.7
49	1.70	.6059	.0077	.0103	.6162	0.4							
50	1.00	.5904	.0162	.0216	.6120	1.4	**XVII.**					.8394	
51*	.50	.5685	.0360	.0480	.6165	0.3	74	3.56	.8323	.0053	.0071	.8394	0
							75	2.60	.8278	.0088	.0117	.8395	0
XII.					.6783		76	1.70	.8147	.0163	.0217	.8364	0.6
52	3.56	.6741	.0030	.0040	.6781	0	77	1.00	.7873	.0322	.0429	.8304	1.6
53	2.60	.6707	.0051	.0068	.6775	0.2	78	.50	.7400	.0667	.0889	.8298	1.7
54	1.70	.6629	.0097	.0129	.6758	0.6							
55	1.00	.6443	.0201	.0268	.6711	1.6	**XVIII.**					.8732	
56*	.50	.6179	.0435	.0580	.6759	0.3	79	3.56	.8651	.0059	.0079	.8730	0
							80	2.60	.8595	.0097	.0129	.8724	0.1
XIII.					.7016		81	1.70	.8467	.0178	.0237	.8704	0.3
57	3.56	.6971	.0033	.0044	.7015	0.1							
58	2.60	.6917	.0056	.0075	.6992	0.6	**XIX.**					.8892	
59	1.70	.6836	.0106	.0141	.6977	0.9	82	3.56	.8809	.0062	.0083	.8892	0
60	1.00	.6658	.0217	.0289	.6947	1.5	83	2.60	.8733	.0101	.0135	.8899	0.1
61*	.50	.6364	.0466	.0621	.6985	0.7	84	1.70	.8593	.0187	.0249	.8842	0.3
							85	1.00	.8293	.0366	.0488	.8781	1.3
XIV.					.7203		86	.50	.7765	.0732	.1003	.8768	2.1
62	3.56	.7153	.0036	.0048	.7201	0							
63	2.60	.7117	.0059	.0079	.7196	0.1	**XX.**					.9384	
64	1.70	.7018	.0113	.0151	.7169	0.7	87	3.56	.9238	.0070	.0093	.9331	0
65	1.00	.6806	.0230	.0307	.7113	1.9	88	2.60	.9181	.0114	.0152	.9333	0
66*	.50	.6412	.0499	.0665	.7077	2.6	89	1.70	.9011	.0209	.0279	.9290	0.7
							90	1.00	.8668	.0406	.0541	.9209	2.0
XV.					.7815								
67	3.56	.7755	.0044	.0059	.7812	0.1	**XXI.**					.9547	
68	2.60	.7718	.0073	.0097	.7815	0	91	3.56	.9443	.0075	.0100	.9543	0.1
69	1.70	.7606	.0137	.0183	.7789	0.5	92	1.70	.9215	.0220	.0293	.9508	0.6
70	1.00	.7376	.0275	.0367	.7743	1.4	93	1.00	.8854	.0426	.0568	.9422	2.0
71	.50	.6922	.0583	.0777	.7699	2.2	94	.50	.8366	.0860	.1147	.9413	2.1

The maximum velocity of approach was in Experiment No. 94, when it was 2.35 ; in this experiment h_a' is not quite one seventh of H.

The final results given in the preceding table, can be compactly stated as follows ;

TABLE XXI.

Fteley and Stearns.— Weir Experiments. Effect of Partial Suppression of Bottom Contraction, less the Loss by Additional Friction and Adhesion between Measuring Point for H and the Crest.
Showing Percentage of Increase of c, due to above Causes combined.

Series	Maximum Value of H	G = hence Depth below Crest.				
		3.86	2.60	1.70	1.00	.50
I.	.193	0		0.2	0.5	0.9
II.	.268	0.1	0	0.1	0.7	0.8
III.	.336	0.1	0	0.3	0.4	2.1*
IV.	.357	0	0.2	0.2	1.0	0.9
V.	.422	0.1	0.2	0.3	1.3	1.3
VI.	.426	0.1	0	0.2	1.1	1.9*
VII.	.493	0.1	0	0.2	1.2	1.5
VIII.	.515	0	0.1	0.4	1.7	3.5*
IX.	.562	0.1	0.3	0.4	1.6	1.5
X.	.600	0.1				1.4
XI.	.614	0	0.1	0.4	1.4	0.3*
XII.	.674	0	0.2	0.6	1.6	0.5*
XIII.	.697	0.1	0.6	0.9	1.5	0.7*
XIV.	.715	0	0.1	0.7	1.9	2.6*
XV.	.775	0.1	0	0.5	1.4	2.2
XVI.	.786	0		0.7		
XVII.	.832	0	0	0.6	1.6	1.7
XVIII.	.865	0	0.1	0.5		
XIX.	.881	0	0.1	0.9	1.9	2.1
XX.	.924	0	0	0.7	2.0	
XXI.	.944	0.1		0.6	2.0	2.1

An asterisk is attached to those experiments, in the preceding tables, which the authors state were imperfectly made, and hence more or less unreliable.

An examination of Table XXI. will show that the percentage of increase in the co-efficient c is fairly uniform for the several series—disregarding the imperfect experiments. With the lower values of Q, and hence lower values of H, the assumed effect of our second and third causes combined, is less than 1 per cent., with the minimum inner depth of .5 below the crest ; with the larger values of H, this effect increases to a little more than 2 per cent.. These percentages of increase in c agree reasonably well with the experiments of Lesbros with orifices, and with the conclusions stated hereafter, in regard to the increased discharge over weirs with partial bottom suppression, compared with full bottom contraction, h being constant in both cases.

Especially with the lower values of H, it must be kept in mind, that we are dealing with very slight differences in the computation of the percentage of increase in c, and therefore cannot expect much greater uniformity—taking into consideration the element of experimental error—than the given results indicate. Somewhat more satisfactory results could have been obtained by making the factor b a variable, having a value of 1.36 for Series I., and gradually decreasing to 1.22 or 1.20 for Series XXI. In fact, from theoretical considerations, it seems to us that b must be a variable, diminishing with $\frac{a_2}{a}$. The given percentages of increased discharge due to partial suppression, for Nos. 78, 86, and 94 appear to be about 1 per cent. too low; for these experiments $\frac{a_2}{a}$ was less than for any of the others.

A careful study of these tables cannot result otherwise, than in impressing us with the definite belief, that in any event $H + h_a$ gives too small a value for h, and that $H + (1\frac{1}{2} \times h_a)$ can be assumed as the value of h in our equation, without very serious error, for weirs with end contractions suppressed.

If we assume that $h = H + h_a$, in $Q = c \frac{2}{3} (2g)^{\frac{1}{2}} h^{\frac{1}{2}} l$, we would have the following percentages of increase in the value of c, to attribute to our second and third causes combined, the inner depth, or G, being constant at .5.

Series	I.			1.5 per cent. with $H = .2$
„	II.			1.9 „ „ $H = .3$
„	V.			3.1 „ „ $H = .4$
„	X.			4.1 „ „ $H = .6$
„	XV.			5.6 „ „ $H = .8$
„	XIX.			6. „ „ $H = .9$
„	XXI.	...	...	6.2 „ „ $H = .9$

It is to say the least most improbable that such an increase in c as is here indicated could be caused by partial suppression; this is especially true for Series I. and II., where H was only about one-half as much as G.

These last computations indicate the general accuracy of our first assumption, that the percentage of increase due to partial suppression, G remaining constant, increases with H. In fact with any other hypothesis, a satisfactory analysis of these 94 experiments would not be possible.

Experiments were also made with the same canal and false bottom, with weirs having the respective lengths of 3, 3.3 and 4 feet, the end contractions being more or less suppressed. These experiments indicate that b should increase as the length of the weir diminishes, the width of the canal remaining constant. We will assume 1.4 as the constant value of b for the shortest weir, with smaller constant values as the length increases.

TABLE XXII.

Fteley and Stearns.—Weir Experiments. Showing effect of increasing e_a, also attended with Partial Suppression of Bottom Contraction.

Weirs of various Lengths ; End Contractions more or less suppressed.

No.	Width of Canal, 5 Feet			G	H	k_a	b	k'_a	h	Percentage of Increase in e_a	b (?)
	L	l	L'								
XXII.									.8732		
95				3.56	.8702	.0020		.0028	.8730	0.1	2.14
96				1.7	.8612	.0056		.0081	.8693	0.7	2.26
97	1.	3.	1.	1.	.8490	.0112	1.4	.0157	.8647	1.5	2.27
98				.5	.8310	.0216		.0309	.8612	2.1	2.00
XXIII.									.8778		
99				3.50	.5765	.0008		.0011	.5776	0.1	2.02
100				1.	.5647	.0057		.0079	.5726	1.4	2.35
101				.5	.5574	.0126		.0175	.5749	0.8	1.65
XXIV.									.8083		
102				3.56	.8062	.0020		.0038	.8090	0.1	2.06
103				2.6	.8036	.0033		.0046	.8082	0.2	1.99
104	0	3.3	1.7	1.7	.7969	.0061	1.39	.0085	.8054	0.7	2.16
105				1.	.7850	.0120		.0167	.8017	1.4	2.09
106				.5	.7677	.0239		.0332	.8009	1.6	1.78
XXV.									.9350		
107				3.50	.9307	.0029		.0040	.9347	0.1	1.99
108				1.	.9024	.0163		.0225	.9249	1.6	2.10
109				.5	.8775	.0309		.0430	.9205	2.4	1.91
XXVI.									.7080		
110	0	4.	1.	3.56	.7048	.0021	1.37	.0029	.7077	0.1	1.81
111				.5	.6639	.0283		.0388	.7027	1.1	1.58

The last column in the table is the value of b, in the equation $Q = c \frac{2}{3} (2g)^{\frac{1}{2}} \left(H + b \frac{v_a^2}{2g}\right)^{\frac{3}{2}} l \left(H + b \frac{v_a^2}{2g}\right)$, the entire variation in H being attributed to velocity of approach. The figures in small type in this column are not the result of experiment, but were assumed by Messrs. Fteley and Stearns, in order to obtain the value of H, with effect of velocity of approach entirely eliminated.

Examining the preceding table, we see results which generally agree with our conception of the causes of the changes in H. Experiments Nos. 100 and 101, however, appear to be flatly contradictory.   More harmonious results could have been obtained by making *b* a variable increasing with *H*, as was also the case for the weir with end contractions suppressed.

From these experiments we can fairly assume, that for weirs having full contraction, *b* can be given a constant value of 1.4, without danger of serious error in the resulting value of *c*.   Hence for such weirs,

$$Q = c\,\tfrac{2}{3}\,(2\,g)^{\frac{1}{2}}\left(H + \frac{(1.183\,v_a)^2}{2\,g}\right)^{\frac{3}{2}} l\left(H + \frac{(1.183\,v_a)^2}{2\,g}\right).$$

Messrs. Fteley and Stearns propose a table of values of *b*, which will doubtless give good results when applied to weirs having the same forms of approach as those employed in their experiments.   But if the form of approach be changed, then appreciable errors may result from using their co-efficients of correction for $\frac{v_a^2}{2\,g}$.   This will inevitably be the case in all formulæ, such as those of Weisbach, et cet., where a distinct conception of the *separate* effects of velocity of approach and suppression of contraction is not kept in view.

J. B. Francis.

Mr. Francis has made a few experiments, from which the effect of v_a in increasing the discharge can be deduced.   They are to be found in his " Lowell Hydraulic Experiments," and will be given in detail in our next chapter.

The experiments embraced in our Weir Nos. 49, 52. and 54 (Francis' Nos. 11-33, 56-61, and 72-78) were with a weir having nearly complete contraction, the inner depth below crest being 4.6, the width of the canal being 13.96, and the length of the weir being nearly 10.00.   The experiments embraced in our Weir Nos. 51, 53, and 55 (Francis' Nos. 36-43, 62-66, and 79-84) were with the same weir, the inner depth, *G*, being 2.014, and the end contractions remaining unchanged.

Mr. Francis adopted, as a correction for velocity of approach, the following formula, $h = \left[\left(H + \frac{v_a^2}{2\,g}\right)^{\frac{3}{2}} - \left(\frac{v_a^2}{2\,g}\right)^{\frac{3}{2}}\right]^{\frac{2}{3}}$.   This formula rests upon the conception that the head h_a which imparts the velocity in the feeding canal at the measuring point for H, accelerates the flow at the weir, considering the vertical section of discharge, having the height H and the length l, as an orifice, with the head h_a above the upper side of the orifice.   This formula gives a smaller value for h, than the equation $h = H + \frac{v_a^2}{2\,g} = H + h_a$.

Mr. Francis' final equation for discharge is hence, $Q = c\,\tfrac{2}{3}\,(2\,g\,h)^{\frac{1}{2}}\,l\,h$, *h* being obtained by his expression $h = \left[\left(H + \frac{v_a^2}{2\,g}\right)^{\frac{3}{2}} - \left(\frac{v_a^2}{2\,g}\right)^{\frac{3}{2}}\right]^{\frac{2}{3}}$.   Although this equation gives an algebraic result agreeing with the stated conception, still its *form* is objectionable, inasmuch as the head due to velocity of approach should only be considered as effective in accelerating the velocity of the escaping sheet, and not in increasing its area.   The expression, $Q = c\,l\,\tfrac{2}{3}\,(2\,g)^{\frac{1}{2}}\left[\left(H + \frac{v_a^2}{2\,g}\right)^{\frac{3}{2}} - \left(\frac{v_a^2}{2\,g}\right)^{\frac{3}{2}}\right]$, is not open to this criticism.

The following table shows the varying values of *c* in the foregoing experiments, with the values of h computed first by Mr. Francis, and then by our corrections deduced from the Fteley and Stearns experiments; i.e., $h = H + b\,h_a$, with b ranging from 1.4 to 1.3.

TABLE XXIII.

J. B. Francis.—Weir Experiments, with nearly full End Contractions.

$$l = 9.997 \qquad (2\,g)^{16} = 8.020 \qquad c = \frac{Q}{l\frac{2}{3}(2\,g)^{16}\,h^{1/2}}$$

No.	No. Weir Experiment.	G	H	Q	Francis.			Smith.	
					h	c	b	h	c
112	49	4.6	.9977	32.580	1.0007	.6089	1.4	1.0019	.6078
113	51	2.014	1.0504	36.002	1.0640	.6137	1.3	1.0697	.6088
114	52	4.6	.7990	23.430	.8007	.6118	1.4	.8014	.6110
115	53	2.014	.8269	25.041	.8347	.6143	1½	.8380	.6107
116	54	4.6	.6238	16.215	.6246	.6146	1.4	.6251	.6138
117	55	2.014	.6493	17.340	.6536	.6140	1½	.6553	.6116

The value of c should be in each of the three series of experiments just given, where H was nearly the same, slightly greater for the inner depth (G) of 2.014, than for the depth of 4.6, on account of the increased discharge to be expected from the partial suppression of bottom contraction with the shallower depth. It will be observed that the values of h given by Mr. Francis in these experiments, more fully satisfy this condition than the values of h given by the author.

These experiments of Mr. Francis are entitled to the greatest respect, as we shall hereafter point out, but in them the corrections for velocity of approach are so minute, that not so much value can be placed upon them in determining the effect of v_a, as upon our conclusions drawn from the experiments of Fteley and Stearns, where very much greater corrections for v_a are required. Should, however, the large value of 2.05 or more for b be applied to the reduction of these Francis experiments, as has been suggested by Messrs. Fteley and Stearns, there would follow palpable inconsistencies in the deduced values of c, which would be altogether incompatible with our belief in the great accuracy of Mr. Francis' work.*

We feel warranted in assuming that these experiments, Nos. 112-117, indicate that b can be but little, if indeed any larger than 1.4 for weirs with full contraction, the section of approach being not much over eight times greater than the area of the weir.

---

* The channel of approach for the Francis experiments with end contraction, was abruptly narrowed to a width of about 12.5 feet, just above the hook-gauges; vide Plate XIV., "Lowell Hydraulic Experiments." It is apparent that it would have been better to have had a channel of uniform width for say 12 or 15 feet up-stream from the weir. This defective arrangement appears to be the only feature open to criticism in the Francis experiments.

Comparing Experiments Nos. 116 and 117, we see that with H about one-third of G, the effect of partial bottom suppression of contraction was not appreciable, even with the small weight given to v_a by Mr. Francis. This fully justifies our primary assumption that when $G = 3.56$ and $H = .944$, the effect of partial suppression is barely sensible.

The last 6 experimental values (Nos. 112-117) represent the means of 55 distinct determinations, in all of which Q was directly measured. This makes them much more reliable than single determinations; if, however, the objectionable form of approach caused a slight error, this error would be constant, and would hence not be eliminated by repetition of the experiments.

The respective heads for the three series somewhat differ ; by reference to our Plate VII. it will be seen that these differences in h produce a slight, but still appreciable, effect upon the values of the co-efficient c. However, the resulting variations in c are not notable enough to change our general conclusions in regard to the effect of v_a.

Castel.

Some experiments made by M. Castel with a weir with end contractions suppressed, should afford data by which the effect of velocity of approach can be determined. An account of these experiments is to be found in the "Mémoires de l'Académie de Toulouse," Tome IV., 1837.

The feeding canal employed was 19.5 long, 2.428 wide, and 1.772 high. The weir was formed by placing a wooden dam at the lower end of the canal, surmounted by a copper bar, having a thickness of .01 ; the height of the weir, or G, was varied from .10 to .74. The supply of water entering the upper end of the canal, was obtained from a vertical pipe-column, the quantity of flow being regulated by a valve. A metallic screen and several wooden boards were placed across the upper portion of the canal, in order to prevent oscillations in the surface at the lower end of the canal where H was measured. In spite of these precautions there must have been some little oscillation, due to the intermittent action of the pumps which furnished the supply. H appears to have been measured at a point in the centre of the canal, 1.61 up-stream from the plane of the weir. The measuring rod, pointed at its lower end, was read to .0001 metre. The length of the dam or weir slightly varied for the different heights of G; how this variation was caused does not appear. The vessel in which q was measured held 113 cubic feet.

In the following table are given the results of these experiments. In the computations for velocity of approach, the width of the canal has been assumed constant at 2.428. The factor b has been given a value of $1\frac{1}{2}$, as in the Fteley and Stearns experiments with end contractions suppressed. The co-efficient C has been obtained by the formula, $Q = C \frac{2}{3} l (2 g)^{1/2} H^{3/2}$, and is given for the purpose of illustrating the great effect in these experiments of v_a. The co-efficient c has been deduced by our formula $Q = c \frac{2}{3} l (2 g)^{1/2} h^{3/2}$. The value of $(2 g)^{1/2}$ has been assumed as 8.023, the value assigned to it by Castel.

TABLE XXIV.

Castel.—Weir Experiments with End Contractions suppressed. Length of Weir nearly constant, and G variable.

$$c = \frac{Q}{l\,\tfrac{2}{3}(\tfrac{2}{3}g)^{\frac12}\,h^{5/4}} \qquad h_\cdot' = \frac{v_a^2}{2g} \times 1\tfrac13.$$

No.	Width Canal l	G	H	v_a	$h_\cdot'$	h	Q	C	c
118	2.438	.1050	.2493	1.587	.0522	.3015	1.3653	.841	.632
119	„		.2001	1.282	.0340	.2341	.9493	.813	.643
120	„	„	.1644	1.030	.0228	.1872	.6866	.790	.650
121	„	„	.1306	.829	.0142	.1448	.4743	.771	.660
122	„	„	.0988	.622	.0080	.1068	.3079	.761	.676
123	2.429	.1345	.2408	1.338	.0371	.2779	1.2188	.794	.640
124	„	„	.1988	1.092	.0247	.2235	.8840	.767	.644
125	„	„	.1634	.885	.0162	.1786	.6378	.750	.650
126	„	„	.1339	.721	.0107	.1446	.4697	.738	.658
127	„	„	.0955	.499	.0052	.1007	.2783	.726	.670
128	2.436	.2461	.2536	.973	.0196	.2732	1.1803	.709	.634
129	„	„	.1965	.740	.0113	.2078	.7953	.701	.644
130	„	„	.1640	.603	.0075	.1715	.6004	.693	.649
131	„	„	.1312	.467	.0045	.1357	.4273	.690	.656
132	„	„	.0988	.332	.0023	.1011	.2778	.687	.663
133	2.434	.3051	.2621	.866	.0155	.2776	1.1930	.683	.627
134	„	„	.1969	.631	.0083	.2052	.7696	.677	.636
135	„	„	.1634	.509	.0054	.1688	.5788	.673	.641
136	„	„	.1306	.389	.0031	.1337	.4121	.671	.648
137	„	„	.0991	.277	.0016	.1007	.2719	.670	.654
138	2.393	.4265	.2687	.713	.0105	.2792	1.2040	.675	.638
139	„	„	.1995	.504	.0053	.2048	.7664	.672	.646
140	„	„	.1673	.407	.0034	.1707	.5870	.670	.650
141	„	„	.1316	.302	.0019	.1335	.4086	.669	.655
142	„	„	.0994	.210	.0009	.1003	.2678	.668	.659
143	2.427	.5578	.2644	.585	.0071	.2715	1.1686	.662	.636
144	„	„	.1923	.398	.0033	.1956	.7247	.662	.645
145	„	„	.1621	.321	.0021	.1642	.5605	.662	.649
146	„	„	.1319	.246	.0013	.1332	.4118	.662	.652

TABLE XXIV.—*continued.*

No.	Width Canal l	v	H	v_a	h_a'	h	Q	C	c
147	2.427	.5578	.1011	.173	.0006	.1017	.2764	.663	.656
148	2.419	.7382	.2415	.429	.0038	.2453	1.0210	.665	.650
149	„	„	.1844	.304	.0019	.1863	.6805	.664	.654
150	„	„	.1644	.262	.0014	.1658	.5742	.666	.657
151	„	„	.1293	.190	.0007	.1300	.4005	.666	.660
152	„	„	.1007	.135	.0004	.1011	.2755	.666	.662

As the several observed heads for each series were not very dissimilar, the foregoing values of c can be arranged for comparison as follows :

H	v						
	.105	.134	.246	.305	.426	.558	.738
.25	.632	.640	.634	.627	.638	.636	.650
.20	.643	.644	.644	.636	.646	.646	.654
.16	.650	.630	.649	.641	.650	.649	.657
.13	.660	.658	.656	.648	.655	.652	.660
.10	.676	.670	.663	.654	.659	.656	.662

In accordance with the views before expressed, with the same head as G increases c should diminish. By reference to the foregoing summary it will be observed that the values of c, H being constant, follow no regular order with variation in v. The results for $G = .738$ appear to be especially irregular, compared with those for smaller values of v. Taken as a whole, the deduced values of c do not sufficiently indicate the effect of partial suppression on the bottom. Assuming that the experiments are trustworthy, this would show that our values of h are placed too high, or, in other words, that b should be somewhat lower than $1\frac{1}{3}$. More satisfactory results would have been obtained by giving b a value of about 1.2 for the smaller values of v.

Although we regard the experiments of Castel as entitled to some weight,[*] still we consider them as much less reliable than the data taken from Fteley and Stearns, and Francis. They possess, however, sufficient value to warrant their insertion here ; with very large values of h_a' in proportion to H, they show results fairly according with our preceding deductions.[†]

---

[*] Note our remarks on Castel's experiments given in Chapter V. in the section treating on short weirs.

[†] In Experiment No. 118 h_a' is 21 per cent. of H, while with the highest velocity of approach in the Fteley-Stearns series—No. 94—h_a' is only 14 per cent. of H.

Other Authorities.

This subject has been investigated experimentally by several savants, in addition to the three authorities we have made use of, but their results are discordant and often self-contradictory. This can be safely attributed to a large range of experimental error.

Conclusions.

It is evident that the force due to velocity of approach, which causes an acceleration of velocity at the weir, must be less than the total kinetic energy of the water as it passes by the measuring point for H. For, with a weir with contraction suppressed on all three sides—or more exactly, a canal of uniform section with the water discharging from an open end—the effective head due to the velocity in the section at H will be diminished by the " frictional " loss between H and the end of the canal, and also further by the changes in velocity of the horizontal layers of water as they feed the escaping sheet, these velocities being in proportion to the respective head for each layer. When there is total or partial contraction – the area at the weir being smaller than that of the canal—there will be in addition losses or absorptions of energy, due to diagonal and eddying movements of the water, as the various fillets make their way more or less directly towards the crest. When there is a weir proper it can be regarded as an obstruction, causing a notable additional variation of direction in the movement of the fillets of water. Any increase of such variation must inevitably result in increased loss of head.

Hence, from theoretical considerations, we can assume that as contraction becomes more and more complete, or in other words as the distance increases of a side of the feeding canal from the respective side of the weir, the greater will be the proportional loss of energy between the measuring point for H and the weir.

Any attempt to compute the allowance to be made for velocity of approach, from the total energy or *vis-viva* of the water, as it passes the measuring point for H, presents so many complicated problems, that we have preferred to use our very simple form, which although not strictly accurate, gives sufficiently satisfactory results, when $h_a{'}$ is not very large in proportion to h. When it is required to measure the flow of water with great accuracy, the velocity of approach should be reduced to the minimum which is practicable. No serious error will result, where v_a is less than one foot, by making for weirs with full contraction $b = 1.4$, and for weirs with end contractions suppressed $b = 1\frac{1}{2}$.

It is doubtless theoretically true that the larger a_e is in proportion to a, the greater should be the value of b in our equation, but this is in a considerable measure compensated for by the fact, as before stated, that as a_e increases, a being constant, the greater is the proportional loss of energy between the point where H is measured and the weir.

In a feeding canal having a considerably larger section than the area of the weir, the various fillets of water in the section at the point for H have widely varying velocities. It seems reasonable to suppose that the additional force produced by the velocity of approach, is chiefly that due to the velocity of the section of the feeding stream nearest or in line with the section of discharge, where the particles of water fairly begin to form into the escaping vein; this velocity is unquestionably greater than v_a. For instance, with a weir with end contractions suppressed, with a very great inner depth below the crest compared with the head, with the measuring point placed the usual distance of 6 feet or so from the crest, the velocity at the bottom of the canal will be practically nil, and v_a will be very low. At the same time, with a considerable value of H, such as 1 foot, there will doubtless be an appreciable surface current at the measuring point, and very much more than is indicated by v_a. With a weir with full contraction, having an unobstructed channel of approach of considerable length, the central surface current, which we conceive is the current which chiefly causes the increased flow at the weir, will be from 1.1 to 1.25 greater than v_a.

Hence we can pretty safely take for weirs with full contraction v_a as the central surface current just above the measuring point; in this case no correction of h_a should be employed. In channels of approach having irregular sections it will probably be safer to take this surface current, rather than to compute h_a' from the mean velocity. Our given value of b of 1.4 for such weirs is expressed in $h_a' = \dfrac{(1.183\ v_a)^2}{2\,g}$, so that assuming the surface current to be 1.2 times the mean velocity, the result is practically the same, whether the surface current be used for v_a without correction, or if the mean velocity be used, and the resulting velocity head increased by the factor b.[*]

The following table gives the values of h_a, $h_a \times 1\frac{1}{2}$, and $h_a \times 1.4$, for velocities of approach, v_a, from .04 to 1.9. The values of h_a, et cet., can be immediately obtained by this table for velocities up to 1., with an error not exceeding .0001. In practice v_a will rarely exceed 1., and as before stated, where much accuracy in the measurement of water is desired, v_a should be maintained at less than 1..

It must be kept in mind that these corrections are intended to compensate solely for additional head due to velocity of approach. Where G or L is so small as to cause partial bottom or side suppression, then further allowance must be made for the increased flow caused by this suppression.

The given corrections are intended especially for weirs; for orifices with full contraction it is probable that, in $Q = C'' \,(2\,g)^{\frac{1}{2}} \left(H + b\,\dfrac{v_a^2}{2\,g} \right)$, b should have a somewhat larger value than 1.4.

---

[*] The ratio of the surface velocity to the mean velocity may be more or less modified by the use of screens, placed near the measuring point for H. Such screens may have the effect of producing a greater velocity near the bottom, than at the surface.

TABLE XXV.

Corrections for Velocity of Approach. $\quad h = H + b\frac{v_a^2}{2g}.\quad 2g = 64.36$

v_a	$\frac{v_a^2}{2g}$	$1\frac12\frac{v_a^2}{2g}$	$1.4\frac{v_a^2}{2g}$	v_a	$\frac{v_a^2}{2g}$	$1\frac12\frac{v_a^2}{2g}$	$1.4\frac{v_a^2}{2g}$	v_a	$\frac{v_a^2}{2g}$	$1\frac12\frac{v_a^2}{2g}$	$1.4\frac{v_a^2}{2g}$	v_a	$\frac{v_a^2}{2g}$	$1\frac12\frac{v_a^2}{2g}$	$1.4\frac{v_a^2}{2g}$
.04	.000 02	0	0	.38	.002 24	.0030	.0031	.71	.007 83	.0104	.0110	.88	.012 03	.0160	.0168
.05	.000 04	.0001	.0001	.39	.002 36	.0032	.0033	.715	.007 94	.0106	.0111	.885	.012 17	.0162	.0170
.06	.000 06	.0001	.0001	.40	.002 49	.0033	.0035	.72	.008 05	.0107	.0113	.89	.012 31	.0164	.0172
.07	.000 08	.0001	.0001	.41	.002 61	.0035	.0037	.725	.008 17	.0109	.0114	.895	.012 45	.0166	.0174
.08	.000 10	.0001	.0001	.42	.002 74	.0037	.0038	.73	.008 28	.0110	.0116	.90	.012 58	.0168	.0176
.09	.000 13	.0002	.0002	.43	.002 87	.0038	.0040	.735	.008 39	.0112	.0118	.905	.012 73	.0170	.0178
.10	.000 15	.0002	.0002	.44	.003 01	.0040	.0042	.74	.008 51	.0113	.0119	.91	.012 87	.0172	.0180
.11	.000 19	.0003	.0003	.45	.003 14	.0042	.0044	.745	.008 62	.0115	.0121	.915	.013 01	.0173	.0182
.12	.000 22	.0003	.0003	.46	.003 29	.0044	.0046	.75	.008 74	.0117	.0122	.92	.013 15	.0175	.0184
.13	.000 26	.0004	.0004	.47	.003 43	.0046	.0048	.755	.008 86	.0118	.0124	.925	.013 29	.0177	.0186
.14	.000 30	.0004	.0004	.48	.003 58	.0048	.0050	.76	.008 97	.0120	.0126	.93	.013 44	.0179	.0188
.15	.000 35	.0005	.0005	.49	.003 73	.0050	.0052	.765	.009 09	.0121	.0127	.935	.013 58	.0181	.0190
.16	.000 40	.0005	.0006	.50	.003 88	.0052	.0054	.77	.009 21	.0123	.0129	.94	.013 73	.0183	.0192
.17	.000 45	.0006	.0006	.51	.004 04	.0054	.0057	.775	.009 33	.0124	.0131	.945	.013 87	.0185	.0194
.18	.000 50	.0007	.0007	.52	.004 20	.0056	.0059	.78	.009 45	.0126	.0132	.95	.014 02	.0187	.0196
.19	.000 56	.0007	.0008	.53	.004 36	.0058	.0061	.785	.009 57	.0128	.0134	.955	.014 17	.0189	.0198
.20	.000 62	.0008	.0009	.54	.004 53	.0060	.0063	.79	.009 70	.0129	.0136	.96	.014 32	.0191	.0200
.21	.000 68	.0009	.0010	.55	.004 70	.0063	.0066	.795	.009 82	.0131	.0137	.965	.014 47	.0193	.0203
.22	.000 75	.0010	.0011	.56	.004 87	.0065	.0068	.80	.009 94	.0133	.0139	.97	.014 62	.0195	.0205
.23	.000 82	.0011	.0012	.57	.005 05	.0067	.0071	.805	.010 07	.0134	.0141	.975	.014 77	.0197	.0207
.24	.000 90	.0012	.0013	.58	.005 23	.0070	.0073	.81	.010 19	.0136	.0143	.98	.014 92	.0199	.0209
.25	.000 97	.0013	.0014	.59	.005 41	.0072	.0076	.815	.010 32	.0138	.0144	.985	.015 07	.0201	.0211
.26	.001 05	.0014	.0015	.60	.005 59	.0075	.0078	.82	.010 45	.0139	.0146	.99	.015 23	.0203	.0213
.27	.001 13	.0015	.0016	.61	.005 78	.0077	.0081	.825	.010 57	.0141	.0148	.995	.015 38	.0205	.0215
.28	.001 22	.0016	.0017	.62	.005 97	.0080	.0084	.83	.010 70	.0143	.0150	1.	.015 54	.0207	.0218
.29	.001 31	.0017	.0018	.63	.006 17	.0082	.0086	.835	.010 83	.0144	.0152	1.1	.018 80	.0251	.0263
.30	.001 40	.0019	.0020	.64	.006 36	.0085	.0089	.84	.010 96	.0146	.0153	1.2	.022 37	.0298	.0313
.31	.001 49	.0020	.0021	.65	.006 56	.0088	.0092	.845	.011 09	.0148	.0155	1.3	.026 26	.0330	.0368
.32	.001 59	.0021	.0022	.66	.006 77	.0090	.0095	.85	.011 23	.0150	.0157	1.4	.030 45	.0406	.0426
.33	.001 69	.0023	.0024	.67	.006 97	.0093	.0098	.855	.011 36	.0151	.0159	1.5	.034 96	.0466	.0489
.34	.001 80	.0024	.0025	.68	.007 18	.0096	.0101	.86	.011 49	.0153	.0161	1.6	.039 78	.0530	.0557
.35	.001 90	.0025	.0027	.69	.007 40	.0099	.0104	.865	.011 63	.0155	.0163	1.7	.044 90	.0599	.0629
.36	.002 01	.0027	.0028	.70	.007 61	.0102	.0107	.87	.011 76	.0157	.0165	1.8	.050 34	.0671	.0705
.37	.002 13	.0028	.0030	.705	.007 73	.0103	.0108	.875	.011 90	.0159	.0167	1.9	.056 09	.0748	.0785

Formulæ.

The famous and much disputed equation of Daniel Bernoulli is, for horizontal orifices ;

$$v = \left(\frac{2\,g\,H}{1 - \left[\frac{a}{a_c}\right]^2} \right)^{\frac{1}{2}}, \text{ or } H = \left(1 - \left[\frac{a}{a_c}\right]^2 \right) \frac{v^2}{2\,g} ; \text{ hence,}$$

(A) $$Q = a \left(\frac{2\,g\,H}{1 - \left[\frac{a}{a_c}\right]^2} \right)^{\frac{1}{2}} = \text{theoretic discharge.}$$

This formula applied to an orifice at the bottom of a prismatic reservoir makes Q infinite when $a = a_c$, thus requiring an infinite supply into the entrance of the reservoir.[*] We believe that the theoretical accuracy of this expression is now generally admitted by mathematicians.[†] In it, however, neither friction or adhesion on the sides of the reservoir, eddying or diagonal movements caused by unequal velocities, nor the phenomena indicated by the form of the escaping vein, are taken into consideration.

The fundamental proposition upon which the equation is based, is that the water moves in one mass with uniform velocity. But in the case of weirs, the escaping horizontal layers of water move at very different speeds, the velocity of the layers being in proportion to the square root of the head, and hence being 0 at the surface, disregarding the effect of velocity of approach. Therefore the equation is especially inapplicable to weirs.

Boileau and others have followed equation (A). Boileau's final equation for weirs with end contractions suppressed, given in his " Traité de la mesure des eaux courantes," is,

(B) $$Q = \frac{\left(1 - \frac{H_o}{H}\right)^{\frac{1}{2}}}{\left(1 - \left[\frac{1}{1 + \frac{a}{H}}\right]^2\right)^{\frac{1}{2}}} \, l\,H\,(2\,g\,H)^{\frac{1}{2}}.$$

With this expression when $a = 0$, Q becomes infinite.

If we propose for vertical rectangular orifices, placed at the end of a feeding canal,

$$\frac{Q}{a_c} = v_a ; \quad h_a = \frac{v_a^2}{2\,g} ; \quad H + h_a = \text{effective head at the orifice, there results,}$$

---

[*] In such a case what will be the proper value of H ? If we could measure it by a piezometric tube attached to the bottom of the reservoir, disregarding effect of frictional and eddy losses, would it not be 0 ?

[†] Vide paper by Weisbach, in the " Allgemeinen Maschinen-encyclopädie," " Ausfluss."

(C) $Q = c \ (2 \, g)^{\frac{1}{2}} \ l \ \frac{2}{3} \ ([H_b + h_a]^{\frac{3}{2}} - [H_c + h_a]^{\frac{3}{2}})$.

Now, as in weirs H_c disappears, and H_b becomes H, we have,

(D) $Q = c \ (2 \, g)^{\frac{1}{2}} \ l \ \frac{2}{3} \ ([H + h_a]^{\frac{3}{2}} - h_a^{\frac{3}{2}})$.

Mr. Francis has adapted equation (D), although in a different algebraic form, in making his corrections for r_a. It is based rigorously upon the conception, that the force causing increased velocity at the weir, is that due to the energy of a section of the feeding stream, having the same area as the area at the weir ($H \, l$), and having the mean velocity, v_a, of the feeding stream.

Mr. Neville proposes a modification of this expression by making $h_a' = \dfrac{r_a^{\ 2}}{2 \, g \, c_d^{\ 2}}$, c_d being the co-efficient of contraction at the orifice. This would approximately result for weirs, in

(E) $Q = c \ \frac{2}{3} \ (2 \, g)^{\frac{1}{2}} \ l \, \{[H + 2.6 \, h_a]^{\frac{3}{2}} - [2.6 \, h_a]^{\frac{3}{2}}\}$

h_a in equation (E) having its usual signification of $\dfrac{r_a^{\ 2}}{2 \, g}$.

The formula which we have adopted is,

(F) $Q = c \ \frac{2}{3} \ (2 \, g)^{\frac{1}{2}} \ (H + b \, h_a)^{\frac{3}{2}} \ l = c \ \frac{2}{3} \ (2 \, g)^{\frac{1}{2}} \ l \ h^{\frac{3}{2}}$,

b having a value of 1.4 for weirs with full contraction, and $1\frac{1}{3}$ for weirs with end contractions suppressed. This mode of expression is faulty, as it assumes that the head representing the effect of velocity of approach not only adds to the measured head, thus imparting additional velocity, but also adds to the section of escape, or area at the weir. We have used this equation on account of its simplicity ; it gives sufficiently satisfactory results when applied to our experimental data.

The expression

(G) $Q = c \ (2 \, g)^{\frac{1}{2}} \ l \ \frac{2}{3} \ ([H + b \, h_a]^{\frac{3}{2}} - [b \, h_a]^{\frac{3}{2}})$,

with b having larger values than in equation (F), would doubtless be more logical, but (G) would give but little, if indeed any closer results than (F), and would involve considerable additional labor in computations.

We think it is apparent that theoretically our factor b should be a variable, increasing with $\dfrac{a_e}{a}$; this view is fairly sustained by the experimental data given in this chapter. The expression $\left(\dfrac{a_e}{a}\right)^{\frac{1}{4}} = b$ would roughly indicate the law of increment in b ; b would hence be unity with suppression on three sides, when $a_e = a$, which is probably not far from the truth. Even if there were at hand data from which an expression could be framed which would accurately state the rate of increment in b, there would result such a complicated formula for weir discharge, as to render it almost fatally objectionable for practical use.

Formulæ, such as that of Boileau, those of Weisbach, et cet., are intended to apply not only for v_a, but also for suppression or partial suppression of contraction. When there is complete suppression on the three sides, these causes are very closely related to each other; still we conceive that even then they should be considered not as a whole, but separately. The two phenomena are essentially distinct, and it seems to us the proper method of investigation to be followed, is to attempt, as we have done, to determine what effect each cause has, in increasing the discharge.

CHAPTER V.
FLOW OVER WEIRS.

It has before been shown that for rectangular weirs, with vertical sides and horizontal crests, the discharge is correctly represented by,

$$Q = c \tfrac{2}{3} (2 g h)^{\frac{1}{2}} a, \text{ and } a = l h; \text{ or, } Q = c \tfrac{2}{3} (2 g)^{\frac{1}{2}} l h^{\frac{3}{2}};$$

c being the co-efficient of discharge to be ascertained by experiment. All the following reductions are made with this formula to determine the varying values of c.

When v_a is of moment, the observed head, H, will be corrected in accordance with the results stated in our last chapter, on Velocity of Approach. It must be kept in mind, however, that H will only be corrected for the additional force due to v_a: H will not be corrected for the increased discharge due to suppression, or partial suppression of contraction; nor, for the diminished discharge caused by increased friction and adhesion on the sides of the feeding canal between the measuring point and the weir, when the canal was of so small a section as to make the effect from these causes sensible.

As the water approaches the crest of a weir, the surface forms a notable curve, extending above the weir an increasing distance as the length of the weir and the depth upon the crest increase; several of the forms of this curve will be shown hereafter. Many experimenters have measured both H—the vertical elevation above the crest to a point where the influence of this curve ceases to be appreciable—and also H_w—the height from the crest to the mean surface line of the water in the plane of the weir—thinking that both heads should be known in order to accurately determine Q.

The exact measurement of H_w adds largely to the labors of the experimenter,[*] and in properly constructed weirs its determination is of no value in obtaining Q. Hence only H will be considered, except in discussing some of Lesbros' experiments, where there was a large loss of head between the measuring point for H and the weir.

Dubuat, Eytelwein, and other early hydraulicians made a large number of experiments with weirs, but beyond proving that the formula $Q = C \tfrac{2}{3} (2 g H)^{\frac{1}{2}} l H$ was

---

[*] Owing to the irregular form of the surface at the weir, H_w can never be measured as exactly as H, where the water presents a very nearly horizontal section, nor can H_w be measured by the hook-gauge, the most delicate appliance yet invented for determining surface heights.

N

approximately correct, or, that the velocity at the weir, v, is very nearly in proportion to $(2 g H)^{\frac{1}{2}}$, these experiments are of no value. They fixed the general value of C at .62 or thereabouts, but gave very contradictory minor results. For instance, with nearly full contraction, some of them showed that as H diminished below .5, C also diminished; while others gave diametrically opposite results.

The later and far more careful experiments of Lesbros, Francis, Fteley and Stearns, and ourselves afford sufficient data to deduce with very considerable accuracy the laws governing the discharge over weirs with the three simplest forms of approach, and with depths up to nearly 2 feet.

It may be remarked at the outset, that the accurate determination of the co-efficient of discharge, c, is far more difficult with weirs than with orifices. With a very small orifice the very exact measurement of its area becomes necessary if it be desired to establish c within close limits, and hence very delicate measuring appliances must be used, but with an orifice with the least side above .5, the dimensions can be obtained with sufficient precision by ordinary scale measurement. The head used for gauging water by orifices is rarely less than .6 and generally 1. or more; with such values of H, a slight error in its measurement is unimportant.

With weirs the great danger of error arises from imperfect determination of H, l being usually of a size where slight errors in its measurement will not sensibly affect c.

Comparing the relative changes in c caused by a small error in H;

Weir.	With $H =$	.1,	an error of .001 will change c	1.30	per cent.
,,	,,	$H =$ .2	,, ,, ,,	.75	,,
,,	,,	$H =$ .6	,, ,,	.25	,,
Orifice.	,,	$H =$ .6	,, ,, ,,	.083	,,
,,	,,	$H =$ 1.	,, ,, ,,	.050	,,
,,	,,	$H =$ 2.	,, ,, ,,	.025	,,
,,	,,	$H =$ 10.	,, ,, ,,	.005	,,

With a vertical gauge-rod pointed at the lower end, such as was used by Lesbros, probably .001 was the limit for accurate measurement of H, although his readings were to .0003 (.1 millimetre). With the much more perfect hook-gauge used by Francis and others, perfectly still water can be measured to .0002 with comparative certainty; there are, however, such fluctuations of the surface of the water in practice, that the experimenter may consider himself very fortunate if his limit of error in H does not exceed .0005, or even more.

We should hence expect to see much greater experimental variations in the value of c for weirs, than was the case with orifices. The weir experiments of Mr. Francis, and Messrs. Fteley and Stearns were, however, executed with such care, and upon such a grand scale, that their resulting values of c for heads above the crest of more than .5, are more reliable, than many of the values of c we have given for orifices. It is apparent that the larger the opening, provided the measuring vessel for q be proportionately large, the more reliable should be the values of c; for in such case, errors of dimension become comparatively smaller.

It is quite likely that with small heads, such as .2 and less, a change in the temperature of the water of 30° or 40° may have a notable effect upon the discharge; it is also possible that with such small heads, there may be variations caused in the flow by unknown changes in the character of the water, such as we have conjectured may occur with very small orifices, or with orifices with very small heads.

EXPERIMENTS.

The weirs we are first about to describe, were all rectangular, with vertical sides, with free or nearly free discharge into the air, and of lengths from .66* to 19. .

They can be divided into three categories :

FIRST. — With both end contractions suppressed, by placing the weir at the end of a feeding canal of rectangular section, whose width was the same as the length of the weir; the plane of the weir being at right angles to the sides of the canal. In all cases the crest was sharp-edged, so that the escaping water only came in contact with the inner corner; the inner side, or end of the canal, below the crest was vertical, and of sufficient depth to give complete or nearly complete bottom contraction.

SECOND.—With complete or nearly complete contraction on the three sides of the weir, the width of the feeding canal therefore being considerably greater than the length of the weir. The discharge into the air was perfectly free. The sides of the weirs were thin (except in one weir of Lesbros), so that the escaping water only came in contact with the square inner corners of the three sides.

THIRD. — With contraction suppressed on the bottom or on one side ; with contraction partially suppressed on one or both sides ; and with various forms of approach as shown by Figs. 4, et cet., Plate I.

FIRST : CONTRACTION SUPPRESSED AT BOTH ENDS.

Lesbros.

Lesbros, using his orifice in a fixed copper plate, of .6562 square as a weir, and with form of approach shown by Fig. 10, Plate I., obtained the following results.

H was measured in still water at a point 11.48 feet above weir, and hence $H = h$.

---

* The discharge over similar short weirs, with l as small as .033, will be discussed hereafter.

TABLE XXVI.

Lesbros.— Weir with End Contractions Suppressed. Perfectly free Discharge into the Air.   Inner Depth below Crest 1.772.

$(2 g)^{\frac{1}{2}} = 8.0227$ Fig. 10, Plate I.

No.	Lesbros No.	H	$H_\bullet$	$H - H_\bullet$	Q	l	c	% Greater than Francis
1	1764-6	.8009	.6549	.1460	1.6158	.6362	.642	3.2
2	1767-9	.5098	.4258	.0840	.8291	,,	.649	4.3
3	1770-1	.3350	.2821	.0529	.4424	,,	.650	4.5
4	1772-4	.1857	.1542	.0315	.1846	,,	.657	5.6
5	1775-7	.0627	.0472	.0155	.03917	,,	.711	14.3

The escaping vein in Nos. 1 to 4 constantly enlarged horizontally; with No. 5 it diminished in length after escape from the weir.

There was a perceptible loss of head caused by primary contraction, as the water entered the narrow feeding canal from the main reservoir, where H was measured.

Francis.

Mr. J. B. Francis made in 1852, at Lowell, Massachusetts, a series of 88 experiments on weirs with various forms of approach, most of them with $l = 10.$, and a few with $l = 4.$. His results are published in "Lowell Hydraulic Experiments," pp. 103-145, New York, 1868.

These experiments were made upon a scale before unknown, and with great care and skill. The use of the Boyden hook-gauge enabled him to determine the surface elevation of the water with much greater accuracy than had before been possible with previous researches, where H was measured by a pointed descending gauge-rod. His measuring vessel or tank had a capacity of 12 138 cubic feet, with a depth of 9.5 feet, thus affording exceptional facilities for the absolute measurement of large volumes of water, with great exactness. The flow of water from the weir was very rapidly connected or disconnected with the flume leading to the measuring tank, by a swinging apron. Times were determined to .1 of a second.

The crest of the weir was of cast iron, $\frac{1}{4}$th of an inch wide on top, and vertical on the up-stream side. The canal was 13.96 feet wide, with its bottom 4.6 feet below crest immediately at the weir, and 5.048 feet below crest at the gauges.

The canal was narrowed to a width of 9.992 feet a distance of 20 feet up-stream from the weir, for experiments with end contractions suppressed.

The head upon the crest was measured by two hook-gauges, placed 6 feet above the weir, on each side of the canal; these were enclosed by small boxes to diminish oscillations in the surface of the water. These boxes were, perhaps, objectionable, as they may have slightly interfered with the normal flow of water to the weirs having full contraction.

To determine effect of r_a, a false bottom was placed in the canal, extending horizontally a distance of 23 feet up-stream from the weir, at an elevation of 2.014 feet below the crest.

To obtain a shorter length, a false piece 2 feet long was placed in the centre of the 10-foot weir, thus forming two weirs each 4 feet long.

He places $(2 g)^{\frac{1}{2}}$ at 8.0202, which value will be used in our reductions from all his experiments.

These weir experiments of Mr. Francis still rank first in reliability, and will be accepted as unquestionable authority ; the only other weir determinations thus far given to the world, which bear comparison with them in regard to accuracy, being those of Messrs. Fteley and Stearns.

The Francis experiments with end contraction suppressed, were with perfectly free discharge into the air—except Nos. (Francis) 51-55 as hereafter noted, with $G = 4.6$. In the following transcript of these experiments, the values of h are given as deduced by the Francis expression of $h = [(H + h_a)^{\frac{3}{2}} - h_a^{\prime\frac{3}{2}}]^{\frac{2}{3}}$; the co-efficient c' is obtained by the formula $Q = c' \frac{2}{3} (2 g)^{\frac{1}{2}} h^{\frac{3}{2}} l$. The length of the weir in all cases was 9.995.

Francis No.	H	h_a	h	Q	c'	Francis No.	H	h_a	h	Q	c'
44	.9867	.0046	.9912	32.909	.6240	51	1.0050	.0049	1.0097	33.818	.6237
45	.9849	.0046	.9893	32.843	.6246	52	1.0060	.0048	1.0106	33.771	.6220
46	.9745	.0045	.9788	32.362	.6254	53	1.0053	.0048	1.0098	33.727	.6219
47	.9762	.0045	.9805	32.430	.6250	54	.9926	.0047	.9971	33.088	.6219
48	.9760	.0045	.9803	32.436	.6253	55	.9924	.0047	.9968	33.069	.6217
49	.9777	.0045	.9821	32.492	.6247	Means	1.0003		1.0048	33.494	.6222
50	.9769	.0045	.9812	32.460	.6249	67	.7362	.0021	.7382	21.153	.6241
Means	.9791		.9834	32.561	.6248	68	.8019	.0026	.8045	24.104	.6251
						69	.8095	.0027	.8121	24.449	.6251
						70	.8149	.0028	.8176	24.688	.6249
						71	.8132	.0027	.8159	24.558	.6236
						Means	.7955		.7979	23.790	.6246

The foregoing means are calculated from $\frac{3}{2}$ power of H and h.

For Nos. 51-55 (Francis) the sides of the canal were extended past the weir, so that the escaping vein was confined to the same length as the weir ; access, however, was given for the entrance of air under the descending sheet of water.

For our purposes the means of the three foregoing series can be taken ; applying our corrections for h_a, we have ;

TABLE XXVII.

Francis.— Weir with End Contractions Suppressed.

Depth on inner side of crest = 4.6. $(2g)^{\frac{1}{2}}$ = 8.0202

No.	Francis No.	Temp. of Water.	H	h_a	b	h_a'	h	l	Q	c
6	44.30	44°	.9791	.0045	1¼	.0060	.9851	9.995	32.561	.6232
7	31.55	44¾°	1.0003	.0048	1¼	.0064	1.0067	„	33.494	.6205
8	67.71		.7955	.0026	1¼	.0035	.7990	„	23.790	.6233

Fteley and Stearns.

Messrs. Fteley and Stearns experimented with two suppressed weirs, one with $l =$ nearly 5., and the other with $l =$ nearly 19. .

The measuring basin for the largest volumes was a section of the new Sudbury conduit for the Boston water supply, and had a capacity of about 300 000 cubic feet, with a change in elevation of 3 feet ; the area was therefore very large in proportion to the depth, and hence this basin was not as favorable for very precise measurements of Q, as would have been one of smaller area but greater depth. The largest volume measured was $Q = 130$ cubic feet, which is probably a larger quantity than has ever before been accurately determined. For Experiments Nos. 9 to 26 inclusive, Q was obtained from this large basin ; for Nos. 27 to 34, Q was obtained from a smaller section of the same conduit. The experiments with the large measuring basin are thought to be freer from errors in Q, than with the small basin.

The crest of the 5-foot weir was a nickel-plated steel straight-edge, having a thickness of .0066 ; the crest of the 19-foot weir was a planed iron bar .02 thick ; the water in these experiments only touched the sharp inner edges of the crests. The length of the shorter weir varied somewhat during the experiments, which was caused by changes in the sides of the canal.

The head upon the crest was measured by a hook-gauge placed below the weir, its lower end or hook determining the elevation of the water in a movable pail. This pail was connected by a rubber hose with an opening in a piece of plate glass, whose face was flush with the side of the feeding canal, and set at a point 6 feet above the weir, and about .4 foot (vertically) below the level of the crest ; the orifice in the glass plate had a diameter of .04, with smooth sides, having its axis normal to the face of the plate.

Those experiments given by these authors, which are stated by them to have been imperfectly made, will not be included in the following tables. All the reductions from the experiments of Fteley and Stearns will be made with our values of h_a'.

TABLE XXVIII.

Fteley and Stearns.— Weir with End Contractions Suppressed. Escaping Vein confined by Prolongation of Sides of Canal, but which did not extend below level of Crest.

Inner depth below crest, 6.55. $h = 1\frac{1}{2}$ $(2g)^{1/2} = 8.020$

No.	F. & S. No.	T of Water.	H	h_a	h_a'	h	l	Q	c	Means h	Means c
9	1	34°	1.6038	.010 97	.0146	1.6184	18.996	130.117	.6223	1.62	.6223
10	2	37	1.4546	.008 44	.0113	1.4659	„	112.066	.6217	1.47	.6217
11	3	37°	1.2981	.006 21	.0083	1.3064	„	94.192	.6211	1.30	.6211
12	5	39°	1.1456	.004 40	.0059	1.1515	„	77.783	.6198	1.15	.6198
13	6	13°	.9873	.002 91	.0039	.9912	„	62.061	.6192	.99	.6194
	7	38°	.9864	.002 91	.0039	.9903	„	62.023	.6197		
14	8	45°	.8191	.001 73	.0023	.8214	„	46.760	.6185	.82	.6185
15	9	44°	.6460	.000 89	.0012	.6472	„	32.685	.6181	.65	.6181
16	10	41°	.4685	.000 35	.0005	.4690	„	20.178	.6186	.47	.6186

TABLE XXIX.

Fteley and Stearns.—Weir with End Contractions Suppressed. Escaping Vein confined by Prolongation of Sides of Canal, but which did not extend below level of Crest. Temperature of Water, 36°.

Inner depth below crest, 3.17. $h = 1\frac{1}{2}$ $(2g)^{1/2} = 8.020$

No.	F. & S. No.	H	h_a	h_a'	h	l	Q	c	Means h	Means c
17	1	.8198	.0064	.0085	.8283	5.	12.750	.6327	.82	.6304
	3	.8118	.0061	.0081	.8199	5.	12.466	.6281		
18	4	.6761	.0037	.0049	.6810	5.	9.430	.6277	.68	.6276
	5	.6713	.0037	.0049	.6762	4.997	9.322	.6275		
19	6	.5203	.0018	.0024	.5227	4.996	6.342	.6283	.52	.6283
20	7	.4810	.0016	.0021	.4831	4.999	5.766	.6425	.48	.6425
21	8	.4761	.0014	.0019	.4780	4.999	5.547	.6280	.47	.6272
	9	.4569	.0013	.0017	.4586	4.999	5.199	.6264		
22	10	.3890	.0008	.0011	.3901	4.999	4.094	.6287	.390	.6287
23	12	.3407	.0006	.0008	.3415	4.994	3.3540	.6294	.341	.6294
24	13	.3114	.0004	.0005	.3119	4.998	2.9355	.6306	.312	.6306
25	14	.2598	.0003	.0004	.2602	4.999	2.2415	.6319	.260	.6319
26	15	.2467	.0002	.0003	.2470	4.999	2.0780	.6333	.247	.6333
27	17	.2190	.0002	.0003	.2193	5.	1.7474	.6365	.220	.6365
28	18	.2182	.0002	.0003	.2185	5.	1.7211	.6304	.218	.6432
	19	.2176	.0002	.0003	.2179	5.	1.7837	.6560		

[Table continued on next page.

TABLE XXIX.—*continued.*

No.	F. & S. No.	H	h_a	h_a'	h	l	Q	c	Means h	Means c
29	20	.1650	.0001	.0001	.1651	4.998	1.1705	.6529	.164	.6504
	21	.1627	.0001	.0001	.1628	4.995	1.1307	.6480		
30	22	.1444	.0001	.0001	.1445	4.993	.9469	.6455	.144	.6455
31	23	.1235	0	0	.1235	4.998	.7495	.6462	.123	.6515
	24	.1225	0	0	.1225	4.998	.7326	.6569		
32	27	.1009	0	0	.1009	4.996	.5816	.6829	.101	.6852
	28	.1008	0	0	.1008	4.996	.5877	.6875		
33	29	.0991	0	0	.0991	4.996	.5498	.6598	.099	.6598
34	30	.0746	0	0	.0746	4.996	.3652	.6710	.075	.6710

It will be observed from the preceding tables that for heads above .3 (with one exception, No. 20) the values of c are very uniform, while for heads below .22 there are variations from one to four per cent. with practically identical values of h. These experimental discrepancies well illustrate our remarks at the beginning of this chapter in regard to the errors which may be expected to attend the weir measurement of water with small heads, even if the experiments are conducted by most careful and experienced investigators with the aid of the most perfect measuring appliances.

Second: Complete, or nearly Complete, Contraction.

Poncelet and Lesbros.

We will first take the results of Poncelet and Lesbros, with a weir .66 long; with inner depth below crest, 1.77, and with width of approach 12.1, thus having absolutely complete contraction on the sides. These experiments were made in 1828.

TABLE XXX.

Poncelet and Lesbros.— *Weir with full Contraction. Discharge free into the Air.*
H measured 11.48 above weir ; hence H = h.   G = 1.77 $(2g)^{1/2} = 8.0227$ Figure 1, Plate I.

No.	H	H_a	H−H_a	l	Q	c
35	.6821	.6273	.0548	.6562	1.1528	.583
36	.5351	.4849	.0502	"	.8098	.589
37	.3376	.2983	.0394	"	.4071	.591
38	.1985	.1686	.0299	"	.1864	.600
39	.1463	.1207	.0256	"	.1194	.608
40	.0771	.0577	.0194	"	.04676	.622

Lesbros.

In 1834 Lesbros obtained the following results, using forms of approach shown by

Figs. 2 and 3, Plate I., and the same length of weir—using same orifice in a thin plate for a weir as in experiments given in Table XXVI.

TABLE XXXI.

Lesbros.—Weir with Full (?) Contraction. Discharge free into the Air.

H measured 11.48 above weir; $H = h$. $(2\,g)^{1/2} = 8.0237$ Length always .6562.

		Fig. 2, Plate 1. $c = 1.77$ $L = 1.77$ $L' = 5.71$						Fig. 3, Plate 1. $G = 1.77$ $L = 1.77$ $L' = 1.77$			
No.	Lesbros No.	H	$H - H_w$	Q	c	No.	Lesbros No.	H	$H - H_w$	Q	c
41	1666	.5955	.0515	.9524		45	1680	.3724	.0397	.4750		
	1667			.9476 .9516	.590		1681			.4791 .4753	.596
	1668			.9547			1682			.4719		
42	1669	.3625	.0393	.4541		46	1683	.1070	.0227	.0774		
	1670			.4588			1684			.0777		
	1671			.4564 .4565	.596		1685			.0767 .0774	.630
	1672			.4567			1686			.0777		
43	1673	.1788	.0279	.1612								
	1674			.1597								
	1675			.1629 .1618	.610							
	1676			.1636								
44	1677	.0958	.0203	.0656								
	1678			.0654 .0653	.628							
	1679			.0649								

With Fig. 2, the vein after its escape from the weir with the higher heads converged a little towards the (prolonged) side of the canal, nearest the weir. With Fig 3, the escaping vein had the same appearance as with Fig. 1.

The head in the plane of the weir, or H_w, in the Lesbros experiments was the mean height above the crest.

It will be observed by looking at the foregoing values of Q, that they vary in one case 2½ per cent. for the same given head. This was probably more due to variations in H, than to errors in measuring q or Q.

Francis.

Mr. Francis made 71 experiments with weirs having nearly full contraction, where he obtained Q by direct measurement of q.[*] We will first give the results of the 10 series, each series having nearly equal heads, with the reductions based upon the Francis formula of $h = [(H + h_w)^{3/2} - h_w^{3/2}]^{2/3}$; c' is obtained by $Q = c' \frac{2}{3} (2\,g)^{1/2} h^{3/2} l$. The width of approach was in all cases 13.96; L and L' were hence each about 2 feet. As before, $(2\,g)^{1/2} = 8.0202$. The discharge was in all cases free into the air. The means are computed from the $\frac{3}{2}$ power of H and h.

[*] Lowell Hydraulic Experiments; 2nd Edition, pp. 122-125.

o

Francis No.	G and l	H	h_a	h	Q	c'	Francis No.	G and l	H	h_a	h	Q	c'
1	G=4.6, l=9.997	1.5243	.0092	1.5330	61.282	.6040	34	G=4.6, l=7.997	1.0102	.0019	1.0121	25.988	.5969
2		1.5504	.0095	1.5594	62.569	.6011	35		1.0262	.0020	1.0282	26.563	.5959
3		1.5593	.0097	1.5684	63.206	.6020	Means		1.0183		1.0202	26.275	.5964
4		1.5691	.0097	1.5783	63.351	.5977	36		1.0280	.0140	1.0410	34.848	.6138
Means		1.5508		1.5598	62.602	.6012	37	l=9.997	1.0372	.0143	1.0504	35.293	.6133
5	l=9.997	1.2369	.0054	1.2421	45.089	.6094	38		1.0445	.0146	1.0580	35.725	.6142
6		1.2419	.0055	1.2472	45.344	.6091	39		1.0449	.0146	1.0584	35.766	.6145
7		1.2479	.0055	1.2532	45.678	.6091	40		1.0460	.0146	1.0595	35.771	.6137
8		1.2508	.0055	1.2561	45.494	.6046	41		1.0513	.0148	1.0649	36.072	.6141
9	G=4.6	1.2529	.0056	1.2582	45.934	.6089	42	G=3.014	1.0794	.0157	1.0930	37.487	.6130
10		1.2549	.0056	1.2602	45.853	.6064	43		1.0711	.0154	1.0853	37.051	.6130
Means		1.2476		1.2528	45.565	.6079	Means		1.0504		1.0640	36.002	.6137
11		.9671	.0028	.9698	31.146	.6101	56	l=9.997	.8186	.0018	.8203	24.319	.6123
12		1.0275	.0033	1.0307	33.942	.6068	57		.8075	.0017	.8092	23.794	.6115
13		1.0339	.0033	1.0372	34.237	.6064	58		.7956	.0017	.7973	23.276	.6117
14		1.0331	.0033	1.0364	34.272	.6077	59		.7769	.0016	.7784	22.480	.6124
15		1.0406	.0034	1.0439	34.655	.6079	60	G=4.6	.8012	.0017	.8029	23.543	.6122
16		1.0373	.0034	1.0406	34.533	.6086	61		.7940	.0016	.7956	23.170	.6108
17		.9632	.0028	.9659	30.957	.6101	Means		.7990		.8007	23.430	.6118
18		.9759	.0029	.9787	31.538	.6094	62	l=9.997	.7711	.0070	.7777	22.527	.6145
19	l=9.997	.9795	.0029	.9823	31.658	.6084	63		.7872	.0073	.7941	23.238	.6149
20		.9888	.0030	.9917	32.144	.6089	64		.8045	.0077	.8118	24.013	.6142
21		.9946	.0030	.9975	32.471	.6097	65	G=3.014	.8796	.0096	.8885	27.498	.6142
22		.9157	.0024	.9180	28.674	.6099	66		.8886	.0099	.8978	27.909	.6138
23		.9280	.0025	.9304	29.193	.6085	Means		.8269		.8347	25.041	.6143
24	G=4.6	.9462	.0026	.9488	30.090	.6091	72		.5919	.0007	.5926	15.000	.6151
25		1.0127	.0032	1.0158	33.300	.6085	73		.5924	.0007	.5931	15.027	.6155
26		1.0116	.0032	1.0147	33.315	.6098	74	l=9.997	.6106	.0008	.6114	15.712	.6149
27		.9949	.0031	.9979	32.554	.6110	75		.6552	.0010	.6562	17.430	.6135
28		1.0336	.0034	1.0368	34.414	.6098	76	G=4.6	.6430	.0009	.6439	16.984	.6149
29		1.0556	.0035	1.0591	35.474	.6089	77		.6379	.0009	.6388	16.760	.6141
30		1.0692	.0037	1.0727	36.175	.6091	78		.6337	.0009	.6346	16.589	.6140
31		.9837	.0029	.9865	31.929	.6096	Means		.6238		.6246	16.215	.6146
32		.9782	.0029	.9810	31.671	.6098							
33		.9670	.0028	.9697	30.994	.6072							
Means		.9977		1.0007	32.580	.6089							

Francis No.	G and l	H	h_a	h	Q	c	Francis No.	G and l	H	h_a	h	Q	c
79		.6515	.0045	.6558	17.425	.6139	85		.6694	.0030	.6723	14.220	.6033
80	l = 9.997	.6559	.0046	.6603	17.613	.6142	86	l = 7.997	.6790	.0031	.6819	14.519	.6030
81		.6598	.0047	.6643	17.772	.6141	87		.6836	.0031	.6866	14.664	.6028
82	G = 2.014	.6313	.0042	.6353	16.618	.6139	88	G = 2.014	.6881	.0032	.6912	14.788	.6019
83		.6425	.0044	.6466	17.058	.6137	Means		.6801		.6830	14.548	.6027
84		.6546	.0046	.6589	17.554	.6140							
Means		.6493		.6536	17.340	.6140							

Experiments (Francis) Nos. 34-35 and 85-88 with $l = 7.997$, were made by dividing the weir with $l = 9.997$ into two weirs, each 3.9985 feet long. This was done by placing a dam or partition in the centre of the long weir, 2 feet in length. These experiments hence should be considered as showing the discharge over a weir with $l = 4.$, with contraction rather imperfect at one end.

Taking the means of the foregoing 71 determinations, we have the following table, with h_a' and h calculated according to our methods, and c deduced from our values of h. The results in this table will be adopted by us as authentic. The greatest difference between our values of c and those of c' deduced by the use of the Francis formula for velocity of approach, is .0049 for No. 51 (Francis Nos. 36-43), where $c = .6088$ and $c' = .6137$; this is a change of about $\frac{1}{120}$th from the co-efficient given by Mr. Francis.

TABLE XXXII.

Francis.—*Weirs with Contraction more or less complete. Width of Approach in all Cases 13.96. Discharge free into the Air. H measured 6. above Weir.* $h = H + b \frac{v^2}{2g}$. $(2g)^{\frac{1}{2}} = 8.0202$

No.	Francis Nos.	T of Water	G	H	h_a	b	h_a'	h	l	Q	c
47	1-4	46.5°	4.6	1.5508	.0095	1.4	.0133	1.5641	9.997	62.602	.5983
48	5-10	48.5°	4.6	1.2476	.0055	1.4	.0077	1.2553	9.997	45.565	.6061
49	11-33	{49.° 41.2°}	4.6	.9977	.0030	1.4	.0042	1.0019	9.997	32.580	.6078
50	34-35	48.°	4.6	1.0183	.0020	1.4	.0038	1.0211	7.997*	26.275	.5956
51	36-43	48.7°	2.014	1.0304	.0148	1.3	.0193	1.0697	9.997	36.002	.6088
52	56-61		4.6	.7990	.0017	1.4	.0024	.8014	9.997	23.430	.6110
53	62-66	48.7°	2.014	.8269	.0083	1.2	.0111	.8380	9.997	25.041	.6107
54	72-78	48.2°	4.6	.6238	.0009	1.4	.0013	.6251	9.997	16.215	.6138
55	79-84	48.7°	2.014	.6493	.0045	1.5	.0060	.6553	9.997	17.340	.6116
56	85-88		2.014	.6801	.0031	1.4	.0043	.6844	7.997*	14.548	.6009

* The co-efficient c, for Nos. 50 and 56, applies to a weir having a length of 4.0.

The channel of approach for the Francis experiments with end contraction, was narrowed to a width of about 12.5 feet, just above the hook-gauges. This change in the section of the feeding canal was objectionable; in our judgment, however, errors, or rather discrepancies, arising from this defect would not be large.

Hamilton Smith, Jun.

In Chapter IX. is given a detailed description of 12 experiments made by the author. They are sufficiently exact when Q does not exceed 10.—Nos. 57-63. After that limit the measuring vessel for q, whose capacity was about 1300 cubic feet, was too small to afford very accurate results. The effect of velocity of approach, which was insensible for volumes less than $Q = 10$., became notable for larger values of Q; for this weir it is difficult to decide what should be the proper correction for v_a; probably C for Nos. 67 and 68 is about $\frac{1}{2}$ per cent. higher than c; on this supposition the correct co-efficients, c, would be for No. 67, $c = .581$, and for No. 68, $c = .578$. The proper correction for v_a for Nos. 64 and 65 would be slight. H was measured at a point 7.6 feet above the weir, *vide* Fig. 8, Plate XV.

TABLE XXXIII.

Smith.— Weir with Complete Contraction.   *Inner Depth below Crest 3.8.*   *No Correction for v_a.*   *Free Discharge into Air.   Temperature of Water from 50° to 60°.*

$(2 g/3 = \times .0177$

No.	H	l	Q	C
57	.5639	2.586	3.582	.6087
58	.6163	„	4.034	.6032
59	.6470	„	4.338	.6030
60	.6703	„	4.567	.6020
61	.7072	„	4.950	.6021
62	1.0681	„	8.988	.5890*
63	1.1063	„	9.455	.5878
64	1.2033	„	10.783	.5910
65	1.3257	2.585	12.241	.5804
66	1.5391	„	15.614	.5918
67	1.7195	„	18.194	.5840
68	1.7327	„	18.318	.5812

* With No. 62 there was some little doubt as to whether termination of experiment was correctly determined or not.

Fteley and Stearns.

Fteley and Stearns indirectly determined the discharge over weirs with nearly complete contraction, and over weirs with contraction suppressed at one end. In one of the weirs of the latter form, contraction was also partially suppressed at the other end, the distance from end of weir to side of feeding canal being only 1. .

For these experiments Q was determined by the flow over the weir 5 feet long, with end contractions suppressed, and whose co-efficients of discharge have been shown by Experiments Nos. 17-34.

To obtain the proper values of c for this 5-foot suppressed weir, we will slightly anticipate by referring to Plates VI. and VII., where the curve of c for this weir has been drawn.

The head, H, for the 5-foot or datum weir, was generally measured at the beginning and ending of each series of experiments, and did not greatly vary ; hence Q was nearly constant for each series. H was in all cases measured at a point 6 feet above weir.

It will be noticed that a number of these experiments belong to our third category, but for convenience they are here included with those belonging to the second.

The discharge was free into the air, except where contraction was suppressed on one or both ends ; in the latter case the side, or sides of the feeding canal were extended, but only above level of crest.

The crest was sharp-edged, being the one employed for Velocity of Approach Experiments, Nos. 1-111. The length was changed by putting in wooden false pieces ; to the irregular swelling of these wooden pieces is due the slightly varying values of l.

The values of c and Q in small type are based upon the 5-foot curve (suppressed weirs) on Plate VII. ; the slightly varying values of Q in ordinary type are interpolated, for each series, in accordance with Table XXVIII. of Fteley and Stearns.

TABLE XXXIV.

Fteley and Stearns.— Weirs with more or less Contraction at one or both Ends.   Temperature of Water about 40°.   Inner Depth below Crest 3.56.

$(2 g)^{\frac{1}{2}} = 8.020$

No.	F. and S. No.	Distances from Ends of Weir to Sides of Canal.		H	$h_{..}$	b	$h_{.}'$	h	l	Q	c	Means.	
		L	L'									h	c
	1	0	0	.1509	.000 05	1½	.0001	.1510	5.0048	1.0127	.6450		
69	2	1.0	0	.1761	.000 05	1.4	.0001	.1762	4.0062	1.013	.6394	.176	.6388
	3	0	1.0	.1763	.000 05	1.4	.0001	.1764	4.0064	1.013	.6383		
70	4	1.0	1.0	.2155	.000 04	1.4	.0001	.2156	3.0080	1.013	.6292	.216	.6292
	5	0	0	.1509	.000 05	1½	.0001	.1510	5.0048	1.0127	.6450		
	6	0	0	.2303	.000 15	1½	.0002	.2305	5.0044	1.8820	.6358		
71	7	1.0	0	.2691	.000 15	1.4	.0002	.2693	4.0063	1.882	.6287	.269	.6285
	8	0	1.0	.2693	.000 15	1.4	.0002	.2695	4.0062	1.883	.6283		
72	9	1.0	1.0	.3301	.000 14	1.4	.0002	.3303	3.0081	1.885	.6168	.330	.6168
	10	0	0	.2304	.000 15	1½	.0002	.2306	5.0044	1.8832	.6358		
	11	0	0	.3368	.000 44	1½	.0006	.3374	5.0045	3.3021	.6297		
73	12	1.0	0	.3942	.000 43	1.4	.0006	.3948	4.0063	3.302	.6214	.395	.6212
	13	0	1.0	.3944	.000 43	1.4	.0006	.3950	4.0061	3.302	.6210		
74	14	1.0	1.0	.4843	.000 41	1.4	.0006	.4819	3.0080	3.303	.6083	.485	.6083
75	15	0	1.7	.4498	.000 42	1.4	.0006	.4504	3.3110	3.303	.6173	.450	.6174
76	16	1.0	1.7	.5824	.000 39	1.4	.0005	.5829	2.3132	3.303	.6001	.583	.6001
	17	0	0	.3369	.000 44	1½	.0006	.3375	5.0045	3.3036	.6297		
	18	0	0	.4244	.000 84	1½	.0011	.4255	5.0043	4.8614	.6277		
77	19	0	1.0	.4978	.000 81	1.4	.0011	.4989	4.0058	4.662	.6177	.499	.6177
78	20	0	1.7	.5678	.000 78	1.4	.0011	.5689	3.3107	4.663	.6139	.569	.6139
	21	0	0	.4245	.000 84	1½	.0011	.4256	5.0043	4.6630	.6277		
	22	0	0	.4308	.000 87	1½	.0012	.4320	5.0049	4.7685	.6275		
79	23	1.0	1.0	.6215	.000 79	1.4	.0011	.6226	3.0070	4.764	.6033	.623	.6032
80	24	1.0	1.7	.7478	.000 75	1.4	.0011	.7489	2.3125	4.764	.5945	.749	.5945
81	25	0	1.7	.5764	.000 81	1.4	.0011	.5775	3.3104	4.763	.6132	.578	.6130
	26	0	1.7	.5766	.000 81	1.4	.0011	.5777	3.3106	4.763	.6129		
	27	0	0	.4302	.000 87	1½	.0012	.4314	5.0049	4.7585	.6276		
	28	0	0	.5115	.001 41	1½	.0019	.5134	5.0046	6.170	.6266		
82	29	1.0	0	.5997	.001 35	1.4	.0019	.6016	4.0063	6.172	.6175	.602	.6177
	30	0	1.0	.5996	.001 35	1.4	.0019	.6015	4.0050	6.172	.6179		

TABLE XXXIV.—*continued.*

No.	F. and S. No.	Distance from Ends of Weir to Sides of Canal. L	L	H	h_a	b	h_c	h	l	Q	c	Means h	c
83	31	1.0	1.0	.7398	.00126	1.4	.0018	.7416	3.0070	6.172	.6011	.742	.6011
84	32	0	1.7	.6860	.00130	1.4	.0018	.6878	3.3101	6.172	.6114	.688	.6114
85	33	1.0	1.7	.8903	.00118	1.4	.0017	.8922	2.3125	6.172	.5924	.892	.5924
	34	0	0	.5117	.00141	1½	.0019	.5136	5.0047	6.172	.6268		
86	35	1.0	1.7	.9548	.00141	1.4	.0020	.9568	2.3126	6.840	.5911	.957	.5911
	36	0	0	.5477	.00170	1½	.0023	.5500	5.0046	6.840	.6267		
	37	0	0	.6010	.00219	1½	.0029	.6039	5.0040	7.870	.6266		
87	38	1.0	0	.7055	.00208	1.4	.0029	.7084	4.0076	7.871	.6161	.709	.6156
	39	0	1.0	.7064	.00208	1.4	.0029	.7093	4.0064	7.871	.6151		
88	40	1.0	1.0	.8714	.00193	1.4	.0027	.8741	3.0104	7.871	.5984	.874	.5984
	41	0	0	.6011	.00219	1½	.0029	.6040	5.0040	7.872	.6268		
89	42	0	1.7	.8062	.00198	1.4	.0028	.8090	3.3112	7.871	.6110	.809	.6105
	43	0	1.7	.8063	.00198	1.4	.0028	.8091	3.3110	7.860	.6101		
	44	0	0	.6002	.00219	1½	.0029	.6031	5.0039	7.854	.6265		
90	45	1.0	1.0	.8702	.00193	1.4	.0027	.8729	3.0098	7.856	.5986	.873	.5986
	46	0	0	.6925	.00322	1½	.0043	.6968	5.0042	9.770	.6278		
	47	0	0	.6924	.00322	1½	.0043	.6967	5.0042	9.768	.6278		
91	48	0	1.7	.9317	.00287	1.4	.0040	.9357	3.3095	9.759	.6093	.935	.8092
	49	0	1.7	.9297	.00285	1.4	.0040	.9337	3.3095	9.722	.6090		
	50	0	0	.6897	.00318	1½	.0043	.6939	5.0042	9.709	.6278		
	51	1.0	0	.9448	.00452	1.4	.0063	.9511	4.0065	12.306	.6193		
92	52	0	1.0	.9432	.00452	1.4	.0063	.9495	4.0043	12.306	.6213	.95	.6196
	[53]	0	0	.8047	.00480	1½	.0064	.8111	5.0038	12.306	.6287		
	54	1.0	0	.9460	.00452	1.4	.0063	.9523	4.0065	12.306	.6182		

Column h_a in preceding table is taken from Fteley and Stearns, and is correct enough for our reductions, although obtained from slightly different values of Q, than those which we have adopted.

As Q was indirectly determined in these experiments, the results cannot be considered to have equal weight compared with the other weir determinations of the same authors, where Q was obtained by absolute measurement. The results, however, shown by this table (XXXIV.) are doubtless near the truth, as the curve on

Plate VII. for the 5-foot weir, which we have used, was deduced from experiments made under very nearly similar conditions.

Lesbros.

Lesbros with a square-edged weir, having a thickness of .164, and approach 12.1 wide (Fig. 1, Plate I.) obtained the following results :

TABLE XXXV.

Lesbros.—Thick-edged Weir with full Contraction.   Discharge free into Air (except the Contact of the Vein on the Crest).

H measured 11.48 above weir; hence $H = h$.   Inner depth below crest, 1.77.

$$(2g)^{\frac{1}{2}} = 8.0227 \qquad \text{Fig. 1, Plate I.}$$

No.	Lesbros No.	H	H_e	$H - H_e$	l	Q	c
93	1840	1.3806	1.2549	.1257	1.9685	10.031	.587
94	1841	1.3796	1.2540	.1256	,,	9.984	.585
95	1842-3	.8711	.7720	.0991	,,	5.031	.588
96	1844-5	.3494	.2877	.0617	,,	1.3193	.607

The escaping vein did not touch vertical sides of weir, but attached itself more and more to the crest, as the heads diminished.

Third : Contraction Suppressed on one Side or Bottom, and Various Forms of Approach.

Lesbros.

A number of experiments in Table XXXIV. belong to this category, and particular reference will be had to them when we discuss the curves representing c.

All the following experiments have been selected from Lesbros.   It will be sufficient to give the means of Q, which were, with one or two exceptions, repeated several times.   This repetition, however, as we have suggested before, does not seem to have checked the accuracy of the measurement of H, and in which generally lies much the greatest danger of error.

The various forms of approach are shown by Plate I., and also by Table VI.

For all these experiments H was measured in still water in the reservoir at a distance of 11.48 above the weir, and hence H is always equal to h.

$(2g)^{\frac{1}{2}}$ is constant at 8.0227.

The discharge was free into the air.   Where there was contraction the edges were sharp, so that the escaping vein only came in contact with the inner corner lines.   These experiments (97-128) were all made by using the orifice, .6562 × .6562 in a fixed copper plate, as a weir.

Temperature of water is not stated; most of the experiments were made in the autumnal months, and a few in August.

TABLE XXXVI.

Lesbros.—Fig. 4, Plate I. Bottom Contraction Suppressed.

No.	Lesbros No.	Distances from Sides of Weir to Sides of Canal.		H	H_e	$H - H_e$	l	Q	c
		Bottom.	Sides.						
97	1687·9			.6732	.6152	.0580	.6562	1.177 7	.607
98	1690·2			.3934	.3383	.0551	,,	.530 4	.613
99	1693·4	0	5.71	5.71	.2198	.1706	.0492	,,	.222 0	.614
100	1695·7			.1014	.0659	.0355	,,	.069 71	.615
101	1698·9			.0528	.0305	.0223	,,	.025 34	.595

The vein in all these experiments (97-101) did not touch the chamfered lower edge of the crest; for heads below .05 the vein attached itself to this chamfer (bevel).

TABLE XXXVII.

Lesbros.—Fig. 8, Plate I. Contraction nearly Suppressed on One Side.

No.	Lesbros No.	Distances from Sides of Weir to Sides of Canal.		H	H_e	$H - H_e$	l	Q	c
		Bottom.	Sides.						
102	1730·2			.6683	.5997	.0686	.6562	1.155 1	.602
103	1733·5			.4885	.4331	.0554	,,	.731 0	.610
104	1736·8	1.77	.066	5.71	.3136	.2733	.0403	,,	.379 1	.615
105	1739·41			.1667	.1444	.0223	,,	.132 1	.637
106	1742·4			.0659	.0591	.0068	,,	.039 73	.669

The surface of the water in the reservoir was elevated a little higher on the side of the canal nearest the weir than on the other side, and the escaping vein converged more or less towards the prolonged direction of this side (.066 from end of weir), according as H was more or less great.

P

TABLE XXXVIII.

Lesbros.— Fig. 5, Plate I.   Contraction Suppressed on Bottom, and nearly Suppressed on One Side.

No.	Lesbros No.	Distances from Sides of Weir to Sides of Canal.			H	H_a	$H - H_a$	l	Q	c
		Bottom.	Sides.							
107	1700-1				.6847	.5984	.0863	.6562	1.215 8	.611
108	1702-3	0	.066	5.71	.3566	.2887	.0679	,,	.454 7	.608
109	1704-6				.0702	.0433	.0269	,,	.037 26	.571

The surface of the water was higher on the side of the canal nearest the weir than on the other side, and the escaping vein converged more or less towards the prolonged direction of this side (.066 from end of weir). For No. 109 the vein attached itself a little to the lower chamfer of the crest, on the end of the weir farthest from the side of the canal.

TABLE XXXIX.

Lesbros.— Fig. 9, Plate I.   Contraction nearly Suppressed on Both Sides.

No.	Lesbros No.	Distances from Sides of Weir to Sides of Canal.			H	H_a	$H - H_a$	l	Q	c
		Bottom.	Sides.							
110	1745				.7133			.6562	1.350 5	.639
111	1746-7				.6857	.6014	.0843	,,	1.267 1	.636
112	1748-50	1.77	.066	.066	.5085	.4446	.0639	,,	.808 0	.635
113	1751-3				.3835	.3379	.0456	,	.525 5	.630
114	1754-7				.3291	.2900	.0391	,,	.417 9	.631
115	1758-60				.1693	.1476	.0217	,,	.156 0	.638
116	1761-3				.0686	.0545	.0141	,,	.041 85	.664

The water in No. 110 covered the upper edge of the orifice, except for a length of .2 in the centre. This experiment, therefore, is just on the dividing line between that for an orifice and that for a weir.

TABLE XL.

Lesbros.— Fig. 6, Plate I.   Contraction Suppressed on Bottom, and nearly Suppressed on Both Sides.

No.	Lesbros No.	Distances from Sides of Weir to Sides of Canal.			H	H_a	$H - H_a$	l	Q	c
		Bottom.	Sides.							
117	1707-9				.8366	.5036	.3330	.6562	1.532 1	.570
118	1710-2				.8340	.5020	.3320	,,	1.525 8	.571
119	1713-4	0	.066	.066	.6460	.3967	.2493	,,	1.046 2	.574
120	1715-6				.5039	.3159	.1880	,,	.722 3	.575
121	1717-20				.3392	.2175	.1217	,,	.396 6	.572
122	1721-3				.2014	.1207	.0807	,,	.176 7	.537
123	1724-6				.1326	.0743	.0583	,,	.089 46	.528
124	1727-9				.0702	.0364	.0338	,,	.031 40	.481

For 117, 118 and 119 there was a strong boiling action at the escape from the weir, caused by the shock of the water against the narrow inner intervals on the ends of .066. See remarks in *Surface Curve at Weir* as to the form of the surface for experiments 117-124.

TABLE XLI.

Lesbros.—Fig. 12, Plate 1. Inclined Sides of Canal.

No.	Lesbros No.	Distances from Sides of Weir to Sides of Canal.			H	H_w	$H - H_w$	l	Q	c
		Bottom.	Sides.							
125	1778-9				.6545	.5873	.0672	.6562	1.124 6	.605
126	1780-2	1.77	.066	.066	.4734	.4200	.0534	,,	.690 5	.604
127	1783-6				.2890	.2506	.0384	,,	.331 8	.608
128	1787-92				.0689	.0594	.0095	,,	.041 30	.654

For Nos. 125 and 126 the appearance of the escaping vein was the same as with Fig. 1.
For No. 127 the horizontal length of the escaping vein was;

At the weir .59
.16 below weir .72, and after this distance, a constant narrowing in length.

The same phenomenon occurred for No. 128.

Lesbros made a number of experiments with other forms of approach, the weir being placed in some of them at the end of a canal of same width as length of weir. All these experiments were more or less complicated by other causes than those we are considering, and are of no value in assisting us to our final conclusions. The reader who desires to investigate them is referred to Lesbros' original volume, *Expériences Hydrauliques sur les lois de l'écoulement de l'Eau.* Paris, 1850.

SURFACE CURVE AT WEIR.

Before comparing the various values of c given for the foregoing selection of experiments, it will be well to discuss the form of the surface curve of the water as it approaches and passes the crest of the weir, and also to investigate the varying values of $H-H_w$, as given by Lesbros and others.

Lesbros.

With Fig. 10, Plate I., the surface on the central axial line of the small feeding canal—shown by vertical section normal to the plane of the weir, and bisecting it—had the following elevations above the crest of the weir; H, measured 11.48 above the crest, was .5089; this weir had both end contractions suppressed, with $l = .6562$.

TABLE XLII.

Lesbros.—Central Surface Curve.

$G = 1.77$ *Fig.* 10, *Plate* I.

Point.	Distance up-stream from Weir.	Elevation Surface above Crest.	Point.	Distance up-stream from Weir.	Elevation Surface above Crest.	Point.	Distance (horizontal) below Weir.	Elevation Surface above Crest.
H	11.18	.5089	l	.984	.4997	H_*	0	.4219
a	6.60	.5036	m	.820	.4954	v	.164	.3773
b	6.40	.5003	n	.656	.4918	w	.328	.2900
c	6.23	.4984	o	.492	.4866	x	.492	.1736
d	3.91	.4990	p_*	.394	.4810	y	.656	.0118
e	3.25	.4990	q	.328	.4754	z	.820	—.1660
f	4.59	.5007	r	.262	.4688			
g	3.94	.5010	s	.197	.4620			
h	3.61	.5026	t	.131	.4518			
i	2.62	.5026	u	.066	.4400			
j	1.97	.5023	H_*	0	.4219			
k	1.31	.5003						

The point b was in the plane of the mouth of the feeding canal, 6.4 feet above the weir.

The irregularities in elevation from a to g seem to have been caused by the contraction at entrance of the canal.

The effect of the surface weir curve extended perceptibly to l, a distance of about 1 foot from the weir; and very slightly—if the above measurements can be relied upon for such small quantities—to the point i, at a distance of 2.6 from the crest.

There was a notable loss of head, caused by contraction at the mouth of the feeding canal. For;

H measured in still water in reservoir   = .5089
H' ,, at l, where surface curve begins = .4997

Apparent loss of head   = .0092

This loss was partly due to head imparting velocity of approach, which can thus be computed : Q for above value of H was about .827, *vide* Experiment No. 2 ; hence $u_c = (.50 + 1.77) \times .656 = 1.49$, and $c_n = \dfrac{.827}{1.49} = .55$; with $b = 1\frac{1}{2}$, $h_a' = b\,\dfrac{v_a^2}{2g} = .0063$, head absorbed in imparting velocity of approach. Therefore $.0092 - .0063 = .0029 =$ head lost by primary contraction ; this loss would, for $H = .5089$, diminish c about 1 per cent..[*]

We will call foregoing experiment No. 129, with $h = (.4997 + .0063) = .5060$; consequently c has a value of .655.

---

[*] A small portion of this loss was due to "frictional" losses as the water passed through the feeding canal.

Poncelet and Lesbros.

For Fig. 1, Plate I., with a weir having full contraction, with $l = .6562$, the central surface curves were as follows, for $H = .5915$, and $H = .0951$.

TABLE XLIII.

Poncelet and Lesbros.—Central Surface Curves.

$l = 1.77$ Fig. 1, Plate I.

Point	Distance upstream from Weir.	Elevation Surface above Crest.	Point	Distance upstream from Weir.	Elevation Surface above Crest.	Point	Distance (horizontal) below Weir.	Elevation Surface above Crest.
H	11.48	.5915 (l)	y	.394	.5840	H_w'	0	.5394
a	.984	5915	h	.328	.5807	m	.066	.5220
b	.853	.5912	i	.262	.5758	n	.131	.4997
c	.722	.5899	j	.164	.5673	o	.210	.4689
d	.591	.5879	k	.098	.5584	p	.295	.4457
e	.525	.5876	l	.033	.5479	q	.394	.3396
f	.459	.3866	H_w'	0	.5394	r	.984	—.307
H	11.48	.0951 (l)	g	.787	.0866	H_w'	0	.0761
a	1.48	.0951	h	.591	.0837	m	.033	.0682
b	1.38	.0945	i	.394	.0810	n	.066	.0608
c	1.28	.0935	j	.295	.0797	o	.098	.0246
d	1.18	.0922	k	.197	.0787	p	.131	—.0115
e	1.08	.0906	l	.098	.0781	q	.164	—.0542
f	.984	.0892	H_w'	0	.0761	r	.295	—.297

The surface sections in plane of the weir for the foregoing heads were as follows :

H	H_w	H_w'	H_w''	H_w'''	H_w''''	H_w^v
.5915	.5410	.5394	.5374	.5682	.5843	.5938
.0951	.0746	.0761	.0722	.0814	.0948	

H_w being mean elevation along weir ; H_w' being in the centre of the weir ; H_w'' being .06 from end ; H_w''' being at end ; H_w'''' and H_w^v being between end of weir and side of canal, H_w^{iv} at a distance of .06 from weir, and H_w^v at .26 from weir.

It will be observed that the effect of the surface (axial) curve is sensible for a greater distance up-stream with small than with large values of H ; for (including two other measurements of this curve, the details of which we do not give) ;

$H = .5915$ Effect of curve is sensible a distance of .98 from crest.
$H = .4311$,, ,, ,, ,, 1.15 ,,
$H = .2369$,, ,, ,, ,, 1.31 ,,
$H = .0951$,, ,, ,, ,, 1.46 ,,

Another remarkable feature is that the surface elevation, with the head of .5915, between the weir and the side of the canal is .0043 greater than H.

Lesbros.

With Fig. 4, Plate I. (bottom contraction suppressed), the form of the surface section in the plane of the weir varied with the head. With H above .42 there was a large descent from the sides of the weir, and a slight rise in the centre; with lower heads there was the same rapid dropping from the sides, then a slight rise, and then a very slight depression in the centre.

The surface axial curve apparently extended back, towards the reservoir, a considerable distance from the crest.

With Fig. 5 (contraction suppressed on bottom, and partially on one side), the surface section in the plane of the weir was highest on the end of the weir nearest the side of the canal, with a very marked depression for considerable heads at a point about two-thirds the length of the weir from this end.

The axial curve, as in Fig. 4, appears to have been sensible a considerable distance up-stream from the crest.

With Fig. 6 (contraction suppressed on bottom and partially on both sides) there was for considerable heads a very notable depression in the axial curve, after the water entered the feeding canal, indicating the loss by primary contraction. With $H = .4649$ the surface elevations above the crest were as follows:

H 11.46	up-stream from crest; in reservoir	.	.	.4649
D 6.40	„	„	at mouth narrow canal . .	.4197
E 5.41	„	„	at lowest depression in canal .	.2615
F 3.28	„	„	highest point between E and H.	.2986
H_c crest at centre .	.	.	. . . .	.2648

For this value of H, Lesbros found $Q = .6391$; hence $c = .5744$, in $c = \dfrac{Q}{l\,H^{\frac{3}{2}}\,(2\,g\,H)^{\frac{1}{2}}}$.

Calling this experiment No. 130, and calculating c from head measured at point F as above (3.28 from crest) we have:

$$H = .2986\; ;\; a_c = .2986 \times .7874 = .2351\; ;\; v_a = \frac{Q}{a_c} = 2.718\; ;$$

$$h_a = \frac{v_a^2}{2\,g} = .1148\; ;\; h = H + h_a = .4134^*\; ;\; \text{hence } c = .685.$$

This indicates a loss of head by primary contraction and friction on sides to the point F, of nearly 20 per cent.

In the above instance the correction for h_a reduces the value of the co-efficient of discharge about 40 per cent. . For:

H being .2986	$C = 1.116$
h „ .4134	$c = .685$

---

* Contraction being suppressed nearly completely on three sides, $h_c = h_c'$.

With this form of approach (Fig. 6) the surface section at mouth of canal—point D—declined from the sides and rose again at the centre ; in the plane of the weir, with $H = .4649$, $H_a{}^i$ (centre) was .2648, and $H_a{}^{iii}$ (at ends) was .3478.

Fteley and Stearns.

Messrs. Fteley and Stearns determined, as shown by following table, the surface curves for a 5-foot suppressed weir, with varying inner depths (G) below crest. The measurements were made on a line 1.5 distant from side of canal, and hence 1. from central axial line. Single observations were taken, which were corrected slightly to form smoother curves by these authors ; we give the original measurements, all made (except H) by point-gauge.

TABLE XLIV.

Fteley and Stearns.—Surface Curve at Suppressed Weir ; l = 5. . Q constant.

Distance from Weir.	1 $G = 3.56$ $v_a = .389$ $h_a = .0034$	2 $G = 2.6$ $v_a = .506$ $h_a = .0040$	3 $G = 1.7$ $v_a = .705$ $h_a = .0077$	4 $G = 1.$ $v_a = 1.022$ $h_a = .0162$	5 $G = .5$ $v_a = 1.529$ $h_a = .0363$
6. up-stream = H	.6143	.6120	.6059	.5904	.5635
3.5 „	.6142	.6117	.6060	.5874	.5635
3. „	.6135	.6126	.6061	.5903	.5631
2. „	.6084	.6102	.6039	.5894	.5621
1.5 „	.6083	.6066	.6013	.5882	.5616
1. „	.6017	.6002	.5939	.5833	.5596
.5 „	.5847	.5827	.5788	.5671	.5474
0 crest = H_a	.5226	.5226	.5182	.5097	.5003
.3 down-stream	.4251	.4261	.4212	.4089	.3914
.38 „	.3845	.3844	.3786	.3677	.3739
$H - H_a$	.0917	.0894	.0877	.0807	.0632

Castel.

In the "Mémoires de l'Académie des Sciences de Toulouse," Tome IV., 1837, M. Castel in his Table VIII., gives the form of 34 surface curves, with various values of l, two widths of feeding canal, and G constant. The curves were determined by 10 vertical rods, pointed at their lower ends, attached to a horizontal bar, which appears to have been placed in the central line of the feeding canals. The results obtained by Castel show, without exception, that G and the width of the feeding canal being constant, the length of the surface curve diminishes as H and l diminish. The maximum length—horizontal ordinate—of the curve was 1.37, with $G = .56$, $H = .38$, $l = 1.18$, and width of canal = 1.18, the weir hence being suppressed ; the minimum length was .32, with $G = .56$, $H = .20$, $l = .10$, and width of canal = 1.18.

Herschel.

Mr. Clemens Herschel, hydraulic engineer of the Holyoke Water Power Company, has been kind enough to give us a section of the surface curve of the Connecticut River, as it passed over the Holyoke dam in 1883, in time of freshet. The form of the curve is shown by the following sketch.

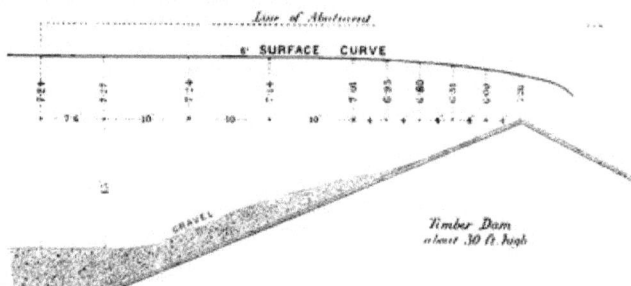

The height of the surface of the sheet above the crest, H_w, was 5.2, with $H = 7.27$; the horizontal length of the curve was about 50 feet. The dam is a long one, perhaps 1200 feet, and the pool formed by it extends a distance of some two miles above the dam.

The section was taken along the southerly abutment of the dam.

Boileau.

M. Boileau found that the horizontal length of the surface curve increased with H; " Traité de la mesure des eaux courants." 1854. p. 55.

$H - H_w$.

For a weir .6562 long, with full contraction, $H - H_w$ was determined to be as follows by Poncelet and Lesbros, and Lesbros, with forms of approach shown by Figs. 1, 2 and 3, Plate I. ; in these experiments H_w was the mean height in the plane of the weir.

H	H_w	$H-H_w$	$\dfrac{H-H_w}{H}$
.6821	.6273	.0548	.0803
.5955	.5440	.0515	.0865
.5351	.4849	.0302	.0938
.3724	.3327	.0397	.1067
.3625	.3232	.0393	.1084
.3376	.2982	.0394	.1167
.1985	.1686	.0299	.1506
.1788	.1509	.0279	.1560
.1463	.1207	.0256	.1750
.1070	.0843	.0227	.212
.0958	.0755	.0203	.212
.0771	.0577	.0194	.252

For weirs with full contraction $H-H_w$ increases with the length of the weir, H being constant. For instance; with weir 1.97 long (Experiment No. 96) $H=.349$ and $H-H_w=.0617$; with weir .066 long (Experiments Nos. 133 and 134) $H=.533$ and $H-H_w=.0026$, also $H=.267$ and $H-H_w=.0017$. Castel's experiments show the same results.

For weirs with end contractions suppressed $H-H_w$ is much larger than for those with full contraction, as will be noticed by reference to Table XXVI. (Fig. 10, Plate 1), where with,

    Experiment No. 1 . . . . . $H=.80$ and $H-H_w=.1460$
               " 2 . . . . . $H=.51$ " $H-H_w=.0840$
               " 4 . . . . . $H=.186$ " $H-H_w=.0315$

In all probability with this form of weir $H-H_w$ also increases with l, although as l increases, losses of head by friction and adhesion on the sides diminish.

By reference to Tables XXXVI., XXXVIII., and XL. (Figs. 4, 5 and 6) it will be observed that $H-H_w$ is often a very large fraction of H, especially with small heads. For instance:

    Experiment No. 117, Fig. 6 . $H=.8366$ and $H-H_w=.3330$
            " 124, " " . $H=.0702$ " $H-H_w=.0338$
            " 109, " 5 . $H=.0702$ " $H-H_w=.0369$
            " 97, " 4 . $H=.6732$ " $H-H_w=.0580$
            " 101, " " . $H=.0528$ " $H-H_w=.0323$

The abnormal value of $H-H_w$ in No. 101 (Fig. 4) is largely due to the friction and adhesion with such a small depth for the distance of 8.2 feet above the weir; hence when these retarding influences cease to be important, as in No. 97, where H is .67,

$H-H_w$ should approach a value similar to that with a weir having full contraction—and which will be observed, is the case.

For No. 117 (Fig. 6) the loss of head, shown by large value of $H-H_w$, is chiefly attributable to the primary contraction at the mouth of the small canal, and to the head absorbed in imparting velocity; in No. 124 the loss of head was chiefly caused by friction and adhesion on the bottom surface.

From the foregoing determinations we can pretty safely draw the conclusion, that the horizontal length of the surface curve—from the plane of the weir to a point where the curve ceases to be sensible—increases directly with l and H; the experiments of Poncelet and Lesbros stand alone in showing that this length decreases as H increases. From the experiments of Fteley and Stearns, Table XLIV., with Q and l constant, and G varying from 3.56 to .50, it is probable that as $\dfrac{G+H}{H}$ increases, the length of the surface curve slightly increases, l and H being constant; this difference in length, caused by wide variation in G, is, however, not very marked in these experiments; with the five given values of G, H could have been safely measured at a point about 3.5 feet from the weir.

The Holyoke curve is of much interest, showing that with very high values of H and l, the length of the surface curve was 50 feet. This, in common with several of the other given curves, indicates that there is a summit or anticlinal point of slight height in the curve, soon after its origin in the feeding channel. H should probably be measured above, or up-stream from this wave summit.

The measurement of H_w as well as H may be of some value in determining in a rough and indirect way, as was the case with the Lesbros experiments, the losses of head between the measuring point for H and the crest. Owing to the mechanical difficulties in the way of the exact measurement of H_w, its given value will always be much less certain than that of H. As we have before remarked, we consider its determination as of no practical value in assisting in the exact reckoning of the discharge over weirs; losses of head between the point for H and the crest, can be more accurately taken account of by other methods.

Lesbros for the reduction of his weir experiments used the formula

$$c' = \frac{Q}{l\,H_w\left(2\,g\left[H-\dfrac{H_w}{2}\right]\right)^{\frac{1}{2}}},$$

thinking that this expression was analogous to the usual formula for orifices. His deduced values of c', it is almost unnecessary to observe, show very misleading results.

EFFECT OF SUPPRESSION.

We can arrive at some general conclusions as to the effect of suppression, or partial suppression of contraction, by a graphic comparison of all the quoted Lesbros experiments with $l = .6562$.

On Plate VIII. are plotted these experiments, including those of Poncelet and Lesbros with Fig. 1; the total head, H, is shown by the vertical lines, and the co-efficient of discharge, c, by the horizontal lines. Our numbers are given on the plate for reference, as are also the numbers of the figures on Plate I., used as forms of approach for the nine series, or experimental curves.

It will be seen by examination of Plate VIII. that the experimental curve for c formed by the series for Figs. 2 and 3, is slightly higher than that for Fig. 1. Lesbros attributed this difference to the effect of partial suppression on one side in Fig. 2, and on two sides in Fig. 3. Arguing from this assumed state of facts, he draws the conclusion that in order to avoid a sensible partial suppression, the width of the feeding canal should be ten times greater than the length of the weir. He therefore considers that, so far as side suppression is concerned, there is entire parallelism between his weir with $l = .656$ and canal 4.2 wide, and a weir with $l = 1.97$ and canal 12.1 wide; the width of canal in each case being about six times the length of the weir, and the inner depth below the crest remaining constant at 1.77.

He also assumes, by comparison of the curves for Figs. 1 and 2, that the increased discharge due to this partial suppression is more notable for small than for large heads, and hence in his table of co-efficients makes c, with $h = .03$, $1\frac{3}{4}$ per cent. larger for Fig. 2 than for Fig. 1; and with $h = .98$, about 1 per cent. larger.

These deductions are manifestly erroneous, as with a weir of great length and small value of h, the effect of even complete suppression at the ends would evidently have no sensible influence upon the value of c. With such a weir, full contraction or full suppression would only affect the particles of water forming into the escaping sheet or vein, for a distance from the end of the weir more or less in proportion to h. Hence with l very much greater than h, the effect of partial *bottom* suppression becomes the factor of importance. In fact Lesbros' own experiments do not warrant the conclusion which he has accepted. For, if with Fig. 3, where each end of the weir was distant 1.77 from the side of the canal, there was an increased discharge compared with Fig. 1, where these distances were 5.71, it seems evident with Fig. 2, where these distances were respectively 1.77 and 5.71, the value of c should be about half way between the values of c for Figs. 1 and 3. But, with $h = .37$, c is the same for Figs. 2 and 3, and perceptibly higher than for Fig. 1. In the other experiment with Fig. 3, h being .1, c is a trifle higher than for Fig. 2; with so small a head the chances of probable experimental error are larger than the discrepancy here shown; and, it will be remembered, that the chances of error with $h = .37$ are much less, than where $h = .1$.

We hence feel warranted in assuming that these discrepancies in the curves for

Figs. 1, 2 and 3 are chiefly due to experimental errors, and that had the investigations been conducted with perfect skill, the curves would have been nearly identical.[*] The experiments of Lesbros in 1834 with Figs. 2 and 3, after his years of experience, appear to be entitled to greater weight than those made in 1828 (Poncelet and Lesbros), and we will therefore adopt the curve of Figs. 2 and 3, as representing with sufficient accuracy the values of c for a weir, with $l = .66$ and h from .1 to .6, having complete contraction.

We will also assume that, with $l = .66$, $h = .66$, the sides of the feeding canal both distant 1.77 from ends of weir, and with an inner depth below crest of 1.77, the effect of partial suppression does not sensibly increase c; i.e., probably not over $\frac{1}{6}$th of 1 per cent.

We are fully confirmed in this assumption by the Francis experiments (Nos. 112-117 of Velocity of Approach). In No. 117 the inner depth below crest, or G, was 2.014, and $h = .65$, while in No. 116, inner depth was 4.6, and $h = .62$; the form of end contractions and l being identical. Using the very low correction for v_a applied by Mr. Francis, c for No. 117 was .6140, while for No. 116 c was .6146; this shows that where G is three times h, there is no sensible increase produced by bottom partial suppression, even when the experimentation is conducted with much greater skill than was the case with Lesbros. By analogy, it is safe to suppose that with a distance between the end of the weir and the side of the canal also three times h, the effect of partial side suppression will not be appreciable.

Complete Suppression.

The curve for Figs. 2 and 3, and that for Fig. 6, Plate VIII., should represent the two extremes for comparing the effect of full contraction, and that of nearly complete suppression on the three sides of the weir. But, as has before been shown in our discussion of the Surface Curve, our values of c for Fig. 6 do not by any means indicate the effect of suppression; c for $h = .41$ being .685, when H was measured in the feeding canal 3.3 up-stream from the weir, while on Plate VIII., with H measured in the reservoir, c for the same head is .574. We might roughly deduce the proper values for c for Fig. 6, showing only the effect of suppression, but it will be preferable to take the curve for Fig. 10, where there was complete suppression on the two ends with full bottom contraction, as our standard for suppression; Experiments Nos. 1 and 2 should be corrected for losses by primary contraction; Nos. 3, 4 and 5 were not largely affected by this loss, and hence need no correction.

The upper heavy dotted symmetrical curve on Plate VIII., which follows the corrected experimental curve for Fig. 10, will be assumed as representing the value of c.

---

[*] With large heads Lesbros states that the vein after its escape from the opening, was perceptibly diverted in a horizontal direction by the form of approach in Fig. 2, both for orifices and for weirs. This very delicate test shows, that for large values of h, this slight partial suppression had some effect.

for a weir with full bottom contraction and end contractions suppressed, and with no retarding influences (losses of head) of much moment between the measuring point for H and the crest. The lower heavy symmetrical dotted curve, on the other hand, represents c for a weir with full contraction; the length of both weirs was the same, being .656.

Assuming that the increased discharge produced by suppression is approximately in proportion to that part of the wetted perimeter suppressed, we have the expression,

$$c_s = c_c \left(1 + x \frac{S}{p}\right), \text{ or } x = \frac{c_s - c_c}{c_c \frac{S}{p}}.$$

c_s being the co-efficient c, for a weir with all or a portion of its perimeter suppressed.

c_c being the co-efficient c, for the same weir with full contraction.

S = length of suppression.

$p = l + 2 h$ = wetted perimeter.

Comparing our two curves of reference we have the following values of x;

$h = .656 = l$; $c_s = .652$ and $c_c = .586$; hence $x = .168$

$h = .4$; $c_s = .650$ and $c_c = .594$; hence $x = .172$

$h = .2$; $c_s = .656$ and $c_c = .611$; hence $x = .194$

$h = .1$; $c_s = .676$ and $c_c = .632$; hence $x = .30$

For the last value of h, it must be remembered that our reference curves are much less reliable than for higher values of h.

Four experiments of Mr. Francis also afford us another opportunity for comparing the values of c for these two forms of approach, the length of both weirs being 10., viz.;

Experiment No. 49.		Full contraction	$h = 1.002$,	$c = .6078$
"	" 6.	Contraction suppressed on both ends	$h = .985$,	$c = .6232$
"	" 52.	Full contraction	$h = .801$,	$c = .6110$
"	" 8.	Contraction suppressed on both ends	$h = .799$,	$c = .6233$

Hence we have,

$l = 10$. $\{ h = 1.0$, and $x = .153$

$\{ h = .8$, and $x = .146$

For Experiment No. 130, with Fig. 6, Plate VIII., with nearly complete suppression on the three sides, it has been shown that with $h = .41$, $c = .685$; for a weir of the same length and head, c_c is .594; hence $x = .15$.

For Experiment No. 97, with Fig. 4, Plate VIII., it has been shown that the given value of c is nearly normal, that is to say, not being considerably diminished by losses of head between the measuring point for H and the crest. In this instance, with complete bottom suppression, $h = .67$ and $l = .656$, $c = c_s = .607$, while c_c for same values of h and l would be about .586; hence $x = .113$.

From the foregoing comparisons it appears to be reasonably safe to give x, in $c_s = c_c \left(1 + x \frac{S}{p}\right)$, a value of about .16 for heads from .3 to 1., when there is complete suppression on one or more sides of a weir.

Weisbach gives the following formula for weirs with end contractions suppressed ; d being depth in canal, or $d = G + H$;

$$\frac{c_s - c_c}{c_c} = .041 + .3693 \left(\frac{H}{d} \right)^2, \text{ or } c_s = c_c \left(1.041 + .3693 \left[\frac{H}{d} \right]^2 \right).$$

Hence

$$Q = c_c \tfrac{2}{3} \left(1.041 + .3693 \left[\frac{H}{d} \right]^2 \right) (2 \, g \, H)^{\frac{1}{2}} \, l \, H.$$

This formula is intended to take into account the effects, both of suppression and velocity of approach. With d very large in proportion to h, c_s would be nearly 1.041 c_c, and would represent the effect of suppression alone, as h_a in this case would be insignificant. In this expression the ratio $\frac{c_s}{c_c}$ does not vary with changes in l or H.

We regard this formula as incorrect in principle ; it certainly is at variance with the experimental data which we have discussed.

Our expression, $c_s = c_c \left(1 + x \frac{S}{p} \right)$, might be changed to $c_s = c_c \left(1 + \frac{S}{K} \right)$, in which K represents the entire perimeter of the water, or $K = 2 \, l + 2 \, h$. With this latter formula, however, x would have a much greater range than with the first.

Partial Suppression.

We have seen that with $l = .66$, if the sides of the feeding canal are distant $3 \, l$ or $3 \, h$ from the ends of the weir, the effect of partial suppression will be insensible. We have now to determine what effect will be produced with smaller distances between the weir and the respective sides of the canal.

With a rectangular feeding canal, having vertical sides, with its axis normal to the plane of a vertical rectangular weir, we will call, as before, the distances from the ends of the weir to the respective sides of the canal, L and L', and the distance from the crest of the weir to the bottom of the canal, G. The least dimension of the weir, whether it be l or h, will be called M, and the distance from any side of the weir to the respective side of the canal, where there is partial suppression, will be called N. The wetted perimeter will be termed $p = l + 2 \, h$, and the length of the side or sides on which there is partial suppression will be termed S'. The co-efficient c for partial suppression will be called c_p.

Experiments Nos. 112-115 (Fig. 9, Plates I. and VIII.), where there was complete bottom contraction, and contraction nearly suppressed on both sides (L and L' both .066), should afford data by which the effect of a small value of N can be ascertained. Experiments Nos. 110, 111 and 116 of the same series (Table XXXIX.) will not be used: No. 110 was almost an orifice, and hence not a fair exponent for a weir; No. 111 was affected perceptibly by primary contraction at the mouth of the feeding canal ; and for No. 116 h was so small as to make c more unreliable than for the other experiments. The comparatively low values of $H - H_a$ for Nos. 112-115 indicate that for them c nearly represents the full effect of partial suppression.

Again, taking the formula $c_s = c_c \left(1 + x \frac{S}{p} \right)$, or $c_p = c_c \left(1 + x \frac{S'}{p} \right)$, we have with $l = .656$;

No.	$H = h$	$c = c_p$	c_r	c_z	x'	x
112	.508	.635	.590	.650	.13	.17
113	.383	.630	.595	.650	.11	.17
114	.329	.631	.598	.651	.11	.18
115	.169	.638	.616	.659	.11	.20

The values of c_r and c_z in the above statement are taken as before, from the two reference curves on Plate VIII. It will be observed that the differences between x and x' increase as the heads diminish. This seems reasonable, as with a very minute head, even with N as small as .066, it is apparent that there should be very nearly complete contraction, and hence that x' should be nearly 0.

The experimental curve for Fig. 8 on Plate VIII., when there was full contraction on the bottom and one side, and $L = N = .066$ on the other side, agrees substantially with the results shown by the series of experiments with Fig. 9, as above given. For, had there been complete suppression on one side, it is apparent that the experimental curve for Fig. 8 on Plate VIII. should have been about half way between our two reference curves on this plate, l being the same for the three weirs; the curve for Fig. 8 (Experiments Nos. 102-106) is, however, quite appreciably nearer the lower than the upper reference curve, thus indicating the effect of partial suppression.

As illustrations of the effect of partial suppression, we have some experiments made by Mr. Francis, comparisons of which have been made in Chapter IV., Velocity of Approach, Experiments Nos. 112-117. For these experiments the length of the weir was constant at 10.; the width of the feeding canal was also constant at 14., there hence being nearly full contraction at the ends. Mr. Francis, it will be remembered, obtained the effective head h by the formula $h = ([H + h_a]^{3/2} - h_a^{3/2})^{2/3}$, and c by the formula $Q = c \frac{2}{3} (2 g h)^{1/2} l h$. We used a larger value for h, making $h = H + b h_a$, with b varying from 1.3 to 1.4. Taking the values of c as given both by Francis and ourselves, we have the following values of x':

l	Velocity of Approach No.	$G = N$	h	c		x'	
				Francis.	Smith.	Francis.	Smith.
10.0	112	4.6	1.00	.6089	.6078	0	0
	113	2.0		.6137	.6088	.010	.002
	114	4.6	.81	.6118	.6110	0	0
	115	2.0		.6143	.6107	.005	—
	116	4.6	.64	.6146	.6138	0	0
	117	2.0		.6140	.6116	—	—

For each of the three comparisons, there was full bottom contraction with the inner depth (G) of 4.6. These results indicate that with $G = 3 h$ there is complete bottom contraction ; that with $G = 2.5 h$ the effect of partial suppression, if felt at all, is very slight ; and that with $G = 2 h$, the effect of partial suppression only increases $c_c \frac{7}{10}$ths of one per cent., even when the very low correction for v_a used by Mr. Francis is employed ; with our assumption of the effect of velocity of approach, the increased discharge caused by partial bottom suppression ($G = 2 h$) only amounts to $\frac{1}{4}$th of one per cent. .

It is probable that partial bottom suppression of a weir with end contractions suppressed, would show larger values of x' than those just given.

From the foregoing comparisons we can roughly place the values of x', in $c_p = c_c$ $\left(1 + x' \frac{S}{l'} \right)$ as follows, for heads from .3 to 1.0, viz. ;

$\frac{N}{M}$	x'
3	0
2	.005
1	.025
$\frac{1}{2}$	.06
0	.16

It will be remembered that M is the least dimension of the weir, whether it be h or l, and N is the distance from the side of a rectangular feeding canal to the respective side of a rectangular weir.

It is evident that the ratio $\frac{N}{M}$ approximately measures the amount of contraction : for, with $l = 1$. and $h = 10$., the distances L and L' need be only 3 l, or 3, in order to obtain full contraction at the ends ; reversing the dimensions, and making $l = 10$ and $h = 1$, L and L' need be only 3 h, or 3. In other words, with an opening of such unequal dimensions, the particles of water as they form into place to feed the escaping vein from **a** to **b** in the following sketches, have their direction changed by the

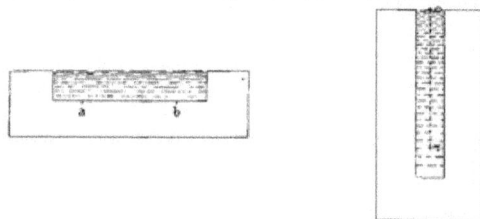

side or sides of the canal parallel to the line **a b**, and are not affected by the sides or side of the canal at right angles to the line **a b**.

The distance from the point **a** to the bottom of the weir in one sketch, or from the points **a** and **b** in the other sketch to the respective end of the weir, can be assumed to be equal to the least dimension of the weir.

In practice, for weirs l is nearly always much greater than h, and hence partial bottom suppression is a more important consideration than partial end suppression, where G, L and L' are equal.

The experimental curve for Fig. 12 on Plate VIII., illustrates the effect of inclined sides ($\beta = 45°$) of the feeding canal, with L and L' both .066, and full bottom contraction ; it will be noticed with this form of approach there is more contraction than with Fig. 9—Nos. 110-116 on Plate VIII.—where l was the same as for Fig. 12, $L = L' = .066$, and sides of approach normal to the weir ($\beta = 0$).

The question of partial suppression is very intricate, and attempts to satisfactorily analyze the effect upon the discharge, due to various modifications of the forms of approach in the feeding canal, are of more interest as curious mathematical and experimental problems, than of practical importance. For, if it be desired to measure the flow of water over a weir with great accuracy, there should be either complete contraction, or complete side suppression ; bottom partial suppression should, if possible, be avoided.

A glance at Plate VIII. is sufficient to warn the experimenter of the danger of taking any other than these two simplest forms of approach. In the nine experimental curves drawn upon this plate, the value of c ranges from .481 to .711 with l and h constant, and the curves cross each other, in what at first sight appears to be a most contradictory manner. We have endeavoured to point out the causes of these irregularities, but without careful and laborious study it is hardly possible for the hydraulic engineer to fully comprehend such apparently anomalous results.

Conclusions.

We consider the following proposition as demonstrated, viz. ;

To secure perfect contraction for a rectangular vertical weir, each side of the feeding canal must be distant from the adjoining parallel side of the weir, at least 3 times the least dimension of the weir. If this distance on all three sides be reduced to 2 times the least dimension of the weir, the discharge will be increased about $\frac{1}{2}$ of one per cent. ; after the last limit, the discharge increases more and more rapidly as the distance diminishes to 0.

This is a very different hypothesis from that adopted by Weisbach and others, of a fixed ratio between a and a_1. For instance, by our proposition full contraction will result

R

by having the following dimensions for the feeding canal, for three weirs, each having the same area of 12.

(A) Let $l = 12, h = 1$; then $(l + 2 [3 h]) \times (h + 3 h) = 72 = a_c = 6a$.
(B) Let $l = 4, h = 3$; then $(l + 2 [3 h]) \times (h + 3 h) = 264 = a_c = 22a$.
(C) Let $l = 1, h = 12$; then $(l + 2 [3 l]) \times (h + 3 l) = 105 = a_c = 8.7a$.

If our views are correct, the channel of approach must be as large as the given value of a_c for each of the three weirs, in order to insure perfect contraction; we therefore have $\frac{a_c}{a}$ varying from 6 to 22 in order to produce this result. With the forms of approach given in (A) and (B) there will be a notable velocity of approach, while with (C) the velocity of approach will be much smaller.

Generally in practice with weirs l is much larger than h; hence making $G = 3 h$, and L and L' each $= 2 h$, will insure almost perfect contraction. For weirs with full contraction on the sides, G can be reduced to 2 h with hardly any appreciable increase of flow, but when end contractions are suppressed it will be well to have G not less than 3 h. For shallow depths on a weir, such as .1 or .2, as a matter of precaution G had better be made not less than 4 or 5 h, as some of Lesbros' experiments appear to show that with such small heads the ratio between h and G should be pretty large in order to secure complete contraction.

Hydraulicians have nearly always confounded the effects of suppression and velocity of approach, by considering them together. As we have hitherto shown, there are three causes to be taken into account when the area a_c is not very large in proportion to the area a. They are:

FIRST.—The head due to velocity of approach, which increases the velocity, v, at the weir.

SECOND.—Suppression or partial suppression, which increases v.

THIRD.—Losses of head by "friction" and adhesion, between the measuring point for H and the weir, which diminish v.

These three causes are entirely distinct, and either one may be the factor of importance in determining the proper value to be given to c. This can well be illustrated by comparing the lower reference curve (c_c) on Plate VIII., with the values of c for Experiments Nos. 130, 129 and 101, l being constant at .656.

In No. 130, with contraction suppressed on bottom and nearly suppressed on both sides, the head, H, was measured 3.28 from the crest, in the narrow canal shown by Fig. 6, Plate I.; the deduced value of C for this experiment, in $Q = C \frac{2}{3} (2 g H)^{\frac{1}{2}} l H$, is 1.116, while, when the effect of v_a is taken into account, c is .685. The value of c_c for the same head is .593, showing an increased discharge due to suppression of about 15 per cent. Therefore in this instance the effect of v_a was extraordinarily great; the effect of suppression, a smaller but still large percentage; while our third cause had but little effect.

In No. 129, with contraction suppressed on both sides and complete on the bottom, the head H was measured .98 from the crest, in the narrow canal shown by Fig. 10,

Plate I. ; the value of C was about .668, while c was .655 ; the value of c_c for same head is .590. Hence in this case the effect of suppression increased c_c about 11 per cent., and the effect of c_a increased c about 2 per cent. .

In No. 101, with contraction suppressed on bottom and full contraction on both sides, the head H was measured in the large reservoir—Fig. 4, Plate I. In this experiment the effect of c_a would have been hardly sensible, had H been measured in the canal. But in this case c was .595, while c_c for same head is somewhere near .660 ; this is doubtless due to our third cause, which much more than counterbalances the effect of suppression. In this instance adhesion of the water to the bottom of the canal, the depth in the canal being quite small, apparently was the chief retarding influence, as the assumption of a very high co-efficient for "frictional" losses in the canal, would only account for the smaller part of the total loss of head.

It is hence apparent that the use of a formula like that of Weisbach,[*] which attempts to express the united effect of these three variables, will give entirely erroneous results, when the conditions are not identical with the experiments from which the numerical values in the expression have been deduced.

DETERMINATION OF c, FOR WEIRS HAVING EITHER FULL CONTRACTION OR END CONTRACTION SUPPRESSED, WITH FULL BOTTOM CONTRACTION ; l NOT LESS THAN .66.

We will now endeavour to ascertain the laws governing the discharge over the two simplest forms of weirs—those with full contraction, and those with contraction suppressed at both ends and with full bottom contraction—.66 being the shortest length, and with heads from .1 to 2. .

The co-efficients of discharge, c, will be denoted as follows ;

c for weirs with full contraction $\quad . \quad . \quad . \quad = c_t$

c „ suppressed weirs $\quad . \quad . \quad . \quad . \quad = c_s$

c „ weir of infinite length and finite depth $\quad . \quad . \quad = c_i$

The experimental data already given for these two forms of approach, can best be contrasted by putting the results in graphic form. This has been done on Plate VI. one of the co-ordinates representing h, and the other c.

We have in all twelve series of experiments, each series forming one experimental curve.

They are :

Series	I.	c_t	Lesbros (Fig. 10)	.	$l =$ .66	Experiments Nos. 1-5.
	II.	„	Fteley and Stearns	.	$l =$ 5.	„	17-34
	III.	„	J. B. Francis	.	$l =$ 10.	„	6-8
	IV.	„	Fteley and Stearns	.	$l =$ 19.	„	9-16

[*] Weisbach for weirs with partial suppression, proposes $\dfrac{c_s - c_t}{c_t} = 1.718 \left(\dfrac{a}{a_t}\right)^4$, or $c_p = c_t \left(1 + 1.718 \left[\dfrac{a}{a_t}\right]^4\right)$.

Series	V	c_e	Poncelet and Lesbros (Fig. 1)	$l = .66$	Experiments Nos. 35-40
,,	VI.	,,	Lesbros (Figs. 2 and 3)		,, 41-46
,,	VII.	,,	,, (thick-edged)	$l = 1.97$	,, 93-96
,,	VIII.	,,	Fteley and Stearns	$l = 2.3$	,, 76, 80, 85 and 86
,,	IX.	,,	Hamilton Smith, Jr.	$l = 2.6$	,, 57-68
,,	X.	,,	Fteley and Stearns	$l = 3.0$	,, 70, 72, 74, 79, 83 and 88
,,	XI.	,,	J. B. Francis	$l = 4.0$	,, 50 and 56
,,	XII.	,,	,,	$l = 10.0$	,, 47-49 and 51-55

Our number for each experiment is given on Plate VI.

An inspection of these twelve experimental curves on Plate VI. shows:

That all values of c_i are larger than those of c_e, for the same heads;

Also that c_i increases as the length of the weir diminishes, and c_e—with some minor exceptions—diminishes as the length of the weir diminishes.

It is apparent that for a weir of infinite length, h being finite and equal, there will be no difference in the values of c_i, c_e, and c_e as the effect of either suppression or contraction will be confined to some finite horizontal distance from the ends of the weir.

In confirmation of this proposition, we observe that as l increases, the curves for c_i and c_e approach each other; those of the longest of each form—19-foot suppressed, and 10-foot with contraction—being quite close together with $h = .6$, and gradually diverging to the maximum head of 1.6. We hence feel warranted in assuming that the curve for c_i will be somewhere between these last two curves.

We can now deduce the general proposition, *That, compared with the curve of c_i for a weir of infinite length, with full contraction c, diminishes as l diminishes, and with contraction suppressed c_i increases as l diminishes.*

This proposition applies where l is greater than .5; for smaller values of l, as we shall see hereafter, it does not strictly apply.

There are minor discrepancies in some of our experimental curves, and hence it will be first necessary to determine what weight should be given to each series, and what experiments in the several series should be rejected, before we draw on Plate VI. our final theoretic curves, representing c_e and c_i for various values of l.

We have already expressed the belief that the Lesbros curve with Figs. 2 and 3, more accurately represents c_i for $l = .66$, than the curve of Poncelet and Lesbros. Hence we will transfer from Plate VIII. the theoretic curve for c_i, $l = .66$, which was based upon results with Figs. 2 and 3.

Series VII. is more or less uncertain, owing to the thickness of the crest (.164). Judging from Lesbros' remarks, Nos. 93 and 94 were almost free from the influence of this thick crest.

Experiments Nos. 57 to 63 of the author, Series IX., will be accepted as determining c_e for $l = 2.6$; the other experiments of this series will only be regarded as approximations.

Series VIII. and X. of Fteley and Stearns, although indirectly determined, will have more weight than Series XI. of Francis, where two 4-foot weirs were separated by an interval of 2 feet, thus complicating form of contraction.

Series XII. of Francis will be accepted as determining c, for $l = 10.$.

Nos. 6 and 8 of Series III. of Francis will be considered a trifle too high, as these experiments were with a perfectly free discharge into the air, while in those of Fteley and Stearns, with which they are to be compared, the escaping vein was slightly confined at the ends of the weir. No. 7 will be assumed as too low, as in that experiment the escaping vein was more confined, than in Series II. and IV.

Experiments Nos. 20, 28, 29 and 32 of Series II. show abnormally large values of c, compared with the other experiments of the same series; they will hence be discarded.

Series I. of Lesbros (Fig. 10, Plate I.) is unreliable, owing to losses by primary contraction and "friction," which we have before described. Hence we will not attempt to form a theoretic curve for c, with $l = .66$. On Plate VIII. will be found this curve, which is approximately correct.

It will be kept in mind that c, represents perfectly free discharge into the air, while c, is based upon the escaping vein being confined by prolongations of the sides of the canal, but which do not extend below the level of the crest; this latter form is adopted as our standard, because Series II. and IV., upon which we chiefly depend for the establishment of c, were made with this form of discharge.

We will first determine the position of the curve for c_i. With $h = .6$ c, is .616 or thereabouts, its position being closely determined by the close approximation of c, for $l = 19.$, and c, for $l = 10.$. For the same head c_i for $l = 2.6$ is .604 or .605, and c, for $l = 5.$ is .627 or .628; c_i is therefore about half-way between c, for $l = 5.$ and c_i for $l = 2.6$; hence we can assume that the $+$ effect of suppression (both ends) is somewhat greater than the effect of full contraction, compared with c_i. This assumption is fully confirmed by a careful contrast of the curves for c, and c_i on Plate VI.

With $h = 1.6$, placing c_i at .614 corresponds with the above hypothesis.

For heads from .1 to .25, c_i should be very close to c, for $l = 5.$ for the smaller value of h, and gradually diverging as the head increases.

Having these points fixed, we can now easily draw on Plate VI. the central heavy line representing c_i with h from .1 to 1.6. It will be observed, that with h from 1. to 1.6, c_i is practically a straight line; hence we can continue c_i with h from 1.6 to 2. with same value of .614 with safety. Whether c_i for heads above 2. can be assumed to continue constant at .614 is uncertain. We will allude to this point again in our comparison of the analogies between orifices and weirs.

For smaller values of h than .1, the experimental data at hand cannot be considered as reliable; in fact c, for heads of less than .2 cannot be considered as being accurately determined. Hence, it is a safe proposition to state that *no weir measurement of water should be made with h less than .2, if accuracy is essential.*

In drawing the curves for c_s and c_c we will be governed, first by the experimental data considered reliable, and next by the proposition, that compared with c_i, the increased discharge produced by suppression and the diminished discharge due to contraction are approximately expressed by the following formulæ;

$$c_s = c_i \left(1 + x \frac{2 h}{l + 2 h}\right), \quad \text{or} \quad c_s = c_i \left(1 + x \frac{S}{p}\right),$$

and

$$c_c = c_i \left(1 - y \frac{2 h}{l + 2 h}\right), \quad \text{or} \quad c_c = c_i \left(1 - y \frac{S}{p}\right).$$

In these expressions x and y will vary so as to best suit the experimental data, x being in general rather larger than y.

By reference to Plate VI. it will be seen that our theoretical curves agree quite closely with the most trustworthy observations, and that there are no striking discordances.

The data for c_s with $l = 2$., and $l = 3$, and for c_c with $l = 1$. are insufficient; hence these curves are shown by dotted lines.

Some experiments of Mr. Francis, given on pp. 88-95 of "Lowell Hydraulic Experiments," will be of service in testing the accuracy of our theoretic curves on Plate VI.

All the experiments were with nearly full contraction, the contraction being less perfect for the larger values of l. Q was constant for each series, but was not determined.

Taking value of c_s from Plate VI. for the first experiment in each series, we can deduce Q, and then c for the other experiments of the same series, so as to make a comparison with values of c_s on Plate VI.

TABLE XLV.

J. B. Francis.—Comparison of Values of H for Various Values of l, Q being Constant for each Series.

$$(2 g)^{1/2} = 8.020$$

Series	Francis No.	l	H	Q	c	Plate VI. c_r	Difference between c and c_r
I.	1	6.987	.3146	4.096		.6214	
	6	3.500	.5085		.601	.609	.005
II.	14	6.987	.3745	5.291		.6180	
	11	3.500	.6049		.601	.606	−.005
III.	17	6.987	.4681	7.347		.6141	
	22	3.500	.7580		.595	.602	−.007
IV.	32	6.987	.6554	12.087		.6098	
	31	3.496	1.0656		.588	.595	−.007
V.	56	6.987	.9554			.6047	
	55	5.487	1.1336	21.096	.596	.599	−.003
	57	8.489	.8361		.608	.609	−.001
VI.	70	8.489	.3662			.6193	
	69	1.829	1.0627	6.229	.581	.587	−.006
	71	3.487	.4936		.612	.612	0

In the foregoing experiments with $l = 7$, there was probably slight partial suppression, which, if taken into account, would have somewhat increased the values of Q, and hence would have increased the deduced values of c.

We can hence conclude that these experiments verify our curves on Plate VI. for c_r.

Some experiments by General Ellis, p. 91 Transactions Am. Soc. of C.E., 1876, where the water first flowed over a short weir, and then over a much longer one, give the following results. Q is deduced from our values of c, for the long weir.

TABLE XLVI.

Ellis.—Orifices used as Weirs.

$(2 g)^{1/2} = 8.020$

	LONG WEIR.					Q	SHORT WEIR.		
l	H	h	c_i Plate VI.	Q'	Leakage.		l	h	c
10.	.6085	.6089	.6133	15.580	.126	15.454	2	1.8372	.5803
„	.6070	.6071	„	15.522	.126	15.396	„	1.8290	.5821
„	.6100	.6104	„	15.637	.126	15.511	„	1.8440	.5793
„	.6072	.6076	„	15.530	.126	15.404	„	1.8326	.5807
6.	.2849	.2849	.6262	3.0547	.0637	2.9910	2.	.5793	.6344
„	.3778	.3778	.6171	4.5970	.0662	4.5308	„	.7849	.6093
„	.4800	.4801	.6130	6.5417	.0702	6.4715	„	1.0079	.5981
4.	.3930	.3930	.6144	3.2373	.1583	3.0790	1.0001	1.0160	.5621

The above values of c are contradictory, and show that there were imperfect methods of observation. This table is inserted as a corroboration of our remarks on the experiments with orifices by the same authority, made at the same time as the foregoing weir determinations.

Weirs with One End Contraction Suppressed.

We can now consider the effect of suppression at one end only.

With this form of approach it is apparent that its co-efficient, c_e', will be very nearly c_i; for, in the equation,

$$c_e' = c_i \left(1 + x \frac{h}{l + 2\,h} - y \frac{h}{l + 2\,h} \right),$$

x and y being nearly equal, c_e' will be practically the same as c_i.

In Table XXXIV. are given experiments by Fteley and Stearns, with contraction suppressed at one end : with $l = 4.0$, L was 1.0 ; with $l = 3.3$, L was 1.7 ; hence with the longer weir there was more or less partial side suppression at the other end of the weir. Comparing the deduced values of c for these experiments, with values of c_i on Plate VI. we have ;

TABLE XLVII.

Fteley and Stearns (Table XXXIV.). Our End Contraction Suppressed.

Comparison of c_i and c_c of Plate VI.

No.	l	h	c	c_c Plate VI.	Differences between c and c_c.
69	4.01	.176	.6388	.6379	+ .0009
71	"	.269	.6285	.6279	+ .0006
73	"	.395	.6212	.6204	+ .0008
77	"	.499	.6177	.6174	+ .0003
82	"	.602	.6177	.6156	+ .0021
87	"	.709	.6156	.6147	+ .0009
92	"	.95	.6196	.6142	+ .0054
75	3.31	.450	.6173	.6187	− .0014
78	"	.569	.6139	.6160	− .0021
81	"	.578	.6130	.6159	− .0023
84	"	.688	.6114	.6149	− .0035
89	"	.809	.6105	.6143	− .0038
91	"	.935	.6093	.6142	− .0050

These differences are not excessive, especially when we remember that c was indirectly determined from our curve for c_i, $l = 5$. It is quite possible that in the determination of this curve for c_i we were not enough influenced by Experiments Nos. 20, 28 and 29 (Plate VI.), and hence placed it slightly too low. If this be the case, c for $l = 3.3$ in foregoing table would be practically identical with c_i, while c for $l = 4.0$ would be somewhat higher than c_i, on account of partial suppression at the other end of the weir; as before remarked in our discussion of Partial Suppression, the greater the head the more appreciable should be the effect from this cause.

Final Conclusions.

Whether or not the curves for c_i, or those for c_c, cross each other with increasing heads, the experimental data thus far given do not indicate, unless we accept Experiments Nos. 93 and 94 of Lesbros as being exact. Should they be so considered, it would show that the curve c_c for $l = 2$ crosses curve c_c for $l = 2.6$ at $h = 1.3$, as will be noticed by reference to Plate VI. This question will again be considered from a theoretic point of view, when we trace the analogies between orifices and weirs.

On Plate VII. will be found a clean copy of the theoretic curves on Plate VI.

It will be observed that for considerable heads the curves for c_c are reversed. For,

s

if c_i continues constant for heads above 1.6, as seems to be indicated by our diagram, then it is evident that the trend of the curves, both for c_s and for c_a, must be towards the straight line c_i; otherwise with great depths, c_s would rise to 1. or more, while c_a would fall below 0.

The following table has been compiled from Plate VII. In applying the co-efficients given, it must be remembered, that:

The weir must have a horizontal crest, with vertical sides, with its plane at right angles to the line of the feeding canal.

The inner face of the bottom and sides of the weir (lower end of the canal) must be vertical.

The edges of the weir must be so thin, that the escaping vein will only come in contact with the inner corner lines. To insure accuracy, these edges should be formed by stiff metallic plates.

The discharge must be free into the air, for full contraction. For suppressed weirs, the escaping vein should be confined by prolongations of the sides of the canal, but which must not extend lower than the level of the crest; in case a suppressed weir has a perfectly free discharge, the following co-efficients are about $\frac{1}{4}$ of 1 per cent. too low. Free access for air must always be provided under the escaping vein.

H should for ordinary cases be measured for suppressed weirs at a point 6 feet upstream from the crest, and the sides of the canal between H and the crest should be of smoothly planed plank. For full contraction H can be measured at any convenient point from 4 to 10 feet from the crest, and the smoothness of the sides is of no importance. When H and l are exceptionally great, the measuring point for H must be taken at a sufficient distance from the weir, to be above (upstream) the origin of the surface curve in the feeding reservoir. It will be remembered that at the Holyoke dam, with $H = 7.24$ and l very great, the proper point for the measurement of H was about 60 feet above the dam or weir.

H should always be corrected for velocity of approach; for this purpose Table XXV. should be used.

When the water enters the feeding canal with a velocity greater than the mean velocity at the measuring point for H, screens or racks should be employed to produce a more uniform velocity; vertical slats, beveled in the direction of the current, are the best; the openings should be so small as to cause a slight, but appreciable, loss of head as the water passes through the racks; the racks should be placed several feet above the measuring point for H.

To obtain perfect contraction, each side of the feeding canal should be distant from the adjoining parallel side of the weir, at least three times the least dimension of the weir. But, as l is generally much larger than h, the distance from the weir end to the canal side, can be fixed at $2h$; the inner depth below the crest (G) can also be placed at

$2 h$, without notable error for weirs with full contraction. These bottom and side distances (G, L and L') had best be made not less than 1 foot.

The surface of the water in the pool below the weir can be brought to the elevation of the crest, without forming any obstacle to free discharge, provided the vein retains its normal form; ample avenues for air to enter under the sheet must be provided, when the level of the lower water is nearly the same as that of the crest, otherwise atmospheric pressure will distort the vein from its normal position.

The formulæ used are;

(A) $\quad Q = c\,\frac{2}{3}\,(2\,g\,h)^{\frac{1}{2}}\,l\,h = c\,\frac{2}{3}\,(2\,g)^{\frac{1}{2}}\,l\,h^{\frac{3}{2}}.$

(B) $\quad Q = c\,\frac{2}{3}\,(2\,g)^{\frac{1}{2}}\,l\,\left(H + b\,\frac{v_a^2}{2\,g}\right)^{\frac{3}{2}} = c\,\frac{2}{3}\,(2\,g)^{\frac{1}{2}}\,l\,\left(H + b\,\frac{\left(\frac{Q}{n_c}\right)^2}{2\,g}\right)^{\frac{3}{2}}.$

Equation (B) is of high degree, but $v_a = \dfrac{Q}{n_c}$ can be easily obtained with sufficient accuracy by one or two approximations.

TABLE XLVIII

Co-Efficient of Discharge, c, in *Formula* $Q = c \frac{2}{3} (\frac{2}{3} g A)^{\frac{1}{2}} L$

For Weirs, either having Full Contraction, or Contraction Suppressed at both Ends, or Contraction Suppressed at one End. There being in all cases Full Contraction at the Bottom. (c, for $l = .66$ from Plate VIII.)

Effective Head — h	c, = Full Contraction. l = Length of Weir.										c, = nearly c', for Contraction at one End.	c, = Contraction Suppressed at both Ends. l = Length of Weir.									h
	·66	1 (f)	2	2·6	4	5	7	10	13	19	19	19	13	10	7	5	4	3 (f)	2 (f)	3 (f) ·66 (f)	
.1	.632	.629	.616	.650	.637	.633	.634	.633	.653	.646	.656	.637	.657	.658	.636	.659	.647	.645	.649	.632	.655	.1
.15	.619	.623	.634	.637	.628	.640	.641	.642	.642	.642	.643	.643	.644	.644	.637	.645	.645	.645	.643	.652	.652	.15
.2	.611	.618	.626	.629	.630	.631	.632	.639	.644	.634	.635	.635	.636	.637	.638	.638	.641	.642	.643	.645	.656	.2
.25	.605	.612	.621	.625	.624	.625	.627	.628	.634	.639	.629	.630	.631	.639	.639	.631	.636	.636	.638	.643	.653	.25
.3	.601	.608	.616	.618	.619	.625	.625	.624	.624	.625	.625	.626	.627	.658	.629	.631	.635	.636	.636	.639	.651	.3
.4	.595	.601	.609	.612	.613	.617	.618	.619	.629	.629	.620	.621	.622	.621	.623	.628	.630	.630	.633	.636	.650	.4
.5	.590	.596	.605	.605	.608	.611	.613	.616	.616	.617	.617	.619	.620	.621	.624	.627	.627	.630	.637	.637	.650	.5
.6	.587	.594	.601	.603	.601	.608	.609	.612	.614	.615	.616	.618	.619	.620	.625	.625	.628	.631	.634	.638	.651	.6
.7	.583	.590	.598	.601	.606	.604	.607	.611	.613	.614	.615	.618	.619	.621	.625	.629	.633	.633	.637	.643	.653	.7
.8			.595	.593	.600	.604	.606	.611	.612	.613	.614	.618	.620	.630	.627	.627	.631	.635	.639	.645	.626	.8
.9			.592	.590	.598	.601	.604	.609	.610	.612	.614	.619	.619	.630	.628	.633	.637	.639	.641	.645		.9
1.0			.590	.593	.595	.598	.601	.606	.605	.611	.614	.619	.621	.624	.639	.635	.637	.639	.644	.648		1.0
1.1			.587	.591	.593	.597	.601	.605	.605	.610	.614	.620	.623	.625	.630	.635	.630	.641	.646			1.1
1.2			.585	.589	.591	.596	.597	.604	.605	.619	.614	.626	.628	.636	.632	.638	.643	.648				1.2
1.3			.583	.586	.589	.596	.596	.603	.604	.609	.614	.628	.624	.628	.633	.638	.643	.644				1.3
1.4			.580	.584	.587	.590	.594	.598	.602	.609	.614	.629	.622	.629	.634	.640	.644					1.4
1.5				.582	.585	.592	.596	.599	.605	.608	.614	.622	.625	.630	.636	.641	.646					1.5
1.6				.582	.587	.591	.590	.600	.604	.607	.614	.623	.625	.631	.638	.642	.647					1.6
1.7				.580	.582	.591	.594	.599	.603	.607	.614	.623	.626	.632	.638							1.7
2.0											.614											2.0

For values of h and l intermediate to those given in the preceding table, c can be obtained by interpolation, or more accurately by reference to Plate VII.

Formulæ in Use.

Lesbros considers the following equation for weirs, analogous with $Q = C a (2 g h)^{\frac{1}{2}}$ for orifices (h for orifices being measured at the centre),

$$Q = c' \, l \, H_w \left(2 g \left[H - \frac{H_w}{2} \right] \right)^{\frac{1}{2}}.$$

For his weir .656 long, with various forms of approach shown on Plate I., and with $H = h$, by this formula c' had the following values :

c' for Figs. 2 and 3	=	.581 – .684
c' ,, Fig. 4	=	.601 – .811
c' ,, ,, 5	=	.622 – .742
c' ,, ,, 6	=	.718 – .756
c' ,, ,, 8	=	.603 – .669
c' ,, ,, 9	=	.638 – .718
c' ,, ,, 10	=	.676 – .797
c' ,, ,, 12	=	.605 – .668

Boileau[*] gives for suppressed weirs the formula,

$$Q = \frac{G + H}{([G + H]^{\frac{3}{2}} - H^{\frac{3}{2}})^{\frac{1}{2}}} \left(1 - \frac{H_w}{H} \right)^{\frac{1}{2}} l \, H \, (2 g H)^{\frac{1}{2}},$$

in which he attempts to take into account the effects of velocity of approach and partial suppression at the bottom. As he makes $\left(1 - \frac{H_w}{H} \right)^{\frac{1}{2}} = .417$, and constant, we can reduce his equation to our form, viz.,

$$Q = \frac{G + H}{([G + H]^{\frac{3}{2}} - H^{\frac{3}{2}})^{\frac{1}{2}}} .625 \tfrac{2}{3} (2 g H)^{\frac{1}{2}} H \, l.$$

M. Boileau afterwards recognised the fact that $\left(1 - \frac{H_w}{H} \right)^{\frac{1}{2}}$ is not constant,[†] and also that his new formula gave results considerably varying from his experimental results. He finally proposes the similar form,[‡] with the measured values of H_w and H,

$$Q = \frac{\left(1 - \frac{H_w}{H} \right)^{\frac{1}{2}}}{\left(1 - \left[\frac{1}{1 + \frac{G}{H}} \right]^{\frac{3}{2}} \right)^{\frac{1}{2}}} l \, H \, (2 g H)^{\frac{1}{2}} ;$$

and gives a table of corrections for his theoretical co-efficient as above, the factor of correction ranging from .96 to 1.07.[§]

We regard this formula as radically wrong in principle, and the experiments of M. Boileau, upon which this expression is more or less based, as inaccurate.

* Jaugeage des cours d'eau, Paris, 1850, par M. P. Boileau, p. 41.
† "Traité de la mesure des eaux courantes." Par P. Boileau. Paris 1854. Page 120.
‡ "Traité de la mesure des eaux courantes," p. 87.
§ "Traité de la mesure des eaux courantes," p. 119.

The experiments of M. Boileau have been frequently quoted as authority, and it will not be out of place for us to here give our reasons for disregarding them.

Taking from his " Traité de la mesure des eaux courantes," pp. 89-97, the four series, with end contractions suppressed, with free discharge, applying our corrections for velocity of approach, and making $(2 g)^{\frac{1}{2}} = 8.0227$, the value assigned by M. Boileau, we have the following results;

I.

No.	G	l	H	v_a	h_a'	h	Q	c_a
1		.9547	.1499	.05	.0001	.1500	.1806	.6088
2		.9449	.2133	.08	.0001	.2134	.3023	.6068
3		.9449	.2690	.12	.0003	.2693	.4303	.6093
4	3.636	.9547	.2936	.13	.0004	.2940	.4978	.6116
5		.9547	.4396	.24	.0012	.4408	.9090	.6162
6		.9547	.5479	.32	.0021	.5500	1.2816	.6153
7		.9449	.9774	.35	.0025	.9799	1.3726	.6181
8		.9547	.9671	.57	.0028	.9999	1.4765	.6223

II.

No.	G	l	H	v_a	h_a'	h	Q	c_a
1			.2067	.10	.0002	.2069	.3029	.6326
2			.2411	.12	.0003	.2414	.3811	.6313
3			.2658	.14	.0004	.2662	.4399	.6294
4	3.018	.9515	.2789	.15	.0005	.2794	.4724	.6286
5			.3970	.25	.0013	.3983	.8040	.6286
6			.4364	.28	.0017	.4381	.9318	.6314
7			.5348	.38	.0030	.5378	1.2773	.6364
8			.5840	.43	.0038	.5878	1.4641	.6371

III.

No.	G	l	H	v_a	h_a'	h	Q	c_a
1			.1870	.09	.0002	.1872	.2542	.6190
2			.2428	.14	.0004	.2432	.3784	.6221
3			.2625	.15	.0005	.2630	.4247	.6250
4	2.690	.9482	.2904	.17	.0006	.2910	.4932	.6195
5			.3396	.22	.0010	.3406	.6334	.6184
6			.4003	.27	.0015	.4018	.7988	.6185
7			.5512	.43	.0038	.5550	1.3160	.6277

IV.

No.	G	l	H	v_a	h_a'	h	Q	c_a
1			.1631	.10	.0002	.1633	.1991	.6249
2			.1847	.11	.0003	.1850	.2384	.6200
3			.2139	.14	.0004	.2143	.2932	.5820
4	2.028	.9482	.2854	.21	.0010	.2864	.4697	.6044
5			.3182	.25	.0013	.3195	.5302	.6106
6			.3543	.29	.0018	.3561	.6873	.6099
7			.5578	.54	.0061	.5639	1.3306	.6195
8			.5742	.57	.0067	.5809	1.4032	.6210

In all these experiments Q was determined by the flow through orifices : H appears to have been measured by placing a glass tube immediately above the weir, with its bottom near the floor of the feeding canal, the surface of the water in this tube being assumed as the height indicating H ; the pressure at the base of the weir is, as we believe M. Boileau was the first to point out, often in excess of the true value of H, but in the experiments we have selected, no serious errors ought to have arisen from this excess of pressure.

Now it seems apparent that, comparing those experiments where the velocity of approach was but slight, our deduced values of c, should be nearly identical for the four weirs with equal heads, no matter whether or not our method of computing h, be correct : that is to say, corrections for v, upon any theory must be so small as not to very largely change the values of c_i.

Contrasting the values of c_i, with h less than 0.3, we have ;

I.		II.		III.		IV.	
h	c_i	h	c_i	h	c_i	h	c_i
.15	.600	.21	.632	.19	.619	.16	.505
.21	.607	.24	.631	.24	.622	.18	.591
.27	.600	.27	.629	.26	.621	.21	.583
.29	.612	.28	.629	.29	.620	.29	.604

We here see that with $h = .21$, c, for the four series has the values of .607, .632, .620, and .583, all dimensions being practically identical, except G, which ranges from 3.6 to 2.0. The thoroughly reliable experiments of Mr. Francis, and Messrs. Fteley and Stearns clearly prove that with such a head, such variations in G will not appreciably affect the discharge, aside from the slight corrections necessary for the respective changes in c_i. If increasing G increases c_i, as Series II., III. and IV. indicate, then c, for Series I. should have a value of .64, instead of .61 as given.

We attribute these discrepancies solely to errors in experimentation ; they are so large as to palpably demonstrate that the experiments have no value except as rough approximations.

We have not selected for illustration M. Boileau's experiments where he directly measured Q, as in these G was comparatively small. Critically examined they present almost equally untrustworthy results, as the ones just discussed.

The Referees, Messrs. Simpson and Galton, for the Metropolitan Drainage, gave the following formula for weirs ;[*]

$Q' = 5.5 \ (d^2 + .8 \ d^2 \ v^2)^{\frac{2}{3}}$.
Q' being discharge in cubic feet per minute for each foot in length on the weir.
d = head H in inches.
v = velocity of approach in feet per second.

Transforming this to our expression, with $(2 g)^{\frac{3}{2}} = 8.02$, $Q = .713 \ \frac{2}{3} / (2 g)^{\frac{3}{2}}$ $(H^2 + .067 \ H^2 \ v_a{}^2)^{\frac{3}{2}}$.

The correction for H^2 is intended to represent effect of velocity of approach.

This formula when used for the reduction of sixty-two experiments made under the auspices of these gentlemen, with l from 3. to 10. and h from .08 to .75, appears to agree very fairly with the measured values of Q. These experiments must assuredly have been made under very peculiar conditions.

Weisbach's formulæ are as follows ;[†]

---

[*] Copy of Letter to Lord John Manners, First Commissioner London, 1858.
[†] Lehrbuch der Ingenieur.

Weirs with contraction ; $Q = c_v \left(1. + 1.718 \left[\dfrac{a}{a_v} \right]^4 \right) \frac{2}{3} (2\,g\,H)^{\frac{1}{2}} l\,H.$

Suppressed weirs ; $Q = c_v \left(1.041 + .3693 \left[\dfrac{H}{H+G} \right]^2 \right) \frac{2}{3} (2\,g\,H)^{\frac{1}{2}} l\,H.$

These expressions take into account both velocity of approach, and partial suppression or complete suppression ; they will give approximately correct results, except with unusual conditions.

Mr. Francis proposes the following formulæ ;

Weirs with both end contractions suppressed ; $Q = 3.33 \, l \, h^{\frac{3}{2}}$.

"   "   one   "   " $Q = 3.33 \, (l - .1\,h) \, h^{\frac{3}{2}}$.

"   "   full contraction ; $Q = 3.33 \, (l - .2\,h) \, h^{\frac{3}{2}}$.

Assuming $(2\,g)^{\frac{1}{2}} = 8.020$, and reducing to our form ;

Weirs with both end contractions suppressed ; $Q = .6228 \frac{2}{3} (2\,g\,h)^{\frac{1}{2}} l\,h.$

"   "   one   "   "   " $Q = .6228 \frac{2}{3} (2\,g\,h)^{\frac{1}{2}} (l - .1\,h)\,h.$

"   "   full contraction ; $Q = .6228 \frac{2}{3} (2\,g\,h)^{\frac{1}{2}} (l - .2\,h)\,h.$

Mr. Francis considers that these expressions will apply when h does not exceed $\dfrac{l}{3}$, and with h from .5 to 2..

Comparing the values of c, in $Q = c \frac{2}{3} (2\,g\,h)^{\frac{1}{2}} l\,h$, deduced from these equations, with our values of c_e, c_i or c_i', and c_v, on Plate VII., we have for the limits prescribed by Mr. Francis ;

	l	h	Francis.	Smith.
Suppressed weirs			$c =$ .623	$c_v = .614-.643$
One end suppressed			.602-.623	$c_i = .614-.617$
	1.5	.5	.581	$c_v = .601$
		.5	.610	.611
	5.	1.	.598	.601
		1.6	.583	.591
Full contraction		.5	.617	.615
	10.	1.	.610	.608
		1.7	.602	.599
		.5	.620	.617
	19.	1.	.616	.611
		1.7	.612	.607

From the above comparison it will be seen that the greatest variation of the Francis formulæ from the values of c as given by us, is in the case of weirs with full contraction, where the difference amounts in one instance to about $3\frac{1}{2}$ per cent. .

Messrs. Fteley and Stearns give for suppressed weirs, $Q = 3.31 \, l \, h^{\frac{3}{2}} + .007 \, l$, with h having a value of not less than .07. Assuming $(2\,g)^{\frac{1}{2}} = 8.020$, and reducing to our form ; $Q = .619 \frac{2}{3} (2 g h)^{\frac{1}{2}} l h + .007 l$. The correction of .007 l is intended to compensate for increased values of c with small heads.

We have shown with a suppressed weir having $l = .66$ and $h = .8$ (Experiment No. 1' of Lesbros, Plate VIII.) that $c_r = .655$; by the above formula c would be about .621, showing a variation of about 5 per cent. from the Lesbros result.

The formulæ of both Francis, and Fteley and Stearns, are based upon the hypothesis that $c_r = c_i$; in our judgment the experimental data we have given[*] prove beyond a doubt that c_r is always greater than c_i, for the same values of h.

M. Graëff, in his "Traité d'Hydraulique," expresses the opinion that the suppression of contraction on the two ends of a weir has the effect of reversing the law of the co-efficients. That is to say, c_r constantly increases with h, while c_i diminishes as h increases ; the curves for c_r and c_i crossing each other. He also thinks that the length of the weir has very little effect upon the co-efficient.

Expressions could be framed for c_c, c_i and c_r which would quite closely agree with the curves on Plate VII., but they would necessarily be complicated and therefore inconvenient for general use. The proper co-efficient for each of these three forms of approach can be obtained immediately from Table XLVIII.

The formulæ of Mr. Francis are from their simplicity convenient for practical purposes, as they can readily be kept in memory. If modified as follows, they will give results sufficiently exact when great accuracy is not required ;

Contraction suppressed on both ends ; $Q = 3.29 \left(l + \frac{h}{7} \right) h^{\frac{3}{2}}$.

" " " one end ; $Q = 3.29 \, l \, h^{\frac{3}{2}}$.

Full contraction ; $Q = 3.29 \left(l - \frac{h}{10} \right) h^{\frac{3}{2}}$.

The limits within which these expressions will apply with reasonable accuracy are, h from .5 to 2. ; l not less than 3 h.

SHORT WEIRS.

The smallest value of l for weirs, thus far considered, has been .656. For rectangular orifices it has been shown that c increases notably, as the least dimension of the orifice diminishes below this distance (.656), and it seems logical to suppose that there should be a similar increase in c for very narrow weirs. The experiments of Lesbros and Castel prove that this supposition is correct.

[*] The experiments of Castel, given in the following section of this chapter treating on Short Weirs, afford an additional proof of the correctness of our conclusions.

T

Lesbros.

Lesbros using his opening in a fixed copper plate .066 × 1.968 as a weir, with form of approach shown by Fig. 1, Plate I., obtained the following results.

TABLE XLIX.

Lesbros.—Short Weir ; Full Contraction ; Free Discharge into Air.   H measured 11.48 above Weir ; hence

$$H = h. \qquad (2\ g)^{1/2} = 8.0237$$

No.	Lesbros' Nos.	H	H_w	$H - H_w$	l	Q	c
131	1829-1830	1.9172	1.9390	.0082	.065 62	.609 0	.639
132	1831-1833	.9892	.9849	.0043		.221 4	.650
133	1834-1836	.5331	.5305	.0026	..	.089 01	.652
134	1837-1839	.2674	.2657	.0017	..	.031 68	.653

Castel.

The only other data for short weirs are the experiments made by M. Castel at Toulouse, the results of which are published in the Mémoires de l'Académie des Sciences de Toulouse, Tome IV., 1837.

The feeding canal employed was a rectangular horizontal box, 19.5 long and 2.428 wide ; this was supplied with water at its upper end by a pipe, with a valve attached, by which the amount of flow was regulated ; a metallic screen and fixed boards were used to prevent undulatory movements in the surface ; the weirs were made by copper plates, .007 thick, screwed to fixed frames. The head was measured by a point-gauge reading to .0001 metre, at the origin of the surface curve in the canal ; the larger the value of v_a, the less accurately was H measured, owing to greater oscillations of the surface in the canal caused by the increase in velocity. The discharge was ascertained by a measuring vessel holding 113 cubic feet ; the times were determined to $\frac{1}{4}''$ by a stop-watch ; the minimum value of t was 70''. There appear to have been frequently several determinations for a given experiment, although no details are given. Castel reduced his experiments by the formula (metrical) of $Q = 2.953\ c\ l\ H^{3/2}$; hence $(2\ g)^{1/2}$ was taken at 8.0232, which will be assumed as its value for our reductions.

In the following table are given the experiments with normal width of canal, various lengths of weir, and c always constant ; it will be noticed that a good many of these experiments were with l greater than .66 ; our reasons for not having hitherto used these determinations will be given hereafter.

TABLE 1.

Castel.—Rectangular Vertical Weirs. Feeding Canal 2.428 wide. C' always .558

$(2 g)^{1/2} = 8.0232$ $h = H + b \frac{v^2}{2g}$, and b constant at 1.1. $Q = c \frac{2}{3} l (2g)^{1/2} h^{3/2}$

No.	l	H	v_a	h_a'	h	Q	c	No.	l	H	v_a	h_a'	h	Q	c
135	.0328	.6526	.02	0	.6526	.061 80	.6681	166	.1637	.3287	.05	.0001	.3288	.101 1	.6122
136		.5899		0	.5899	.053 43	.6720	167	„	.2608	.04	0	.2608	.071 38	.6119
137		.5249		0	.5249	.044 92	.6730	168	„	.1959		0	.1959	.046 44	.6118
138		.4626		0	.4626	.037 26	.6748	169	„	.1624		0	.1624	.035 18	.6137
139		.3947		0	.3947	.029 52	.6785	170	„	.1303		0	.1303	.025 25	.6134
140	.0653	.7868	.05	.0001	.7869	.155 68	.6386	171	.3291	.7887	.22	.0011	.7898	.734 6	.5940
141		.7218	.04	0	.7218	.136 92	.6394	172		.7228	.21	.0010	.7238	.643 5	.5931
142		.6421		0	.6421	.114 99	.6400	173		.6519	.19	.0008	.6527	.550 6	.5926
143		.5889		0	.5889	.101 18	.6411	174	„	.5912	.17	.0006	.5918	.475 7	.5931
144	„	.5217		0	.5217	.084 44	.6418	175	„	.5207	.15	.0005	.5212	.392 0	.5913
145	„	.4623		0	.4623	.070 36	.6429	176	„	.4551	.13	.0004	.4555	.320 2	.5911
146	„	.3911		0	.3911	.055 09	.6451	177		.3934	.11	.0003	.3937	.256 4	.5890
147		.3255		0	.3255	.042 03	.6482	178		.3297	.09	.0002	.3299	.197 1	.5904
148		.2608		0	.2608	.030 34	.6521	179		.2618	.07	.0001	.2619	.139 8	.5919
149		.2008		0	.2008	.020 70	.6587	180		.1991	.05	.0001	.1992	.093 13	.5945
								181		.1660	.04	0	.1660	.071 13	.5968
150	.0988	.6549	.06	.0001	.6550	.175 95	.6285	182		.1401		0	.1401	.055 84	.6044
151		.5925	.05	.0001	.5926	.151 30	.6279	183		.0991		0	.0991	.033 94	.6176
152	„	.5236		0	.5236	.125 69	.6289								
153		.4597		0	.4597	.103 33	.6278	184	.6542	.6785	.388	.0052	.6817	1.164 7	.5914
154		.3947		0	.3947	.082 18	.6274	185	„	.5837	.335	.0024	.5861	.928 1	.5911
155		.3294		0	.3294	.062 63	.6274	186		.5233	.300	.0020	.5253	.787 9	.5914
156		.2615		0	.2615	.044 36	.6280	187	„	.4615	.263	.0013	.4628	.649 8	.5898
157		.1975		0	.1975	.029 10	.6276	188	„	.3921	.22	.0011	.3932	.508 9	.5899
158		.1627		0	.1627	.021 83	.6294	189	„	.3268	.18	.0007	.3275	.387 4	.5907
								190	„	.2633	.11	.0004	.2635	.280 8	.5933
159	.1637	.7959	.12	.0003	.7962	.382 5	.6118	191	„	.1962	.10	.0003	.1965	.185 3	.6022
160	„	.7326	.11	.0003	.7329	.337 5	.6113	192	„	.1690	.08	.0001	.1691	.118 8	.6102
161	„	.6595	.10	.0002	.6597	.287 9	.6135	193	„	.1299	.06	.0001	.1300	.101 4	.6184
162	„	.5821	.09	.0002	.5826	.238 7	.6129	194	„	.0994	.04	0	.0994	.068 44	.6240
163	„	.5266	.08	.0001	.5267	.204 9	.6122								
164	„	.4564	.07	.0001	.4565	.165 6	.6133	195	.9849	.4528	.394	.0034	.4562	.968 0	.5963
165	„	.3963	.06	.0001	.3964	.133 7	.6120	196	„	.3953	.342	.0025	.3978	.791 1	.5985

TABLE L.—*continued.*

No.	l	H	v_a	h_a'	h	Q	c	No.	l	H	v_a	h_a'	h	Q	c
197	.9849	.3264	.277	.0017	.3281	.5937	.5996	213	1.6483	.1650	.21	.0010	.1660	.3740	.6272
198		.2602	.21	.0010	.2612	.4231	.6016	214	„	.1335	.16	.0006	.1341	.2738	.6324
199		.1985	.16	.0006	.1991	.3843	.6074	215	„	.1027	.12	.0003	.1030	.1863	.6392
200		.1663	.13	.0003	.1666	.2202	.6147								
201		.1342	.09	.0001	.1343	.1602	.6180	216	1.9689	.3251	.586	.0074	.3325	1.2576	.6229
202		.1037	.07	.0001	.1038	.1109	.6296	217	„	.2654	.465	.0048	.2702	.9281	.6275
								218	„	.1975	.335	.0022	.1997	.5961	.6343
203	1.3117	.4068	.483	.0050	.4118	1.1315	.6103	219	„	.1696	.268	.0015	.1711	.4736	.6354
204		.3448	.405	.0035	.3483	.8829	.6122	220	„	.1273	.18	.0007	.1280	.3083	.6392
205	„	.2641	.296	.0020	.2661	.5805	.6131	221		.1020	.14	.0004	.1024	.2236	.6479
206	„	.1962	.21	.0010	.1972	.3797	.6179								
207	„	.1591	.16	.0006	.1597	.2788	.6226	222	2.2323	.3054	.632	.0087	.3141	1.3257	.6298
208		.1309	.13	.0003	.1312	.2100	.6290	223	„	.2612	.526	.0060	.2672	1.0450	.6337
209	„	.1011	.09	.0001	.1012	.1434	.6348	224	„	.1988	.378	.0031	.2019	.6940	.6406
								225	„	.1644	.298	.0020	.1664	.5216	.6436
210	1.6483	.3192	.471	.0048	.3240	1.0037	.6173	226	„	.1358	.23	.0011	.1369	.3920	.6482
211		.2641	.379	.0031	.2672	.7565	.6212	227	„	.0945	.14	.0004	.0949	.2287	.6553
212	„	.1991	.269	.0015	.2006	.4951	.6251								

In Experiment No. 204 there may be a typographical error in the given values of Castel. If this be so, *c* for this experiment should be about .6220.

The same feeding canal was then narrowed to a width of 1.184, by placing a longitudinal vertical partition in the canal, having a length from the lower end of 7.35. At the lower end of this new canal weirs were placed, being formed as before of thin copper plates fastened to fixed frames. There was some loss of head by primary contraction as the water entered the narrow section of the canal, but this loss should not have affected the accuracy of the results, as M. D'Aubuisson appears to think, as H was measured not more than 1.4 from the weir, and hence about 6 feet below the mouth of the narrow section. These experiments should only be less accurate than those given in Table L., on account of the greater values of v_a, which caused, as before stated, greater surface oscillations.

TABLE LI.

Castel.—Rectangular Vertical Weirs. Feeding Canal 1.184 wide.

G always .558. $(2g)^8 = 8.0232$ $h = H + b\,\dfrac{v_a^2}{2g}$, and b constant at 1.4. $Q = c\,\tfrac{2}{3}l(2g)^8 h^{1/2}$

No.	l	H	v_a	b_a'	h	Q	c	No.	l	H	v_a	b_a'	h	Q	c
228	.0328	.7710	.05	.0001	.7741	079 60	.6660	258	.1637	.2631	.07	.0001	.2632	072 47	.6129
229	,,	.6578	.04	0	.6578	062 41	.6665	259	,,	.1959	.05	.0001	.1960	046 55	.6126
230	,,	.5262		0	.5262	044 75	.6679	260	,,	.1657	.04	0	.1657	036 20	.6130
231	,,	.3944		0	.3944	029 31	.6745	261		.1303	.03	0	.1303	025 25	.6134
232	.0653	.7917	.11	.0003	.7920	159 00	.6460	262	.2582	.7530	.36	.0028	.7558	559 8	.6169
233	,,	.6542	.09	.0002	.6544	119 19	.6447	263	,,	.7195	.34	.0025	.7230	520 2	.6140
234		.5906	.07	.0001	.5907	102 03	.6435	264	,,	.6545	.31	.0021	.6566	448 5	.6104
235		.5230	.07	.0001	.5231	085 04	.6437	265	,,	.5929	.28	.0017	.5946	384 6	.6074
236		.4603	.06	.0001	.4601	070 35	.6449	266	,,	.5259	.25	.0014	.5273	320 0	.6052
237		.3950	.05	.0001	.3951	056 08	.6467	267	,,	.4610	.22	.0011	.4621	261 7	.6032
238		.3301	.04	0	.3301	042 91	.6480	268	,,	.3931	.18	.0007	.3938	205 1	.6010
239		.2572		0	.2572	029 81	.6543	269	,,	.3182	.14	.0004	.3186	148 8	.5991
								270	,,	.2625	.11	.0003	.2628	111 3	.5983
240	.0988	.7841	.14	.0004	.7845	230 9	.6290	271	,,	.1946	.08	.0001	.1947	071 06	.5989
241	,,	.7244	.13	.0004	.7248	204 2	.6365								
242	,,	.6565	.12	.0003	.6568	175 9	.6335	272	.3012	.7238	.40	.0035	.7273	607 4	.6079
243		.5902	.11	.0003	.5905	149 8	.6249	273	,,	.6401	.35	.0027	.6428	502 2	.6049
244		.5226	.10	.0002	.5228	124 70	.6245	274	,,	.5922	.33	.0024	.5946	445 0	.6025
245		.4590	.08	.0001	.4591	102 49	.6237	275	,,	.5249	.29	.0018	.5267	369 8	.6005
246		.3934	.07	.0001	.3935	081 30	.6233	276	,,	.4597	.25	.0014	.4611	301 6	.5979
247	,,	.3271	.06	.0001	.3272	061 63	.6234	277	,,	.3940	.21	.0010	.3950	238 6	.5966
248	,,	.2615	.04	0	.2615	044 07	.6240	278	,,	.3284	.17	.0006	.3290	181 4	.5968
249	,,	.1952	.03	0	.1952	028 54	.6263	279	,,	.2628	.13	.0004	.2632	129 6	.5958
								280	,,	.1988	.09	.0002	.1990	085 68	.5991
250	.1637	.7700	.24	.0012	.7712	368 7	.6218	281	,,	.1627	.08	.0001	.1628	063 85	.6034
251	,,	.7211	.22	.0011	.7222	333 6	.6189	282	,,	.1306	.05	.0001	.1307	046 44	.6101
252	,,	.6585	.20	.0009	.6594	289 0	.6163	283	,,	.1004	.04	0	.1004	032 03	.6251
253	,,	.5906	.18	.0007	.5913	244 7	.6146								
254	,,	.5262	.16	.0006	.5268	205 4	.6134	284	.3294	.7815	.476	.0049	.7864	754 0	.6137
255	,,	.4577	.14	.0004	.4581	166 5	.6132	285	,,	.7231	.438	.0042	.7273	665 0	.6086
256	,,	.3947	.12	.0003	.3950	133 3	.6131	286	,,	.6601	.400	.0035	.6636	577 1	.6059
257	,,	.3268	.10	.0002	.3270	100 4	.6130	287	,,	.5869	.355	.0027	.5896	481 7	.6039

TABLE LI.—*continued.*

No.	l	H	v.	h.'	h	Q	c	No.	l	H	v.	h.'	h	Q	c
288	.3294	.5249	.32	.0021	.5270	.405 4	.6015	306	.6542	.0965	.09	.0001	.0966	.0663	.6309
289	,,	.4603	.27	.0016	.4619	.331 1	.5986								
290	,,	.3957	.23	.0012	.3969	.262 7	.5963	307	.9849	.4603	.927	.0187	.4790	1.1160	.6390
291	,,	.3304	.19	.0008	.3312	.200 1	.5960	308	,,	.3937	.769	.0129	.4066	.8667	.6345
292	,,	.2625	.15	.0005	.2630	.141 8	.5967	309	,,	.3264	.616	.0083	.3347	.6442	.6315
293	,,	.1959	.10	.0002	.1961	.091 65	.5990	310	,,	.2608	.472	.0049	.2657	.4581	.6348
294	,,	.1650	.08	.0001	.1651	.071 59	.6057	311	,,	.1919	.324	.0023	.1942	.2885	.6399
295	,,	.1319	.06	.0001	.1320	.051 81	.6132	312	,,	.1568	.25	.0014	.1582	.2135	.6441
296	,,	.0948	.04	0	.0948	.032 28	.6275	313	,,	.1309	.20	.0009	.1318	.1632	.6473
								314	,,	.0988	.14	.0004	.0992	.1073	.6521
297	.6542	.6037	.736	.0124	.6161	1.039 7	.6144								
298	,,	.5249	.652	.0093	.5342	.835 9	.6119	315	1.1844	.5780	.929	.0188	.3968	1.0298	.6504
299	,,	.4613	.567	.0070	.4683	.683 7	.6097	316	,,	.3333	.791	.0136	.3469	.8349	.6450
300	,,	.3898	.470	.0048	.3946	.527 6	.6083	317	,,	.2638	.594	.0077	.2715	.5774	.6443
301	,,	.3317	.391	.0033	.3350	.412 5	.6080	318	,,	.1991	.420	.0038	.2029	.3765	.6303
302	,,	.2644	.30	.0020	.2664	.293 2	.6093	319	,,	.1634	.327	.0023	.1657	.2790	.6529
303	,,	.1946	.21	.0010	.1956	.183 3	.6122	320	,,	.1289	.24	.0013	.1302	.1959	.6582
304	,,	.1591	.16	.0006	.1597	.137 9	.6174	321	,,	.1004	.17	.0006	.1010	.1348	.6628
305	,,	.1339	.13	.0004	.1345	.107 1	.6217								

Of the foregoing experiments made by M. Castel, Nos. 135-261 and 313-321 have been plotted on Plate IX., with h and c as co-ordinates. If his weirs were placed in the centre of each canal, so that in all cases $L' = L$, it is apparent that when $L = 3$ l for either width of canal, l and h being constant, c should have similar values. We therefore have two series of the Castel experiments for four values of l, with complete end contraction, viz.; .033, .065, .099 and .164. By reference to Plate IX. it will be seen that while each separate series for these four weirs forms a curve of nearly perfect symmetry, the two curves often vary considerably with the largest heads; for instance, with $l = .065$ and $h = .79$, c for No. 140 is .6386, while for No. 232 it is .6460, showing a variation of over 1 per cent.. This variation could not have been caused by erroneous assumptions of the value of h_v', as in both these experiments v_a was quite small. These variations are perhaps due to experimental error, or perhaps L may not always have been the same as L', for the narrow canal, so that there was in same cases partial suppression on one side.

On Plate IX. have been drawn five curves in heavy solid lines, representing the most probable values of c for lengths of weir from .329 to .033, all with full end contraction. For the greatest length, $l = .329$, there was a slight partial bottom suppression for heads from .20 to .79 ; for the other four lengths there was complete bottom contraction. In the determination of these curves, greater weight has been attached to the experiments with the broad canal, than to those with the narrow canal. The Lesbros Experiment No. 134, with $l = .066$, agrees closely with the curve for $l = .065$; Nos. 133 and 132 are about $1\frac{1}{2}$ per cent. too high.

The series of experiments with the broad canal, where end contraction was partially suppressed, are represented by light dotted lines, except that for $l = 2.23$, when there was nearly complete end suppression, which is shown by a heavy dotted line. The series with $l = 1.18$ for the narrow canal, when there was complete end suppression, is also shown by a heavy dotted line.

For all the Castel experiments b, in $h_a' = b \dfrac{v_a^2}{2\,g}$, has a constant value of 1.4 in our reductions ; this value is in all probability excessive for the larger values of v_a, and consequently our deduced values of c are probably in some cases too low, and especially so for the series with $l = 1.18$, Experiments Nos. 315-321. For No. 315, where v_a was the greatest, c as calculated by us appears to be abnormally high ; this would seem to show that b should have had a greater value than even 1.4 ; on the other hand the Castel experiments given in Chapter IV. (Velocity of Approach, Experiments Nos. 118 to 152) indicate that the smaller value of b of $1\frac{1}{2}$ is too great. Taking them as a whole, the experiments given in Tables L. and LI. show that our correction for v_a is too small, while those in Table XXIV. show that it is too great ; hence we can assume that the Castel experiments do not contradict our conclusions in regard to the effect of velocity of approach.

Comparing the five heavy symmetrical curves on Plate IX., where there was practically complete contraction on all three sides, it will be observed that as the length diminishes from .329, the co-efficient c constantly increases for equal heads. With $h = .8$, c has the following values ;

$l = .329$	$c = .591$
$l = .164$	$c = .615$
$l = .099$	$c = .628$
$l = .065$	$c = .642$
$l = .033$	$c = .666$

For heads diminishing from $h = .3$, with l constant, c constantly increases ; for heads increasing from $h = .3$, c for two of the curves slightly diminishes, and for the three other curves slightly increases. Taking into consideration the chances of experimental error, it cannot be said that these Castel experiments clearly prove, that with complete contraction, as h increases above .3, c does not remain constant.

The irregular curve for $l = .654$ (Nos. 184-194) quite closely agrees with the curve for $l = .329$; for this series (Nos. 184-194) there was a very slight partial suppression of contraction, but not enough to notably alter the form of the curve. We have seen before that for weirs with full contraction, with h constant, c, or c_e, increases as l increases above .66; we can hence assume that the law in regard to the effect of l upon c_e changes when l is about .5. Therefore, *for weirs having full contraction, the co-efficient of discharge for equal heads, increases as l increases above .5, and also increases as l diminishes below .5.* When h is a small fraction of l, the effect upon c_e of variation in l will doubtless be small.

In the curves on Plate IX., for the lengths .654, .985, 1.31, 1.65, 1.97 and 2.23, c for equal heads constantly increases with the length; this is partly due to the effect of changes in l, but more largely to the effect of partial suppression; the weir with $l = 2.23$ being nearly suppressed on both ends.

For the purpose of comparing the results of Castel with the data discussed in the first part of this chapter, there have been drawn on Plate IX. in light solid lines curves for c, with $l = .66$, and for c_e with $l = 1$, and $l = 2$; these curves have been taken either directly, or by interpolation, from the curves on Plate VII. It will be seen that the curve for c_e with $l = .66$, is for heads less than .5, higher than the irregular curve for Castel's Nos. 186-194 with a weir of nearly the same length; Castel's experiments with this weir, agree more closely with Experiments Nos. 35-40 of Poncelet and Lesbros, Plate VIII., than with those of Lesbros from which this curve of c_e was determined. The irregular curve for Castel's Nos. 315-321, with complete end suppression and $l = 1.18$, agrees closely with the curve for c_e with $l = 1$, taken from Plate VII., except No. 315, which appears to be abnormally high. The irregular curve for Castel's Nos. 222-227, with end contraction nearly suppressed and $l = 2.23$, is somewhat lower than the curve for c_e with $l = 2$, taken from Plate VII.; had there been full end suppression for this series of Castel, the two curves would apparently have been nearly identical. Velocity of Approach Experiments Nos. 143 and 147 of Castel were with $l = 2.43$ and end contraction suppressed, with very nearly complete bottom contraction; by comparing the values of c for these experiments with the curves for c_e on Plate VII., it will be seen that they agree very fairly; for instance with $l = 2.4$ and $h = .26$, c_e by Plate VII. is .638, while for Experiment No. 143 (Velocity of Approach) with the same length and head, c is .636. Nos. 148-152, in the same Table No. XXIV., do not, however, agree so well with Plate VII.; the values of c in these latter experiments, when compared with the other experiments given in the same table, appear to be unreliable.

From the preceding comparisons, we are warranted in the statement, that the experiments of Castel with weirs having a greater length than .50, agree as closely as could be expected with the deductions we have drawn from the data furnished by Lesbros, Francis, Fteley and Stearns, and ourselves.

The experiments made by Castel with orifices having convergent mouth-pieces, and with weirs, form wonderfully smooth curves, when each series is plotted by itself. This symmetry for his weir experiments is somewhat more perfect, when H and C are used for co-ordinates, than when h and c as deduced by us are the co-ordinates.* When, however, the various curves for different weirs are critically compared, they present discordances, not very great to be sure, but still considerably larger than one would expect with such apparently accurate experimentation. It may perhaps be uncharitable, but we cannot refrain from the conjecture, that M. Castel devoted a good deal of time in his study in adjusting his experimental data, so that they would show a very high degree of accuracy when presented to the world. To one familiar with the very great physical difficulties in the way of the exact determination of C for weirs with small values of H, the exceedingly smooth curves shown by the Castel experiments, appear most suspicious. His experimental appliances were in every way inferior to those employed by Mr. Francis, and by Messrs. Fteley and Stearns; hence the remarkable accuracy of M. Castel inferentially shows that these later experimenters did their work carelessly ; this is an inference which we are by no means prepared to accept. We are frank to say that with the apparatus employed by Castel, with so small a measuring vessel that t was often as short as 80″, and with heads as low as .1, we cannot conceive it possible that such smooth curves could have resulted, unless each experiment represented the mean of many separate determinations, and which does not appear to have been the case. It may be, that seemingly incongruous results were rejected, and only those were given which seemed to be harmonious ; such suppression is never justifiable ; an experiment made under normal conditions, which at first sight seems to be discordant, may upon closer study have an important bearing upon the problem to be solved. The perfect frankness of Messrs. Fteley and Stearns, who faithfully give *all* their results, affords an example which should be always followed.

We are of the opinion that M. Castel, considering the appliances which were at his disposition, made his experiments with much skill, and that his results are fairly accurate, but that they bear such evidences of having been "doctored," as to render them more or less open to doubt. Had it not have been for this opinion, we should have used many of his determinations in the first part of this chapter.

SUBMERGED WEIRS.

A submerged or incomplete weir, is one where the surface of the water down-stream from the weir is higher than the crest.

Dubuat proposed the following formula for such weirs, when there is no velocity of approach ;

(A) $$Q = c' \, l \, (2 \, g \, h)^{\frac{1}{2}} \, (H_t + \tfrac{2}{3} \, h) .$$

H_u = head measured up-stream in still water, above the crest.
H_t = ,, ,, ,, down-stream ,, ,, ,,
h = $H_u - H_t$, effective head producing the flow over the weir.

This formula is deduced by the following reasoning :

Dividing the section, $l \, H_u$, at the weir into two horizontal sections $l \, h$ and $l \, H_t$, it can be assumed that the discharge through the submerged section $l \, H_t$ is represented by

(B) $$Q' = c'' \, l \, H_t \, (2 \, g \, h)^{\frac{1}{2}} .$$

Also that the discharge through the unsubmerged section $l \, h$ is represented by the usual weir formula, or,

(C) $$Q'' = c''' \, l \, \tfrac{2}{3} \, h \, (2 \, g \, h)^{\frac{1}{2}} .$$

---

* Castel in his reductions did not take into account the effect of v_0.

If it be assumed that c'' and c''' are equal, by adding equations (B) and (C) together we have,

(A) $\qquad Q = Q' + Q'' = c' l (2 g h)^{\frac{1}{2}} (H_t + \tfrac{2}{3} h).$

When there is an additional head due to velocity of approach at the measuring point for H, it should be added to h.

Lesbros, very erroneously it seems to us, assumed that the entrance of the water from his reservoir into the feeding canal in Fig. 6, Plate I., and in a similar canal shown by his Fig. No. 19 which we have not given, should be considered as belonging to this class of weirs, and constructs his Table No. XXIV. from experiments similar to that from which we obtained the data for our No. 130. In this case (Lesbros' No. 2015) he takes the head H measured in the still water of the reservoir as H, (the canal being horizontal H_u is the same as H for the weir at the lower end of the canal); for H_t, or what he assumes as the submerged portion, he takes the point E (p. 110) being the lowest depression in the canal, and .99 down-stream from its upper end. He hence has $H_u = .4649$, $H_t = .2615$, $h = .2034$; measured value of $Q = .6391$, and $l = .7874$; then by formula $Q = c'' l (2 g h)^{\frac{1}{2}} H_u,$[*] he obtains $c'' = .4835$.

Strictly speaking there was in this case no weir, there being no dam at the head of the canal. The loss of head between the points H and E being caused, as we have before pointed out, by the velocity which the water in the canal had acquired at E, by contraction as the water entered the mouth of the canal at D, and slightly by friction on the sides between D and E (vide p. 110).

These experiments of Lesbros cannot therefore be considered of value in giving direct data for submerged weirs. They are especially alluded to here, because they have been quoted by Mr. John Neville, in his Hydraulic Tables, which is generally regarded as the standard text-book upon Hydraulics in the English language.

Fteley and Stearns.

The experiments by Messrs. Fteley and Stearns give us valuable data for this form of weir. The weir used by them had a length of 5 feet, end contractions suppressed, with inner depth below crest, G, of 3.17. The lower canal was about 8.7 wide. The head H_u was measured 6 feet above the weir; the head H_t was measured 6 feet below the weir. Q was determined by the flow over a sharp-crested weir, with free discharge, and then the level of the water in the lower canal was raised above the crest by stop-gates, the same volume of water passing over the now submerged weir. We have calculated Q for the following experiments by curve for c, with $l = 5$. on Plate VII.

The correcting factor h, in $h = H + b \dfrac{\left(\dfrac{Q}{a_u}\right)^2}{2 g}$, is placed at 1.33.

For corrections for velocity of approach for H_u the same expression is adopted, hence $h'' = \left(H_u + b \dfrac{v_u^2}{2 g} \right) - H_t$. The value of C'' is then obtained by the formula of Dubuat, $Q = C' l (2 g h'')^{\frac{1}{2}} (H_t + \tfrac{2}{3} h'')$.

Very likely our correction for v_u with the submerged weir is not exactly logical

---

[*] This is an incorrect formula for submerged weirs.

but these corrections are not large, and any error from this cause will have but slight effect upon the values of C'.

The velocity in the lower canal at the measuring point for H_l is not given, although were it large it might possibly be an important factor ; judging from the sketches of the apparatus, the section of the lower canal was large, and hence the effect of this lower velocity must have been slight.

All these experiments were with the weir entirely submerged ; *i.e.*, no air under the escaping sheet.

The value of the co-efficient x, given in the last column of the table, has been obtained by $Q = x \, l \, (2 \, g \, h'')^{\frac{1}{2}} \, (.915 \, H_l + \frac{2}{3} \, h'')$.

TABLE LII.

Fteley and Stearns.—Comparison Flow over Sharp-crested Weir, with Free Discharge, and the same Volume passed over the same Weir Submerged; $l = 5$. End Contractions Suppressed; full Bottom Contraction.

$$(2g)^{1/2} = 8.020 \qquad Q = C l (2g h^2)^{1/2} (H_1 + \tfrac{2}{3} h^2)^{1/2} = z l (2g h^2)^{1/2} (.915 H_1 + \tfrac{2}{3} h^2)$$

No.	F. & S. No.	Determination of Q							Plate VII. c_i	Q	Submerged.										
		H	z_i	h_o	h_v	h	c_i				H_v	c_i	h_o	h_v	h_{vo}	h_o	H_v	h'	C	z
322	9	.3053	.3641	.0004	.0005	.3058		.6308	2.8517	.3251	.163	.0004	.0005	.3456	.0956	.2300	.596			
323	19	.3033	"	"	.0005	.3058			2.8517	.4382	.158	.0004	.0005	.4287	.3373	.0914	.591			
324	21	.3033	"	"	.0005	.3058			2.8517	.6246	.150	.0004	.0005	.6251	.3894	.0357	.614			
325	22	.3033	"	"	.0005	.3058			2.8517	.8819	.144	.0004	.0005	.8151	.7917	.0207	.611			
326	8	.3957	"	.0009	.0012	.3966		.6257	4.226	.4157	.235	.0009	.0012	.4169	.1183	.2984	.608	.628
327	14	.3953	"	"	.0012	.3965			4.225	.4394	.231	.0009	.0012	.4316	.2193	.2321	.584	.615
328	15	.3952	"	"	.0012	.3964			4.223	.4857	.231	.0008	.0011	.4868	.3078	.1790	.583	.621
329	18	.3953	"	"	.0012	.3965			4.225	.5464	.227	.0008	.0011	.5465	.4186	.1279	.585	.629
330	6	.5074	.331	.0017	.0023	.5097		.6280	6.098	.5080	.331	.0017	.0023	.5101	.0735	.4366	.631	.642
331	12	.5069	"	"	.0023	.5092			6.049	.5462	.327	.0017	.0023	.5485	.2093	.3390	.599	.624
332	20	.5062	"	"	.0023	.5085			6.077	.7193	.318	.0015	.0020	.7213	.5731	.1482	.586	.632
333	7	.5788	.357	.0024	.0032	.5820		.6267	7.439	.5922	.396	.0024	.0032	.5951	.1122	.4832	.614	.625
334	10	.5790	"	"	.0032	.5822			7.443	.6098	.394	.0024	.0032	.6130	.1859	.4271	.603	.624
335	13	.5782	"	"	.0032	.5814			7.427	.6328	.391	.0024	.0032	.6360	.2614	.3746	.592	.619
336	17	.5780	"	"	.0032	.5812			7.423	.7423	.340	.0023	.0031	.7456	.5186	.2250	.580	.621
337	11	.6655	.473	.0035	.0047	.6702		.6275	9.204	.7165	.473	.0035	.0047	.7213	.2663	.4517	.598	.622
338	16	.6650	"	"	.0047	.6697			9.194	.8153	.462	.0033	.0044	.8167	.5157	.3010	.583	.621

Francis.

Mr. J. B. Francis has published, in the Transactions of the Am. Soc. of C.E., September, 1884, an account of a number of experiments made with submerged weirs at Lowell, Massachusetts, in 1883.

The feeding water was drawn through a turbine water-wheel at rest, under a nearly constant head, from an upper canal into a lower canal. The amount of flow was regulated by the gates for this wheel; the size of the opening presumably remained constant for each series of experiments; the discharge into the lower canal was submerged. In this lower canal were placed two weirs, the feeding stream entering about midway between the weirs into the pool formed by them. The weirs were sharp crested, with end contractions suppressed, 11.22 and 10.98 feet long respectively, and having horizontal crests at the same elevation. The discharge was first gauged for each series by allowing the water to pass over the weirs with free discharge into the air; the water level in the canal below the two weirs was then raised a certain distance above the crests for each experiment. This diminished the head of the feeding stream or submerged jet, and hence decreased the flow; there were also fluctuations in the elevation of the surface in the upper canal; for each experiment the amount of head for the submerged feeding jet was determined.

There were 5 series of experiments made, the gates regulating the supply not being changed during the experiments constituting one series. Now, calling A the area of the opening for the feeding supply, Y the head actuating the flow through this opening, being the observed difference in level between the surface of the water in the two canals, and Z the co-efficient of discharge for this particular opening, we have $Q = Z A (2 g Y)^{\frac{1}{2}}$. The head Y for any one series did not vary largely, and was never less than 12 feet; we have seen for orifices that with slight variations in such large heads the co-efficient of discharge is practically constant; we can hence assume that Z for each series was constant. Taking 42 experiments made under normal conditions where there was perfectly free discharge, from the given values of Z and H we can determine the most probable value of $Z A$ for each of the 5 series, Q being computed from the interpolated curve for c, with $l = 11.1$ on Plate VII. The depth of the water in the feeding lower canal below the crest of the weirs was about 5.8, so that the velocity of approach was never very great, and the effect of v_a can be neglected without serious error. The deduced values of $Z A$ for each series do not vary much, showing that A was practically constant. These computations need not be given here; the results slightly vary from those indicated by Mr. Francis, who adopted a somewhat different method of computation.[*]

[*] Mr. Clemens Herschel in the Transactions of the Am. Soc. of C.E., May, 1885, has discussed these experiments of Mr. Francis. Mr. Herschel has calculated Q by the mean values of $Z A$, using the Francis formula for suppressed weirs, and making allowances for v_a.

We can now readily calculate Q for the experiments with submerged or " drowned " discharge, as we have $Z A$ and Y, in $Q = Z A (2 g Y)^{\frac{1}{2}}$.

H_u was measured at a point 6 feet from each weir. H_l was determined by the height of water in a vessel, which was connected with the back-water below each weir by a pipe having openings at its ends, distant 18 feet from the crests ; H_l was hence the mean elevation of the back-water in the two lower pools. There was so much commotion in the water near the weirs, that it was necessary to measure H_l at this considerable distance of 18 feet below the two crests.

Using the same notation employed for the preceding experiments of Fteley and Stearns, and not attempting to make any corrections for velocity of approach, we have the following results.

TABLE LIII.

Francis.—Submerged Weirs. $l = 11.22 + 10.98 = 22.20$; *hence representing Flow over a Weir, with* $l = 11.10$.
End Contractions Suppressed and Full Bottom Contraction. $(2 g)^\frac{1}{2} = 8.020$

$$Q = C \tfrac{2}{3} (2 g H_a)^\frac{1}{2} H_a l, \text{ when } H_i \text{ is minus.}$$
$$Q = C' \, l (2 g h'')^\frac{1}{2} (H_i + \tfrac{2}{3} h'') = x \, l (2 g h'')^\frac{1}{2} (.915 H + \tfrac{2}{3} h'')$$

No.	Y	Q	H_a	H_i	h''	Co-Efficients			Remarks
						C	C'	x	
339	13.959	73.82	1.000	-.144	1.000	.6219			**Series I.**
340	13.956	73.81	1.000	-.066	1.000	.6219			$Z A = 2.4636$.
341	13.918	73.71	.999	.052	.947		.623	.627	For Nos. 339 and 340, c, (free discharge) = .6232.
342	13.921	73.72	1.002	.136	.866		.624	.634	At $H_i = 0$, air under escaping
343	13.858	73.55	1.037	.263	.774		.603	.621	veins began to disappear; at
344	13.800	73.40	1.091	.448	.643		.586	.613	$H_i = +.17$, air had all disappeared.
345	13.719	73.18	1.156	.657	.499		.588	.623	
346	13.706	73.15	1.149	.636	.513		.587	.621	
347	13.545	72.72	1.328	1.015	.313		.596	.642	
348	13.901	95.79	1.185	-.238	1.185	.6256			**Series II.**
349	13.903	95.79	1.183	-.155	1.183	.6272			$Z A = 3.2034$.
350	13.938	95.91	1.186	-.009	1.186	.6256			For Nos. 348-351, c, (free discharge) = .6254.
351	13.949	95.95	1.186	-.022	1.186	.6239			At $H_i = +.01$ air under escaping
352	13.945	95.94	1.183	.096	1.087		.630	.636	veins began to disappear; at
353	13.943	95.93	1.182	.089	1.093		.630	.636	$H_i = +.09$, air had all disappeared.
354	13.930	95.89	1.203	.207	.996		.620	.631	
355	13.903	95.79	1.227	.309	.918		.610	.628	
356	13.860	95.64	1.277	.478	.799		.594	.619	
357	13.785	95.38	1.391	.860	.531		.606	.644	
358	13.663	94.96	1.491	1.039	.452		.592	.634	
359	13.136	161.53	1.673	-.083	1.673	.6289			**Series III.**
360	13.128	161.48	1.670	.025	1.645		.630	.632	$Z A = 5.5573$.
361	13.133	161.51	1.670	.022	1.648		.630	.632	For Nos. 359 and 368, c, (free discharge) = .6298.
362	13.123	161.46	1.670	.075	1.595		.631	.634	At $H_i = -.10$, air under escaping
363	13.063	161.08	1.720	.466	1.254		.621	.640	veins began to disappear; at
364	13.043	160.96	1.740	.465	1.275		.609	.628	$H_i = +.06$, air had all disappeared.
365	13.016	160.80	1.743	.483	1.260		.608	.628	
366	13.147	161.60	1.804	.792	1.012		.615	.645	
367	13.060	161.06	1.917	.996	.921		.585	.618	
368	13.265	162.32	1.679	-.143	1.679	.6286			

TABLE LIII.—*continued.*

No.	Y	Q	H_a	H_l	h"	Co-Efficients			Remarks.
						C	C'	z	
369	12.597	206.77	1.965	−.127	1.965	.6324			Series IV.
370	12.596	206.77	1.968	−.133	1.968	.6310			Z A = 7.2642
371	12.591	206.72	1.963	−.069	1.963	.6333			For Nos. 369-372, c, (free discharge) = .6318.
372	12.588	206.70	1.962	−.049	1.962	.6337			
373	12.570	206.55	1.965	.029	1.936		.632	.633	At $H_l = -.14$, air under escaping veins began to disappear; at $H_l = +.17$, air had all disappeared.
374	12.573	206.58	1.965	.049	1.916		.632	.634	
375	12.558	206.45	1.976	.128	1.848		.627	.632	
376	12.549	206.38	1.976	.173	1.803		.628	.635	
377	12.523	206.16	1.994	.327	1.667		.624	.636	
378	12.493	205.92	2.034	.528	1.506		.615	.634	
379	12.537	206.28	2.092	.730	1.362		.606	.630	
380	12.553	206.41	2.090	.732	1.358		.608	.632	
381	12.450	205.56	2.188	1.054	1.134		.599	.630	
382	12.449	205.56	2.190	1.071	1.119		.601	.632	
383	12.086	228.92	2.212	.727	1.485		.615	.637	Series V.
384	11.953	227.66	2.319	1.111	1.208		.607	.639	Z A = 8.2105.
385	11.955	227.67	2.318	1.102	1.216		.606	.638	c, with free discharge, and H = 2.12, has a value of .6327. At $H_l = -.30$, air under escaping veins began to disappear; at $H_l = +.10$, air had all disappeared.

It will be observed from an inspection of Tables LII. and LIII., that for each series of experiments, as H_l increases, the co-efficient C', in the Dubuat formula* of $Q = C' l (2 g h'')^{i} (H_l + \frac{2}{3} h'')$, steadily diminishes, the only exception being the first series of Fteley and Stearns—Nos. 322-325. The maximum variation in C' for any one series, is .047 for Series IV. of Fteley and Stearns, or about 8 per cent. . As Series I. of Fteley and Stearns gives values for C' differing from all the other series, it is probable that these anomalous results are due to some experimental error.

In examining the experiments of Fteley and Stearns we found that in the equation (D) of $Q = x l (2 g h'')^{i} (.915 H_l + \frac{2}{3} h'')$, x would agree fairly well with c, disregarding the series considered faulty. Upon examination of Mr. Francis' paper, we find that he has proposed substantially the same expression; he gives, $Q = 3.33 l (h'')^{1/2} + 4.5988 l$

---

* Formula (A) as given before, with $C' = c$, and $h'' = h$.

$H_i(h^n)$; reducing this to our form, with $(2\,g)^{\frac{1}{2}} = 8.0202$, we have $Q = c\,l\,(2\,g\,h^n)^{\frac{1}{2}}$ $(.9207\,H_i + \frac{2}{3}\,h^n)$.

In the preceding tables we have given the value of the co-efficient c in equation (D), and it will be seen that it approximates very much closer to the value of c_i, than the co-efficient c'' in equation (A). Messrs. Fteley and Stearns propose variable co-efficients, based upon the ratio of $\dfrac{H_i}{H_a}$; their co-efficients, however, are irregular, and when applied to our experimental data give but little closer results than the general expression (D). Equation (D) is founded upon the assumption that the co-efficient c' in equation (B) is $\frac{915}{1000}$ths of c'' in equation (C). There seems to be no valid reason for supposing that c' and c'' should have values considerably differing, so equation (D) must be considered as purely empirical; it gives fairly satisfactory results with the data given by Fteley and Stearns, and Francis, but it by no means follows that it will give equally satisfactory results when applied to experiments made under different conditions.

When the sheet passes over the crest of a submerged weir, there is a depression in the surface immediately below the weir, the surface of the water gradually rising for some distance down-stream from this depression, as shown by the following sketch.

In the preceding experiments H_i was measured in one case 6 feet, and in the other case 18 feet, below the crest of the weir, and hence probably near the summit of the surface in the lower pool, or back-water. Owing to the boiling of the water at the point of greatest depression, shown in the sketch at **a**, it is practically impossible to measure H_i' with any reasonable degree of accuracy; probably $H_i - H_i'$ increases with $\dfrac{H_i}{G}$; that is to say, the shallower the depth of water in the canal below the weir, the greater will be the depression, other things remaining constant.

If the section of the contracted vein after its escape from the weir be assumed to be at H_i', the "piling-up" of the water down-stream from the point **a**, can be considered as being due to the energy of the water after its proper escape from the weir; any work done by a jet after this escape will in all probability not have any appreciable effect upon the quantity of discharge; hence with this assumption, the lower head should be measured at **a**, in the trough of the wave. If, on the other hand, the section of the

x

contracted vein be supposed to be at **X** in the sketch, the depression from **b** to **a** is caused by the energy of the water after its full escape from the weir, and in this case H_t' would be less than the true lower head. The problem cannot be satisfactorily solved, as it is impossible with such a submerged discharge to determine the precise position of the contracted vein.

Bornemann.—Herr K. R. Bornemann has made a large number of experiments with submerged weirs, of which he has published detailed accounts in "Der Civilingenieur," Vol. XVI., 1870, and Vol. XXII., 1876. In these experiments the height of the dam forming the weirs, or G, was quite small in proportion to H_u, so that there was often a high velocity of approach, and also a high velocity in the canal below his weirs. The upper head, or H_u, was measured about 3 feet above the weirs, and the lower head, or H_t, about 9 feet below the weirs; H_t was hence probably not far from the summit of the surface curve below the weirs. End contractions were suppressed for all these experiments.

We have reduced the 103 experiments given by Herr Bornemann, by the formula $Q = c' \, l \, (2 \, g \, h)^{\frac{1}{2}} \, (H_t + \frac{2}{3} \, h)$; in which $h = H_u + b \cdot \frac{v_u^2}{2 \, g} - H_t$; as v_u was generally not very large in proportion to v_t, we gave b a constant value of 1.25, which in all probability is amply great; we did not take into consideration the effect of the velocity in the canal below the weir. The resulting values of c' range from .596 to .869; in general c is greatest, when $\frac{H_t}{H_u}$ is nearest unity. The results, however, are often contradictory, which was to have been expected, as the experimental apparatus of Herr Bornemann was very far from being perfect. We think that the abnormally high values of c' in many of these experiments were due to the fact that the lower head, or H_t, was measured on the summit of the wave below the weir. These results confirm the suggestions we have already made, in regard to the proper point for the measurement of H_t.[*]

Submerged weirs afford a most imperfect method of gauging the flow of water; this is due to the impossibility of correctly measuring the lower head, and also to the fact that slight errors in the determination of h, or $H_u - H_t$, may largely affect the deduced values of Q. When, as was the case with the Fteley-Stearns and Francis experiments, the depth in the canal is quite large in proportion to h, so that the velocity below the weir is not great, then Q can be roughly determined by a submerged weir; but, when the velocity below the weir is excessive, the estimation of Q becomes wild guess-work. We therefore advise, that when it is not practicable to obtain a free

---

[*] We have in Chapter III. alluded to the experiments made by Bornemann, Boileau, et cæt., with orifices under sluice-gates (Schutzen). The very large values of the co-efficient of discharge in some of these experiments, were in a great measure due to the incorrect measurement of the lower head, which was taken on the summit of the lower wave. Experiments with such orifices, are analogous to those with submerged weirs.

discharge over a weir, the weir be abandoned, and the quantity determined by measurements of the mean velocity in a canal of uniform section ; even if it be only practicable to obtain the central surface current in such a canal, multiplying this velocity by $\frac{7}{10}$ to obtain the mean velocity, will give better results than can usually be obtained by the use of a submerged weir.

A singular phenomenon occurs for suppressed weirs when H_l is very small, and when therefore the escaping vein clings more or less perfectly to the lower vertical side or wall of the weir, the sheet being thus diverted from its normal position by atmospheric pressure : in this case with Q and l constant, as the lower back-water is raised slightly above the level of the crest, the elevation of the water up-stream from the weir instead of being increased by the submergence of the crest, is slightly diminished. Mr. Francis[*] found with $H = .8525$, being the head for free discharge, that with the same value of Q for the same weir, when H_l was .063, H_u was .8485, thus showing a lowering in the surface above the weir of .0040. Messrs. Fteley and Stearns[†] with Q and l constant, found that with free discharge H was .5793, but when H_l was .0463, H (which then became H_u) was .5743, thus showing a lowering in the surface above the weir of .0050 ; in one of these experiments it is stated that there was no air under the escaping sheet. Some of the experiments given in Table LIII. indicate a similar slight decrease in the upper head, with Q nearly constant.

When the vein escaping from a weir is suddenly enclosed on all sides, so that no fresh supply of air can enter underneath the sheet, the inner particles of moving water carry with them more or less air ; this process is continued until all, or nearly all, the air is abstracted, and the sheet is hence pressed against the outer wall of the weir by atmospheric pressure. Mr. Francis thinks that the increased value of the co-efficient of discharge under such circumstances, is analogous to the increased flow through a tube, produced by the addition of a divergent adjutage. Very likely it may be due simply to the change in the form of the contracted vein, caused by the pressing in of the sheet.

When the contracted vein retains its normal position, it is evident that with Q and l constant, the surface of the water in the lower pool can be raised until it reaches the lower edge of the contracted section, without notably interfering with the flow ; this lower edge is higher than the crest, and hence the surface in the lower pool may be raised a slight distance above the crest, without affecting the height H or H_u in the upper pool. The experiments given in Table LIII. show that this supposition is correct.

It may be remarked in conclusion, that the Francis experiments clearly show,

---

[*] Lowell Hydraulic Experiments ; 2nd ed., p. 102.
[†] Transactions Am. Soc. of C.E. ; 1883, p. 104, Table XXV.

when the air is allowed to freely enter under the escaping sheet, the surface of the water in the lower pool can be brought to the level of the crest, without interfering in the slightest manner with free discharge.

In the experiments made with orifices with great heads (Orifice Nos. 7-14), the escaping jets very soon after emerging from the nozzles struck the buckets of a "hurdy-gurdy" wheel, which were moving with a velocity much less than that of the water; in these experiments c was practically unity. This confirms our opinion that any work done, or resistance encountered, by a jet after it has passed the plane of the contracted vein, has no effect upon the discharge.*

Broad and Rounded Crests.

Lesbros experimented with both broad and rounded crests, but as Messrs. Fteley and Stearns have investigated the same subject more fully, we will avail ourselves of their experiments.†

A fixed sharp-crested weir, made of hard pine, was placed across a canal 5 feet wide, end contractions hence being suppressed. The length of this canal was 6 feet; the inner depth below crest, or G, was 3.17; H was measured at a point 6 feet above the weir, near the bottom of the channel. A constant flow of water was admitted into the canal, and H was determined for the flow over the sharp crest; a false piece of wood of the desired width was then slid into place on the lower side of the fixed crest, and drawn into close contact by several fastenings; the height of the water in the canal, or H', was then measured, Q remaining constant. The quantity of flow into the canal was then increased and another set of comparisons made; Q being always constant for each set of comparisons. The upper sides of the false or broad crests were horizontal; the total widths of these crests were respectively 2, 3, 4, 6 and 10 inches.

In the following table are given the results of these comparisons, except for three experiments which were thought to be faulty.

---

* If, however, an unyielding obstacle should be placed before the escaping jet, at a very short distance from the plane of the contracted vein, so that the rebounding particles of water come in contact with the inner portion of the contracted vein, the discharge will probably be notably changed.

† Transactions Am. Soc. of C.E., 1883, pp. 89-101.

TABLE LIV.

Fteley and Stearns.—Comparison of Heads with Weirs, having Sharp and Broad Crests, Q being Constant for each Companion.

Crest = 3"			Crest = 3'			Crest = 4'			Crest = 6'			Crest = 10"		
H	H'	H − H'	H	H'	H − H'	H	H'	H − H'	H	H'	H − H'	H	H'	H − H'
.0956	.1158	−.0172	.1088	.1307	−.0219	.1095	.1318	−.0223	.1092	.1390	−.0298	.1120	.1359	−.0432
.1190	.1360	−.0170	.1112	.1329	−.0217	.1114	.1331	−.0217	.1115	.1349	−.0334	.1153	.1394	−.0241
.1579	.1722	−.0150	.1375	.1621	−.0246	.1139	.1368	−.0229	.1394	.1666	−.0272	.1449	.1713	−.0363
.1939	.3047	−.0108	.1728	.1974	−.0245	.1390	.1659	−.0269	.1896	.2237	−.0341	.2432	.2872	−.0450
.2297	.2363	−.0066	.1882	.2111	−.0229	.1642	.1929	−.0287	.1899	.2237	−.0338	.2877	.3403	−.0526
.2140	.2486	−.0046	.1883	.2119	−.0237	.1741	.2036	−.0295	.2519	.2928	−.0409	.2924	.3449	−.0525
.2960	.2956	+.0040	.2122	.2348	−.0226	.2136	.2445	−.0309	.2832	.3355	−.0423	.4224	.4907	−.0683
			.2491	.2694	−.0203	.2349	.2665	−.0316	.3023	.3474	−.0451	.4634	.5350	−.0716
			.2810	.2978	−.0168	.2559	.2814	−.0294	.3121	.3564	−.0443	.5155	.5861	−.0706
			.3096	.3237	−.0141	.2680	.3121	−.0294	.3456	.3912	−.0456	.5544	.6307	−.0763
			.3127	.3512	−.0085	.3504	.3745	−.0241	.3904	.4352	−.0448	.6017	.6794	−.0777
			.3480	.3961	−.0081	.3902	.4106	−.0197	.4492	.4913	−.0421	.6581	.7323	−.0742
			.3870	.3891	−.0021	.4123	.4273	−.0190	.4984	.5371	−.0387	.7061	.7776	−.0715
			.3912	.3928	−.0016	.4185	.4616	−.0131	.5473	.5717	−.0344	.7561	.8322	−.0661
			.3991	.3985	−.0004	.4983	.5034	−.0051	.5838	.6131	−.0293	.8129	.8752	−.0623
			.4062	.4033	+.0009	.5369	.5382	−.0013	.6386	.6617	−.0231			
			.4081	.4073	+.0008	.5456	.5447	+.0009	.6739	.6929	−.0199			
			.4157	.4143	+.0024	.5640	.5612	+.0028	.6855	.7036	−.0171			
			.4293	.4356	+.0037	.5815	.5750	+.0065	.7347	.7440	−.0093			
			.4668	.4684	+.0084	.6028	.5961	+.0067	.7716	.7760	−.0044			
			.4730	.4619	+.0111	.6012	.5912	+.0100	.7592	.7907	−.0015			
						.6329	.6168	+.0161						
						.6368	.6368	+.0190						
						.6716	.6484	+.0232						

From an examination of the foregoing table it will be seen, that when H' is small in proportion to the width of the crest, H' is larger than H. Also, that as H' increases for any particular width of crest, the relative difference, or $\dfrac{H'}{H}$, diminishes; for the three narrowest crests, H' is less than H, when H' is considerably greater than the width of the crest.

These results are probably due to the following causes, which can be best illustrated by sketches showing the forms* of the escaping sheet; Fig. 1 represents the normal

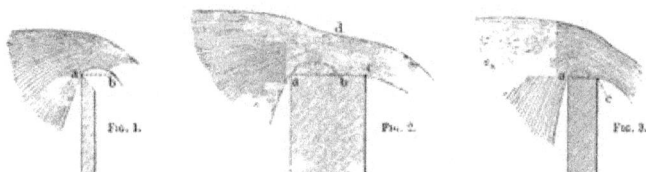

Fig. 1.   Fig. 2.   Fig. 3.

shape of the sheet over a sharp crest; Fig. 2, its shape when H' is small in proportion to the width of the crest; Fig. 3, its shape when H' is large in proportion to the width.

In Fig. 1 there was free admission of air under the sheet to the point a at the sharp crest; the sheet hence preserved its normal form.

In Fig. 2 there was a loss of head by what may be termed primary contraction as the water passed over the crest at a; this is indicated by the depression in the surface curve at d; there was also an increased velocity† due to suppression of bottom contraction, but the loss caused by primary contraction was greater than the gain by suppression; hence H' is greater than H. This view is supported by the experiments of Lesbros, with a weir having the form of approach shown by Fig. 6, Plate I.; this form of approach was with a weir having end contractions nearly suppressed and bottom contraction completely suppressed; the head H was measured in the deep reservoir of supply, and hence this weir may be considered as one with an exceedingly broad crest, with end contractions nearly suppressed; the co-efficients of discharge, as shown on Plate VIII. (Experiments Nos. 117-124), were abnormally low, the cause of this being chiefly due to the loss by primary contraction as the water entered the canal of approach, and which loss was much greater, especially for low heads, than the gain by suppression.‡ With quite small values of H, some of the Lesbros experiments show that there is a

---

* The forms of the surface curves in these sketches are partly taken from the diagrams given by Messrs. Fteley and Stearns.

† In the section measured by $H' \times l$.

‡ This is strikingly shown by the experimental point for No. 130 on Plate VIII., which is very much higher than the curve formed by Nos. 117-124. The head, h, for No. 130 was obtained by the measurement of H in the feeding canal, while for the other experiments H &h was measured in the deep reservoir.

comparatively large loss of head with very broad crests, by friction (adhesion ?) of the thin sheet with the floor of the crest ; possibly such losses may be in part due to atmospheric pressure.

In Fig. 3 the form of the surface curve is but slightly changed, showing that there is but little loss from what we term primary contraction ; the increased velocity, due to partial suppression of bottom contraction is greater in its effect than the loss by primary contraction, and hence H' is less than H ; it will be noticed that $\dfrac{H'}{H}$ is, in these experiments, never much less than unity.

If the escaping vein only touches the inner edge at a, there will be of course no appreciable difference between H and H', no matter what may be the width of the crest. When, however, there is no access provided for the entrance of air on the sides of the crest, if the lower side of the sheet in its normal position nearly touches the point c, the current produced by the lower particles of water in the sheet will gradually abstract the air from a to c, and the atmospheric pressure will then force the sheet in a downward direction, thus producing bottom suppression. This phenomenon occurred in several of the given experiments.

Judging from Lesbros' experiments with the form of approach shown by Fig. 4, Plate I. (which can be considered a weir with an exceedingly broad crest, with full end contraction), with minute heads H' will be larger than H, but with larger heads such as .7, H' will be considerably less than H.

The comparisons in the foregoing table of the varying differences between H and H', can be applied with safety to the measurement of water over broad crests, when the conditions are similar to those which obtained in the experiments made by Messrs. Fteley and Stearns. It will probably be dangerous to use them, when much accuracy is essential, in cases where the conditions are widely different, such as the value of l, et cet. .

The following experiments were made with inner rounded crests, with the same canal and sharp-crested weir used for the preceding experiments with broad crests. The curves were quadrants of a circle, with the respective radii of $\frac{1}{4}$, $\frac{1}{2}$ and 1 inch ; the quadrants were placed on the inner side of the wooden sharp-crested weir, which had a horizontal breadth of .035 ; these curves were formed on the upper side of a plank, placed against the fixed weir, so that the inner face below the curve was vertical.

Comparisons were made by admitting a constant flow of water into the feeding canal and measuring the respective heads H and H' ; H being the head for the sharp crest, and H' the head for the rounded crest.

When the depth on the weir was above a certain limit, the lower side of the escaping sheet detached itself from a portion of the rounded edge and from the hori-

zontal crest, in the manner shown by the following sketch; this limit was reached
when the depths on the weir were .17, .26 and .45, for the ¼, ½ and 1 inch radii, respectively.

The weir, as before, had end contractions suppressed, with $C = 3.2$, and $l = 5.0$.

TABLE LV.

Fteley and Stearns.—Comparison of Heads on Weirs with Sharp and Rounded Crests. Q being Constant for each Comparison.

¼-inch Radius.			½-inch Radius.			1-inch Radius.		
H	H'	$H - H'$	H	H'	$H - H'$	H	H'	$H - H$
.1159	.1096	.0063	.1158	.1118	.0040	.1157	.1127	.0030
.1662	.1524	.0138	.1663	.1535	.0128	.1665	.1569	.0096
.2837	.2723	.0114	.2166	.1955	.0211	.2166	.1986	.0180
.3556	.3408	.0148	.2835	.2547	.0288	.2835	.2554	.0280
.4096	.3953	.0143	.3555	.3273	.0282	.3555	.3164	.0391
.4889	.4742	.0147	.4096	.3813	.0283	.4096	.3636	.0140
			.4889	.4594	.0295	.4888	.4371	.0517

Inner rounded crests or sides will always increase the co-efficient of contraction, and hence the co-efficient, c, of discharge.   Broad crests will sometimes increase c, and sometimes diminish it.   Both forms are to be avoided, if it be desired to make accurate measurements of the flow of water.

MEASUREMENT OF H.

M. Boileau[*] with suppressed weirs, measured the head H by placing a vertical glass tube just above the weir; the tube was open at both ends, and when held in position, the bottom end was near the base of the dam forming the weir.   The

---

[*] *Jaugeage des cours d'eau,* Paris, 1859. And *Traité de la mesure des eaux courantes.* Paris, 1854.

height of the water in the tube was observed, and assumed to be the head H. M. Boileau found that generally the head indicated by the tube, was greater than the head measured at the origin of the surface curve.

Mr. J. B. Francis[*] made a number of experiments, comparing the heights indicated by a piezometric column, having its lower end near the base of the weir, with the heights shown by another piezometric column, having its lower end 6 feet up-stream from the weir. The observed differences of these heights were quite minute, and Mr. Francis came to the conclusion that the depth on a weir, could be determined with sufficient accuracy " by means of a pipe opening into the dead water, near the bottom of the canal on the up-stream side of the weir."

Messrs. Fteley and Stearns[†] have lately investigated this subject in great detail, and we will give the salient features developed by their experiments.

The head above the origin of the surface curve was measured by the height of water in a pail placed below the experimental weir; this pail was connected by a rubber pipe with an opening in a piece of plate glass; the face of the glass was exactly flush with the side of the feeding canal; the opening was true and normal to the side of the canal; it was placed slightly below the crest of the weir, and distant from it 6 feet up-stream. The head near the weir was measured by the height of water in another pail, connected by a rubber pipe with an iron pipe of 1 inch diameter, stopped at both ends, and perforated on its upper side with 8 $\frac{1}{4}$-inch holes, .63 apart; this pipe rested on the bottom of the canal, and was placed parallel to the weir, quite near its inner base. The heights in the two pails were determined by two hook-gauges, supported by the same post; the zeros of the two gauges could hence be readily and accurately compared.

The first experiments were with a weir, having both end contractions suppressed, and 5 feet in length, with various inner depths below the crest. We will call H the head at the point 6 feet above the weir, and H' the indicated head at the base of the weir; h_a is, as usual, the head due to velocity of approach, or $\frac{v^2}{2g}$. Of the 129 determinations given, we will select a sufficient number of those which represent fairly the law of variation between H and H'. The feeding canal was necessarily 5 feet wide, rectangular in section, with its bottom horizontal.

---

[*] *Lowell Hydraulic Experiments,* 2nd ed., p. 141, and also p. 183.
[†] Transactions Am. Soc. of C.E., 1883; pp. 24-51.

Y

TABLE LVI.

Fteley and Stearns.—Comparison of Heads for a Suppressed Weir, with $l = 5.0$.

H = head 6 feet up-stream from weir.
H' = head at base of weir.

G = 3.56			G = 2.60			G = 1.70			G = 1.00			G = .30		
H	$H'-H$	h_a	H	$H'-H$	h_a	H	$H'-H$	h_a	H	$H'-H$	h_a	H	$H'-H$	h_a
.1509	.0002	.0001	.2685	.0004	.0004	.1926	.0004	.0004	.1914	.0005	.0009	.1884	.0011	.0027
.1935	-.0001	.0001	.3359	.0005	.0008	.2676	.0005	.0009	.2649	.0012	.0022	.2594	.0026	.0060
.2304	.0002	.0002	.4206	.0009	.0015	.3551	.0008	.0019	.3496	.0022	.0044	.3404	.0046	.0114
.3360	.0007	.0004	.4925	.0015	.0022	.4183	.0012	.0030	.4155	.0035	.0068	.3974	.0064	.0165
.4222	.0006	.0008	.5598	.0021	.0032	.4887	.0021	.0044	.4784	.0046	.0097	.4585	.0090	.0230
.4926	.0011	.0013	.6119	.0031	.0040	.5551	.0031	.0062	.5409	.0065	.0132	.5186	.0110	.0302
.5615	.0017	.0018	.6703	.0046	.0051	.6057	.0043	.0077	.6113	.0108	.0201	.5509	.0126	.0345
.6143	.0022	.0024	.7117	.0034	.0059	.7018	.0060	.0113	.6906	.0120	.0230	.6926	.0224	.0583
.6897	.0025	.0032	.7716	.0043	.0073	.7695	.0085	.0142	.7873	.0160	.0322	.7761	.0238	.0752
.7747	.0028	.0044	.8284	.0049	.0088	.8467	.0103	.0178	.8293	.0179	.0366	.8369	.0296	.0860
.8336	.0041	.0053	.8595	.0067	.0097	.9009	.0116	.0209	.8854	.0205	.0426			
.8812	.0045	.0062	.9180	.0069	.0114	.9223	.0134	.0321						
.9449	.0047	.0075												

From the preceding table it will be observed that H' was—with one exception doubtless due to experimental error—higher than H, and that this excess is nearly in proportion to the value of h_a; h_a is shown to be larger than $H'-H$, the ratio between these quantities, or $\dfrac{h_a}{H'-H}$, increasing as G diminishes.

With a suppressed weir, 19 feet long and $l = 6.55$, the following results were obtained :

$$H = 1.16 \qquad\qquad H'-H = .0114 \qquad\qquad h_a = .0045$$
$$H = 1.61 \qquad\qquad H'-H = .0274 \qquad\qquad h_a = .0110$$

In this instance $H'-H$ was about 2.5 h_a; for these two experiments H was measured 6 feet above the weir ; which, for such large values of l and H, was perhaps at a point below the origin of the surface curve ; this may partly be the reason of the apparently abnormal excess of $H'-H$ above h_a ; the screens for equalizing the current in the feeding canal for this weir were, however, placed a comparatively short distance above the weir, and quite possibly this may have had a notable effect upon H'.

The following results were obtained with end contraction more or less perfect, the width of the feeding canal being in all cases 5 feet.

TABLE LVII.

Fteley and Stearns.—Comparison of Heads for Weirs with End Contraction more or less perfect.

$$L + L' + l = 5.00$$

H = head 6 feet up-stream from weir.
H' = head at base of weir.

	$G = 3.56$										
$L=1.0,\ L'=0,\ l=4.0$			$L=0,\ L'=1.7,\ l=3.3$			$L=1.0,\ L'=1.0,\ l=3.0$			$L=1.0,\ L'=1.7,\ l=2.3$		
H	$H'-H$	h_e	H	$H'-H$	h_e	H	$H'-H$	h_e	H	$H'-H$	h_e
.1761	.0002	.0000	.4498	.0004	.0004	.2155	.0031	.0000	.5824	.0003	.0004
.2691	.0002	.0001	.5678	.0008	.0008	.3301	.0002	.0001	.7478	.0011	.0008
.3942	.0003	.0004	.6869	.0015	.0013	.4843	.0003	.0004	.8905	.0014	.0012
.5997	.0015	.0014	.8062	.0024	.0020	.6215	.0010	.0008	.9548	.0013	.0014
.7055	.0023	.0021	.9317	.0026	.0029	.7398	.0015	.0013			
.9460	.0029	.0046				.8714	.0024	.0020			
	$G = 1.00$										
			.5647	.0017	.0058	.8490	.0042	.0112			
			.7850	.0031	.0120						
			.9024	.0037	.0162						
	$G = .50$										
.6639	.0064	.0283	.5574	.0001	.0126	.8310	−.0009	.0217			
			.7677	−.0023	.0239						
			.8775	−.0041	.0309						

The results shown by the preceding table appear to be contradictory; they clearly illustrate the danger of error in obtaining the head for a weir, by measuring it by means of a piezometric column placed near the weir.

From the foregoing experiments, it is apparent that the head for a weir should be measured at a point above the origin of the surface curve, if accuracy is essential. Where there is full contraction, the hook-gauge had better be placed directly in the feeding canal, rather than below the weir as was done by Messrs. Fteley and Stearns. When end contractions are suppressed the hook-gauge should be placed on the outside of the canal, with a number of small openings connecting the water in the canal and that in the box in which the gauge is placed; these openings should be large enough to permit the surface in the box to quickly respond to the fluctuations of level in the canal. The gauge should be read at regular intervals of time; when the oscillations are minute, the arithmetical mean of the observed heights will represent the proper value of H, but when the heights vary considerably, the means should be deduced from the $\frac{3}{2}$ power of the readings.

The method followed by Messrs. Fteley and Stearns of placing the gauge in a movable pail, has the advantage that the heights can be read with greater convenience by the observer, than if the gauge is placed in the canal.

Messrs. Fteley and Stearns also made a large number of experiments of the velocity of approach in canals in various points of the cross-section, and at various distances from the weir; the velocity was measured by current-meters.

From their investigations they conclude, that H should be measured "at a distance from the weir equal to $2\frac{1}{2}$ times its height above the bottom of the channel." In our opinion, when G is greater than $3\,H$, any increase in G will have no appreciable effect upon the length of the surface curve. In discussing the surface curve, we have shown that its length is nearly altogether governed by the values of l and H; when G is much less than $3\,H$, then doubtless changes in G will affect its length.

With full contraction, the surface curve is produced by the changes in velocity of the various horizontal layers in the escaping vein, and by the contraction of the vein; its form is simply modified by changes in l and H.

CHAPTER VI.

THE FLOW OVER WEIRS AND THROUGH ORIFICES COMPARED.

In our discussion of the flow over weirs we have assumed that the quantity of flow can be most accurately determined by the measurement of the total head H, which has been ascertained at a point just above the origin of the surface curve; the measurement of the height H_o in the plane of the weir being of no practical advantage. This assumption is analogous to our proposition in regard to the flow through orifices, where we show (p. 26) that the co-efficient of discharge for orifices can only be approximately determined by the measurement of the section of the contracted vein.

We have also assumed that the surface curve at a weir represents the normal contraction of the upper side of the escaping sheet. We will now attempt to determine how nearly this contraction corresponds with full contraction from an orifice; the experiments of Lesbros afford approximate data from which some conclusions can be drawn in regard to this question.

It will be remembered that Lesbros used his square aperture in a fixed copper plate, .66 × .66, first as an orifice and then as a weir, with all the forms of approach shown on Plate I., except Figs. 7 and 11. For all these orifice experiments there was necessarily contraction on the upper side of the escaping vein, although when the surface of the water in the feeding reservoir was only a small distance above the top of the aperture, the resulting form of contraction somewhat approached the form of contraction for the upper side of a sheet escaping from a weir.

It may be here remarked, that as the surface in the feeding reservoir was gradually lowered from a height slightly above the top of the aperture, just at the moment of transition from an orifice to a weir, the surface in the canal of supply shown by Fig. 6, Plate 1, detached itself abruptly from the upper edge of the opening and fell at once a notable distance. In all probability the same phenomenon occurred with the other forms of approach, but in a less marked degree.[*]

Now, if we draw the curve for the orifice experiments formed by using H_b—H_b being the height from the bottom of the orifice to the surface of the still water—and c as co-ordinates, and also the curve for the weir experiments by using H and c as co-ordinates,

---

[*] We have frequently noticed similar phenomena.

the form of approach being identical, it is apparent that if there be an abrupt break at the junction of the two curves, this will indicate that the two forms of contraction are not alike. In the following statement this comparison is made; it will be remembered that for both orifices and weirs the actuating head was measured in the still water of the deep reservoir, 11.48 feet up-stream from the aperture. The size of the orifice was always .66 × .66; the length of the weir always .66; and the discharge in all cases free into the air.

Form of Approach. Plate I.	Orifices.		Weirs.		Remarks.
	H_1	c	H	c	
Figs. 1, 2 and 3.	5.0	.604	.70	.585	Full contraction.
	3.0	.605	.60	.587	These quantities are obtained from the most probable curves for c on Plates II. and VIII.
	1.5	.602	.50	.590	
	1.2	.600	.40	.594	
	1.0	.599			
	.70	.594			
Fig. 8.	3.3	.612	.67	.602	Contraction nearly suppressed on one side.
	1.8	.610	.49	.610	
	1.2	.607	.31	.615	
	.74	.607			
Fig. 4.	4.0	.624	.67	.607	Contraction suppressed on bottom.
	2.8	.624	.39	.613	
	1.6	.624	.22	.614	
	1.3	.624			
	1.0	.621			
Fig. 9.	6.1	.627	.71	.639	Contraction nearly suppressed on two opposite sides.
	3.3	.629	.69	.636	
	1.6	.633	.51	.635	
	1.4	.633	.38	.630	
	.75	.649	.33	.631	
Fig. 5.	5.6	.637	.68	.611	Contraction suppressed on bottom, and nearly suppressed on one side.
	3.2	.637	.36	.608	
	1.4	.636			
	1.0	.635			

Form of Approach. Plate I.	Orifices.		Weirs.		Remarks.
	H_s	c	H	c	
Fig. 10.	5.6	.637	.80	.642	Contraction suppressed on both sides.
	3.3	.639	.51	.649	
	1.6	.641	.33	.650	Some loss of head by primary contraction at mouth of canal.
	1.1	.647			
	.87	.658			
Fig. 6.	4.5	.663	.84	.570	Contraction suppressed on bottom and nearly suppressed on both sides.
	2.9	.666	.65	.574	
	1.7	.673	.50	.575	
	1.2	.684			Large loss of head by primary contraction at mouth of canal.
	1.1	.695			
Fig. 12.	3.3	.612	.65	.605	Sides of approach inclined at 45°, vide Plate I.
	1.2	.609	.47	.604	
	.73	.610	.29	.608	

It will be observed by plotting the two curves for each form of approach, that there is in all cases an abrupt break where they join, c for the weir being always lower than c for the orifice. This difference for the form of approach shown by Fig. 6 is very large, and is chiefly due to the loss of head by primary contraction at the mouth of the feeding canal being proportionately greater for the weir experiment with $H = .84$, than for the orifice experiment with $H_s = 1.1$; the same observation applies to the comparison for Fig. 10.

The comparisons for Figs. 1 and 12 probably indicate pretty fairly the difference between full contraction from an orifice, and the contraction given by the surface curve for the weir; c for the first being say $1\frac{1}{2}$ per cent. greater than for the latter. Now, as the change in form of contraction was confined to one side of a square, we can roughly assume that the contraction given by the surface curve is 6 per cent. ($\frac{1}{16}$th) greater than full contraction from the side of a square orifice, when both H and l are about .7.[*]

Comparing in the same manner the orifice experiments of Lesbros, for the forms of approach shown by Figs. 1, 2 and 3 (full contraction), l being constant at .66 and w from .066 upwards, with the curve on Plate VIII. for a weir with full contraction having the same length, we see considerably greater differences between the values of c for orifices and those for the weir. We are inclined to attribute this result to the reason that the co-efficients for these orifices with w less than .66 are placed at too high values. It will be remembered that these values are considerably higher than those indicated by our Holyoke experiments; vide p. 56.

[*] Very likely this ratio may be somewhat different when H and l differ considerably.

In the case of a weir having H very much in excess of l, it is apparent that the increased contraction caused by the surface curve will not notably diminish c. This proposition is experimentally proved by the experiments of Lesbros with his other fixed orifice in a copper plate, having $l = .066$ and $w = 1.97$ (Table VII.) ; this same aperture was afterwards used as a weir (Table XLIX.).

Comparing the respective values of c, we have ;

Orifice No. 190 $H_b = 2.06$ $c = .638$

Weir No. 131 $H = 1.95$ $c = .639$

The value of c, as indicated by Plate VII., for a weir with full contraction, having l and H both 1 foot, is about .583 ; hence from the preceding deductions from Lesbros' experiments, c for an orifice 1. × 1. with a head slightly above the top of the opening should be about .592. For our orifice experiment No. 2 (which we consider quite trust-worthy) having nearly that size and a head above top of opening (H_t) of .5, c is .599 ; c for this orifice with H_t = say .2, would probably have a value of about .595. This shows a fair degree of agreement between our respective co-efficients for weirs and orifices, the chances being that c for the weir is placed at a value slightly too low.

The value of c for a weir with full contraction, having $l = 2$ and $H = 1$, is by Plate VII. .590 ; hence for an orifice with $H_b =$ say 1.2, $l = 2$. and $w = 1$., c should have a value of about .602—i.e., $.590 \times \left[1. + \left(.06 \dfrac{2}{(2+1) \times 2}\right) \right] = .602$. This agrees reasonably well with the Ellis experiments given in Table XIII. ; these experiments, however, are not sufficiently reliable to permit of satisfactory comparisons.

For a weir of infinite length (of course having full bottom contraction) with H from .5 to 2. we assume (Plate VII.) that c will be about .615 ; hence for an orifice of infinite length, with widths from .5 to 2., and with a comparatively small head above the top of the opening, c should be about .633 ; this is probably not far from the truth. For a weir of infinite length, with $H = .1$, c is about .656 (Plate VII) : hence for an orifice of infinite length, with $w = .1$ and $H_b = .12$, c should be about .676 ; it will be seen by an examination of Plate III. that some such value is probable.

It will be observed by reference to Plate VII. that our curves for weirs with full contraction (c_i) do not cross each other, but continually diverge as H increases. The curve c_i for $l = 2$. indicates that with $H = 2$. c_i will be nearly .570 ; hence for an orifice 2. × 2., with H_t small, c should be about .579. This last value is, in all probability, considerably too low ; probably it should be somewhere from .590 to .605. From this comparison we may infer that for this weir c_i should be about .585, which is about its value as indicated by Lesbros' experiments, Nos. 93-95 on Plate VI.

If the curves for c_i continually diverge as H increases, for a weir having $l = .66$ and $H = 2$, c_i will be .560 or less ; hence for the similar orifice c would not be over .570. This latter value we conjecture is altogether too low.

From the foregoing comparisons we are inclined to draw the conclusion that the curves for c_c on Plate VII., if extended for increasing values of H, instead of diverging will approach each other, and very likely may cross each other. That is to say, for instance with $l = .66$, c_c for a weir may have a minimum value when H is about .7; the value of c_c increasing both as H diminishes below this value, and as H increases above this value. And for a weir with $l = 2.$, c_c may have a minimum value when H is about 1.6; c_c increasing as before when H either diminishes or increases from this stated value.

The interesting question presents itself; What will be the value of c_c for a weir of great size, such as one having a length of 20 feet with a head of 20 feet? If the curve for c_c remains constant at .614 for heads increasing from 2 feet, as is indicated by Plate VII., it is almost certain that c_c for such a weir will be less than .614. It may be that for all weirs with lengths above 1 or 2 feet, c_c will practically have the same minimum value of about .585 when H and l are nearly equal; assuming that this is the case, c_c for an orifice 20×20, with H_t small, should be about .595. This last value does not seem to us to be an improbable one; hence inferentially the constancy of c_c does not appear to be improbable.

We hope that the next experimenter, who has at his command a measuring vessel of proper size, will carefully determine the value of c_c for weirs with l from 1 to 3, and with maximum heads of 4 or 5; some accurate experiments with orifices 2×2, and 3×3, would also be of advantage. The possession of such data in addition to the data we have discussed in the preceding pages, would enable one to generalize with safety in regard to the laws governing the flow over weirs with full contraction.

It is hardly worth while to pursue this subject still further into spaces where we have absolutely lost the guidance of experimental light. Such speculations, though possibly ingenious and plausible, when tested with facts generally prove to be very wide of the truth.*

---

* Very likely this remark may apply to some of the suggestions given in this chapter.

CHAPTER VII.

FLOW THROUGH OPEN CONDUITS.

It is not proposed to discuss the flow of water through open conduits at great length. This chapter is introduced partly for the purpose of deducing from the careful experiments of Darcy and Bazin, and Fteley and Stearns with open uniform conduits of considerable size, data which will be of advantage in elucidating the laws governing the flow through circular pipes of large diameter.

Dubuat showed that the force imparting velocity to a stream is due to its surface fall or descent, and that this accelerating force in a channel of uniform section and inclination is equal to the sum of the forces which retard the flow; thus uniform motion is maintained.

We will assume that the expression $v = n \, (r \, s)^{\frac{1}{2}}$, known as the Chezy formula, approximately represents the law governing the flow in uniform channels, after the regimen of flow has been established. In this equation;

$v =$ mean velocity in channel, $\dfrac{Q}{a}$.

$n =$ a co-efficient, now known to be a variable with a wide range.

$r =$ hydraulic mean radius,* being the area divided by the length of the wetted

surface, or $\dfrac{a}{p}$.

$s =$ sin of inclination of the surface of the stream, being the surface fall in a certain

distance along the course of the stream, divided by that distance, or $\dfrac{h}{l}$.

It involves the hypotheses that the mean velocity is in direct proportion to the square root of the inclination, and also in direct proportion to the square root of the hydraulic mean radius. The values of the co-efficient n, which will be shown to vary with v, r and Δ, are to be determined by the experimental data now to be considered.

---

* Dubuat appears to have first introduced this factor.

Experimental Data.

Darcy and Bazin.

By far the most valuable experiments that have been made of the flow through open conduits of uniform section, are those which were begun by M. H. Darcy in 1855, and completed after M. Darcy's death in 1858 under the direction of M. H. Bazin. The results of these experiments have been published by M. Bazin in his " Recherches Hydrauliques," Paris, 1865.* This work appears to be remarkably free from errors, either of the author or of the printer ; in fact we have only discovered one error, and that an insignificant one.

Fifty series of experiments are given, of which thirty were made with experimental conduits of various sizes, forms and inclinations, and twenty with canals and sluice-ways. In most of the experiments the regimen of flow had been fully established ; where this was not the case, the sign ɵ attached to the value of *n* indicates that the co-efficient for this experiment has been deduced by M. Bazin by his formula for variable flow.

In Series I., with sections of the canals of Marseilles and Craponne, *r* was determined by numerous measurements with a Pitot tube and a current meter, under the direction of M. Baumgarten.

For No. 1 the section was nearly rectangular ; the surface was very uniform, the bottom being of cement and the sides of brick. No. 2 ; section rectangular, surface of cut-stone and smooth. Nos. 3, 4, 5 and 6 ; section nearly rectangular, surface of hammered stone, quite rough. No. 7 ; section an irregular trapezoid, bottom covered with mud and some vegetation.

TABLE LVIII.—Series I.

Canals of Marseilles and Craponne.—Baumgarten.

$$v = n (r \ s)^{\frac{1}{2}}$$

No.	*v* †	*d* ‡	*a*	Q	*r*	*s*	*v*	*n*	Δ
1	7.4	2.5	17.8	182.73	1.504	.003 72	10.26	137.1	Very smooth.
2	8.5	3.0	25.9	143.74	1.774	.000 84	5.55	123.0	Quite „
3	3.5	1.2	3.9	43.93	.708	.029	11.23	78.4	
4	3.5	0.9	3.1	43.93	.615	.060	13.93	72.5	Hammered stone.
5	3.9	1.6	5.8	43.93	.881	.012 1	7.58	73.5	Rather rough.
6	3.6	1.5	5.3	43.93	.835	.014	8.36	77.3	
7	19.7	4.5	66.1	167.65	2.871	.000 43	2.34	72.2	Mud & vegetation.§

* Also in Tome XIX. Savants étrangers, l'Académie des Sciences.

† In this column *v* indicates mean surface width.

‡ In this chapter *d* always indicates mean depth in deepest part of channel.

§ Grass and weeds.

Series II. to XXVII. inclusive were made by the aid of a wooden rectangular flume or canal, 6.6 wide × 3.1 deep, 1956 feet long, having three inclinations respectively of .0049, .0020 and .0084. In this outer flume were placed interior linings, forming conduits of various forms and of various surfaces. The original flume in the course of the experiments did not exactly retain its inclination, thus forming for some of the series a notably undulatory bottom profile. Q was determined by the flow through orifices, whose co-efficients of discharge had been determined by direct measurement, the experimental conduit serving as a measuring vessel.* There were 12 of these orifices or gates, and in general, experiments were made for each series by first opening completely one gate, and then another. The mean area was determined by many measurements for each experiment. All the experimental data appear to have been obtained with great care.

The bottom longitudinal profiles for Series II. to VIII. inclusive, XX., XXII., and XXIV. to XXVII. inclusive, were very nearly straight; for Series IX. to XVII. inclusive there were small angles, necessarily causing undulations in the surface of the water, thus making it difficult to correctly measure mean depths, and causing some uncertainty in regard to the inclinations; for Series XVIII., XIX., XXI. and XXIII. the bottom was slightly irregular.

When the conduit was lined with gravel, the mean area was measured from the outer surfaces of the pebbles, the interstices between them hence not forming part of a; the same was the case when the wooden conduit was lined with strips of wood.

---

* These experiments have been referred to in Chapter III. on Orifices.

TABLE LIXa.

Darcy and Bazin.—Rectangular Conduits with various Surfaces, and of various Proportions.

$$v = n (r s)^{1/2}$$

	II. Δ = pure cement. s=.0049, w=3.94, T=54°					III. Δ = brick (not very smooth). s=.0049, w=6.27, T=about 66°					IV. Δ = small gravel (.03 to .07 dia.) fixed in cement. s=.0049, w=6.01, T=58°					V. Δ = large gravel (.10 to .13 dia.) fixed in cement. s=.0049, w=6.11, T=about 59°			
No.	d	r	v	n	No.	d	r	v	n	No.	d	r	v	n	No.	d	r	v	n
8	.18	.168	3.34	116.5	20	.20	.192	2.75	89.7	32	.27	.230	2.16	61.7	44	.32	.291	1.79	47.5
9	.28	.251	4.39	125.1	21	.31	.284	3.66	98.3	33	.41	.357	2.95	70.5	45	.48	.417	2.43	53.8
10	.36	.322	5.04	126.9	22	.41	.365	4.18	98.8	34	.53	.450	3.40	72.5	46	.61	.510	2.90	58.0
11	.43	.375	5.68	132.4	23	.49	.424	4.72	103.7	35	.63	.520	3.84	76.1	47	.73	.587	3.27	61.1
12	.50	.430	6.08	132.4	24	.57	.481	5.10	105.1	36	.73	.588	4.14	77.2	48	.84	.656	3.56	62.8
13	.56	.474	6.51	135.1	25	.66	.540	5.33	103.7	37	.82	.644	4.43	78.8	49	.93	.712	3.85	65.2
14	.63	.518	6.83	135.5	26	.71	.582	5.68	106.3	38	.91	.700	4.64	79.3	50	1.03	.772	4.03	65.5
15	.69	.558	7.12	136.2	27	.77	.620	6.01	109.0	39	.99	.746	4.88	80.7	51	1.13	.823	4.23	66.6
16	.76	.595	7.41	137.3	28	.85	.668	6.15	107.4	40	1.06	.785	5.12	82.6	52	1.21	.867	4.43	68.0
17	.80	.632	7.63	137.2	29	.90	.697	6.47	110.8	41	1.15	.832	5.26	82.4	53	1.29	.909	4.60	69.0
18	.86	.665	7.86	137.8	30	.97	.739	6.60	109.7	42	1.23	.871	5.43	83.1	54	1.37	.946	4.78	70.3
19	.91	.696	8.07	138.2	31	1.04	.779	6.72	108.7	43	1.30	.910	5.57	83.4	55	1.46	.987	4.90	70.4

	VI. Δ = unplaned plank. s=.00208, w=6.53, T=45°					VII. Δ = uplaned plank. s=.0049, w=6.53, T=47°					VIII. Δ = unplaned plank. s=.00824, w=6.53, T=47°			
No.	d	r	v	n	No.	d	r	v	n	No.	d	r	v	n
56	.26	.240	2.06	93.2	68	.20	.188	2.71	89.3	80	.15	.147	3.53	101.4
57	.41	.363	2.69	97.8	69	.30	.272	3.70	101.2	81	.25	.231	4.42	101.4
58	.53	.453	3.16	102.8	70	.38	.342	4.35	106.2	82	.32	.289	5.23	107.1
59	.63	.528	3.53	106.5	71	.46	.402	4.85	109.4	83	.38	.341	5.83	109.8
60	.73	.601	3.78	106.9	72	.53	.453	5.39	112.2	84	.45	.393	6.24	109.7
61	.81	.648	4.13	112.5	73	.60	.504	5.61	113.0	85	.50	.431	6.74	113.1
62	.90	.704	4.34	113.5	74	.66	.547	5.93	114.5	86	.54	.466	7.17	115.6
63	.99	.759	4.51	113.5	75	.72	.587	6.23	116.1	87	.60	.506	7.44	115.2
64	1.06	.801	4.72	115.8	76	.78	.628	6.45	116.4	88	.65	.541	7.73	115.8
65	1.14	.846	4.88	116.3	77	.83	.662	6.71	117.8	89	.69	.572	8.03	116.9
66	1.20	.880	5.09	119.0	78	.89	.698	6.90	117.9	90	.74	.604	8.26	117.1
67	1.28	.922	5.21	118.9	79	.94	.727	7.15	119.8	91	.78	.630	8.57	119.0

IX. Δ = unplaned plank. s=.0015 w=6.51 T=58°					X. Δ = unplaned plank. s=.0059 w=6.52 T=58°					XI. Δ = unplaned plank. s=.008 39 w=6.50 T=about 64°				
No.	d	r	v	n	No.	d	r	v	n	No.	d	r	v	n
92	.30	.276	1.80	88.3	99	.18	.172	2.99	93.7	106	.15	.146	3.54	101.1
93	.46	.406	2.37	96.3	100	.28	.255	3.98	102.7	107	.24	.224	4.57	105.4
94	.72	.590	3.10	104.2	101	.43	.376	5.23	111.1	108	.37	.334	6.00	113.4
95	.92	.720	3.63	110.4	102	.55	.472	6.06	114.8	109	.49	.424	6.89	115.6
96	1.10	.824	4.05	115.1	103	.67	.554	6.69	117.0	110	.59	.500	7.57	116.8
97	1.27	.912	4.41	119.1	104	.77	.623	7.24	119.3	111	.68	.565	8.19	118.9
98	1.44	.998	4.66	120.4	105	.87	.686	7.71	121.1	112	.77	.621	8.74	121.0

XII. Δ = laths of wood, .09 wide x .03 deep, nailed on sides and bottom of flume. Placed .03 apart (.12 centre to centre) at right angles to line of flume. Not butted squarely together. s=.0015 w=6.43 T=about 46°					XIII. Δ = as in Series XII. s=.0059 w=6.43 T=about 43°					XIV. Δ = as in Series XII. s=.008 86 w=6.40 T=57°				
No.	d	r	v	n	No.	d	r	v	n	No.	d	r	v	n
113	.33	.302	1.65	77.4	120	.22	.205	2.50	71.8	127	.19	.182	2.85	70.8
114	.51	.442	2.17	84.5	121	.33	.302	3.34	79.0	128	.30	.273	3.75	76.4
115	.73	.634	2.86	916	122	.51	.442	4.40	86?	129	.46	.403	4.92	82.4
116	1.06	.775	3.33	949	123	.67	.552	5.08	89.0	130	.59	.499	5.77	86.6
117		.889	3.68	976	124	.80	.643	5.63	91.4	131	.71	.582	6.38	83.9
118		.946	3.98	990	125	.92	.716	6.14	94.5	132	.83	.658	6.86	89.9
119		1.046	4.19	990	126	1.05	.790	6.48	94.8	133	.94	.726	7.26	90.5

XV. Δ = laths of wood as in Series XII., except that they were placed .16 apart (.25 centre to centre). s=.0015 w=6.43 T=about 59°					XVI. Δ = as in Series XV. s=.0059 w=6.44 T=45°					XVII. Δ = as in Series XV. s=.008 86 w=6.40 T=about 59°				
No.	d	r	v	n	No.	d	r	v	n	No.	d	r	v	n
134	.43	.378	1.28	53.7	141	.29	.264	1.91	48.3	148	.25	.232	2.21	48.7
135	.66	.550	1.68	58.6	142	.44	.384	2.56	53.7	149	.39	.350	2.85	51.2
136	1.02	.777	2.21	64.8	143	.67	.553	3.37	59.0	150	.60	.509	3.75	55.8
137	1.33	.942	2.55	67.8	144	.87	.686	3.88	61.0	151	.78	.628	4.37	58.6
138	1.61	1.073	2.81	70.1	145	1.05	.791	4.51	63.1	152	.94	.730	4.85	60.5
139	1.91	1.197	2.97	70.0	146	1.21	.882	4.65	64.5	153	1.09	.812	5.22	61.5
140	2.18	1.289	3.11	70.5	147	1.38	.965	4.91	65.1	154	1.22	.885	5.57	63.9

	XVIII.					XIX.					XX.			
	Δ = unplaned plank.					Δ = unplaned plank.					Δ = unplaned plank.			
	s = .0049					*s* = .0043					*s* = .006			
	w = 3.93					*w* = 2.625					*w* = 1.575			
	T = 46°					*T* = 13° to 31°					*T* = about 57°			
No.	d	r	v	n	No.	d	r	v	n	No.	d	r	v	n
155	.27	.235	3.37	99.1	167	.26	.214	2.85	94.0	178	.34	.237	3.57	94.5
156	.41	.341	4.43	108.3	168	.39	.299	3.47	96.9	179	.44	.281	4.00	97.3
157	.35	.428	5.05	110.2	169	.50	.364	4.14	104.6	180	.50	.304	4.20	98.3
158	.67	.498	5.54	112.3	170	.60	.412	4.54	107.8	181	.53	.317	4.23	97.1
159	.78	.558	5.94	113.7	171	.71	.461	4.91	110.4	182	.62	.347	4.67	102.3
160	.89	.612	6.26	114.3	172	.81	.499	5.12	110.5	183	.70	.372	4.94	104.6
161	1.00	.661	6.50	114.2	173	.90	.535	5.41	112.7	184	.79	.393	5.11	105.3
162	1.10	.703	6.76	115.3	174	.99	.563	5.60	113.9	185	.87	.112	5.26	105.7
163	1.19	.741	7.00	116.1	175	1.17	.618	5.92	114.9	186	.95	.431	5.49	107.9
164	1.29	.777	7.20	116.7	176	1.33	.662	6.33	116.7					
165	1.37	.808	7.42	118.0	177	1.50	.700	6.48	118.1					
166	1.46	.839	7.59	118.1										

TABLE LIXa.

Darcy and Bazin—Trapezoidal and Triangular Plank Conduits.

$$v = n\,(r\,s)^{\frac{1}{2}}$$

	XXI.					XXII.					XXIII.			
	For depth of 1.64 sides inclined at 45°, and then vertical.					One side vertical; the other side inclined at 45°.					Sides inclined at 45°; section being a right-angled triangle with vertex on bottom.			
	Bottom width = 3.28					Bottom width = 3.10								
	Top width = 6.56													
	Δ = unplaned plank.					Δ = unplaned plank.					Δ = unplaned plank.			
	s = .0015					*s* = .0049					*s* = .0049			
	T = about 43°					*T* = 76°					*T* = 63° to 72°			
No.	d	r	v	n	No.	d	r	v	n	No.	d	r	v	n
187	.40	.331	2.39	107.0	199	.30	.237	3.58	100.7	211	.92	.327	4.13	103.1
188	.63	.485	2.93	108.5	200	.46	.361	4.71	112.1	212	1.19	.422	5.02	110.4
189	.79	.586	3.35	113.0	201	.60	.450	5.29	112.7	213	1.40	.494	5.56	113.0
190	.95	.673	3.62	113.8	202	.72	.517	5.79	115.1	214	1.55	.549	6.03	116.2
191	1.08	.744	3.85	115.4	203	.83	.570	6.25	116.2	215	1.69	.597	6.36	117.6
192	1.21	.809	4.03	115.7	204	.94	.624	6.51	117.7	216	1.82	.643	6.59	117.3
193	1.32	.864	4.20	116.7	205	1.03	.665	6.85	120.0	217	1.93	.685	6.83	118.0
194	1.41	.911	4.39	118.9	206	1.12	.707	7.05	119.8	218	2.03	.719	7.03	118.4
195	1.51	.959	4.51	119.0	207	1.20	.740	7.37	122.4	219	2.13	.752	7.23	119.0
196	1.60	1.002	4.64	119.7	208	1.28	.773	7.57	122.9	220	2.22	.783	7.40	119.5
197	1.69	1.047	4.76	120.2	209	1.36	.807	7.76	123.4	221	2.30	.814	7.54	119.4
198	1.77	1.097	4.87	120.1	210	1.44	.837	7.93	123.8	222	2.37	.839	7.75	120.9

TABLE LIXc.

Darcy and Bazin.—Semicircular Conduits.

$$v = n \, (r \, s)^{\frac{1}{2}}$$

XXIV. Δ = pure cement. $s = .0015$ Top width (also diameter) = 4.10 $T = 63°$					XXV. Δ = cement mixed with ⅓rd sand (very fine). $s = .0015$ Top width (also diameter) = 4.10 $T = 63°$					XXVI. Δ = partly planed plank. $s = .0015$ Top width (also diameter) = 4.59 $T = 64°$ to $73°$					XXVII. Δ = small gravel (.03 to .07 dia.) fixed in cement. $s = .0015$ Top width (also diameter) = 4.00 $T = 54°$				
No.	d	r	v	n	No.	d	r	v	n	No.	d	r	v	n	No.	d	r	v	n
223	.59	.366	3.02	128.9	235	.61	.379	2.87	120.5	247	.63	.390	2.61	107.8	260	...	.454	2.17	78⁰
224	.83	.503	3.72	135.6	236	.88	.529	3.43	122.0	248	.88	.537	3.23	113.8	261	...	.546	2.50	82⁰
225	1.03	.605	4.16	138.0	237	1.09	.635	3.87	125.3	249	1.07	.632	3.71	120.6	262	...	.619	2.69	82⁰
226	1.18	.682	4.60	143.7	238	1.24	.706	4.30	132.1	250	1.24	.717	4.04	123.0	263	...	.681	2.93	84⁰
227	1.34	.750	4.87	145.1	239	1.41	.787	4.51	131.3	251	1.40	.796	4.25	123.2	264	...	.731	3.05	84⁰
228	1.47	.809	5.12	147.1	240	1.54	.839	4.80	135.3	252	1.53	.856	4.51	125.8	265	...	.784	3.22	85⁰
229	1.61	.867	5.29	146.7	241	1.69	.900	4.94	134.5	253	1.68	.921	4.64	124.7	266	...	.826	3.33	84⁰
230	1.72	.913	5.51	148.8	242	1.80	.941	5.20	138.3	254	1.79	.964	4.87	128.2	267	...	.900	3.54	85⁰
231	1.83	.949	5.75	152.5	243	1.92	.983	5.38	140.1	255	1.93	1.015	5.00	128.2	268	...	.968	3.73	85⁰
232	1.94	.992	5.91	153.3	244	1.98	1.006	5.48	141.0	256	2.02	1.054	5.18	130.3	269	...	1.012	3.95	88⁰
233	2.05	1.029	6.06	154.2	245	2.04	1.022	5.55	141.7	257	2.14	1.096	5.29	130.4					
234	2.08	1.034	6.11	155.1	246	2.09	1.038	5.66	143.5	258	2.24	1.129	5.45	132.3					
										259	2.29	1.148	5.54	133.5					

Series XXVIII. to XXXI. inclusive were made with a rectangular trough cut in four longitudinal solid timbers, nicely joined, and having a total length of 62 feet. Q was measured directly, with much accuracy. Series XXVIII. and XXIX. were made with the planed and very uniform wooden surface of the trough. A strip of strong cloth was then tacked to the sides and bottom for Series XXX. and XXXI.; this lining somewhat rounded the lower corners, and caused notable undulations in the surface of the flowing water.

TABLE LX.

Darcy and Bazin.—Small Rectangular Conduit.

$$v = n (r s)^{1/2}$$

XXVIII. Δ = wood, very smooth. s = .0047 w = .328 T = 50°					XXIX. Δ = as in XXVIII. s = .0152 w = .328 T = 50°				
No.	d	r	v	n	No.	d	r	v	n
270	.036	.029	.90	76.5	277	.037	.030	1.87	87.5
271	.076	.052	1.30	83.0	278	.058	.043	2.30	90.0
272	.111	.066	1.58	89.4	279	.078	.053	2.68	94.4
273	.138	.075	1.74	92.7	280	.097	.061	3.00	98.5
274	.173	.084	1.94	97.6	281	.134	.074	3.56	106.4
275	.204	.091	2.11	102.1					
276	.215	.093	2.16	103.2					

XXX. Δ = cloth. s = .0081 w = .312 T = 50°					XXXI. Δ = cloth. s = .0152 w = .312 T = 50°				
No.	d	r	v	n	No.	d	r	v	n
282	.048	.038	.73	40.7	288	.036	.031	.69	31.8
283	.062	.046	.89	46.1	289	.051	.040	.82	33.5
284	.081	.055	1.11	32.5	290	.072	.051	1.19	42.8
285	.111	.067	1.33	56.9	291	.077	.054	1.25	43.7
286	.147	.078	1.51	59.8	292	.106	.066	1.55	49.0
287	.269	.102	1.88	65.3	293	.111	.067	1.62	50.7
					294	.150	.079	1.91	55.0
					295	.193	.089	2.12	57.6
					296	.226	.095	2.23	58.7

Series **XXXII.** and **XXXIII.** were made with a sluice-way, 865 feet long, of hammer-dressed masonry laid in cement, the bottom being covered with a slight deposit of adhering slime; the bottom was flat, and the sides had a batter of $\frac{1}{10}$: the sluice-way had two inclinations. These experiments are notable for the very high values of *v*.

Series **XXXIV.** and **XXXV.** were made with another sluice-way, revetted with dry hammer-dressed paving on sides and bottom; its form was a flat trapezoid, with bottom width of 5.9 and side slopes of $1\frac{1}{2}$ to 1. Both sides and the bottom were covered

A A

with moss and grass (there appears to have been a deposit of mud on the bottom and on one side); No. XXXIV. was made with this condition of Δ.   The vegetation and mud were then carefully removed, and No. XXXV. made; the section was very slightly changed, but the discharge for same depths increased nearly one half.

In these four series, Q was measured by orifices, whose co-efficients of discharge, C, were assumed to be from .60 to 67.

TABLE LXI.

Darcy and Bazin.—Masonry Sluice Ways.

$v = n \, (r \, s)^{\frac{1}{2}}$

	XXXII.					XXXIII.			
Δ = hammer-dressed masonry, with some adhering slime. Form nearly rectangular. $s = .101$ Bottom width = 5.91 $T = 64°$					Δ = as in XXXII. Form as in XXXII. $s = .037$ Bottom width = 5.91 $T = 64°$				
No.	d	r	c	n	No.	d	r	v	n
297	.36	.324	12.29	67.9	301	.49	.424	9.04	72.2
298	.55	.467	16.18	74.5	302	.77	.620	11.46	75.7
299	.71	.580	18.68	77.2	303	.97	.745	13.55	81.6
300	.84	.662	21.09	81.6	304	1.16	.852	15.08	84.9

	XXXIV.					XXXV.			
Δ = dry hammer-dressed masonry, covered with moss and grass (some mud). Form, flat trapezoid. $s = .0146$ Bottom width = 6.50 $T = 39°$					Δ = dry hammer-dressed masonry. Form nearly as in XXXIV. $s = .0142$ Bottom width = 5.87 $T = 50°$				
No.	a	r	v	n.	No.	n	r	v	n
305	8.9	.856	4.19	37.5	310	6.6	.703	5.66	56.6
306	13.9	1.087	5.75	45.7	311	10.0	.930	7.36	61.0
307	19.3	1.383	7.20	50.7	312	15.5	1.227	8.94	67.7
308	24.4	1.586	8.27	54.3	313	19.3	1.394	10.12	71.9
309	27.5	1.694	8.99	57.2	314	21.6	1.491	11.96	77.4

The experiments embraced in Series XXXVI. to L. inclusive, were made with two feeding sluices or canals, each several miles in length, and having along their courses Δ varying from earth covered with vegetation to smooth masonry. There were many curves, but of large radii. The measurements of a, and hence r, were necessarily more or less inaccurate when the canals were in earth. Q was measured by the flow over a weir, with allowances for water received and lost along the length of the canals.

These experiments are hence much rougher than the preceding ones.

TABLE LXII.

Darcy and Bazin.—Canals with various Sections and Surfaces.

$$v = a\,(r\,s)^{\frac{1}{2}}$$

XXXVI. Δ = earth covered with vegetation at many points. Form, trapezoidal. Bottom width = 3.7						XXXVII. Δ = earth (stony); but little vegetation. Form, trapezoidal. Bottom width = 3.9						XXXVIII. Δ = as in XXXVII. Form, trapezoidal. Bottom width = 4.1					
No.	a	r	s	v	n	No.	a	r	s	v	n	No.	a	r	s	v	n
315	13.0	1.14	.000 678	.91	33	319	9.5	.96	.000 792	1.23	45	323	9.3	.96	.000 937	1.24	41
316	19.9	1.42	.000 633	1.28	43	320	14.9	1.20	.000 808	1.67	53	324	14.1	1.18	.000 929	1.70	51
317	25.2	1.61	.000 644	1.45	45	321	19.4	1.41	.000 858	1.81	52	325	18.8	1.41	.000 993	1.80	48
318	29.1	1.74	.000 622	1.65	50	322	22.9	1.56	.000 842	2.00	55	326	22.2	1.54	.000 986	1.96	50

XXXIX. Δ = smooth masonry. Form nearly rectangular. Bottom width = 3.9						XL. Δ = rough stone; very little vegetation. Form, trapezoidal. Bottom width = 4.2						XLI. Δ = as in XXXVII. Form, trapezoidal. Bottom width = 4.4					
No.	a	r	s	v	n	No.	a	r	s	v	n	No.	a	r	s	v	n
327	2.0	.406	.0081	5.73	100	331	10.5	1.05	.000 936	1.08	34	335	11.3	1.04	.000 445	.96	45
328	3.2	.571	do.	7.52	111	332	17.2	1.37	do.	1.37	38	336	18.1	1.38	.000 450	1.27	51
329	4.0	.680	do.	8.19	110	333	21.1	1.52	.000 957	1.56	41	337	22.9	1.57	.000 455	1.40	52
330	4.9	.766	do.	8.75	111	334	24.6	1.64	.000 964	1.71	43	338	27.2	1.71	.000 441	1.51	55

XLII. Δ = Bottom earth; sides of masonry and dry paving. Form, trapezoidal. Bottom width = 7.1						XLIII. Δ = as in XXXVI. Form, trapezoidal. Bottom width = 4.3						XLIV. Δ = masonry in rather bad order; some mud and stones on bottom. Sides nearly vertical; flat circular invert on bottom. Bottom width = 6.6					
No.	a	r	s	v	n	No.	a	r	s	v	n	No.	a	r	s	v	n
339	10.5	1.00	.000 525	1.01	44	343	11.6	1.06	.000 420	.89	42	347	9.7	1.07	.000 30	1.12	63
340	16.3	1.36	.000 450	1.38	56	344	19.0	1.41	.000 470	1.18	46	348	14.3	1.38	.000 35	1.69	77
341	20.0	1.54	.000 462	1.58	59	345	24.0	1.60	do.	1.31	47	349	18.0	1.57	.000 33	1.92	84
342	23.1	1.67	.000 487	1.74	61	346	28.7	1.76	.000 450	1.39	49	350	21.1	1.71	.000 30	2.18	96

| XLV. |
| Δ = masonry in better order than XLIV. Bottom clean. Form as in XLIV. Bottom width = 6.2 |

No.	a	r	s	v	n
351	8.2	.98	.000 303	1.32	77
352	12.7	1.29	.000 308	1.90	95
353	16.3	1.49	.000 331	2.12	96
354	18.6	1.60	.000 347	2.47	105

| XLVI. |
| Δ = as in XLIV. Form as in XLIV. Bottom width = 6.6 |

No.	a	r	s	v	n
355	7.4	.88	.000 648	1.47	62
356	12.0	1.23	.000 671	2.02	70
357	14.7	1.40	.000 683	2.34	76
358	16.6	1.50	do.	2.78	87

| XLVII. |
| Δ = earth ; some vegetation. Form, nearly arc of circle. |

No.	a	r	s	v	n
359	11.8	1.09	.000 464	.82	36
360	17.2	1.38	.000 450	1.32	53
361	23.0	1.63	.000 479	1.43	51
362	26.8	1.71	.000 493	1.68	58

| XLVIII. |
| Δ = as in XLVII. Form, nearly arc of circle. |

No.	a	r	s	v	n
363	10.1	.99	.000 555	.96	41
364	15.4	1.30	do.	1.48	55
365	20.9	1.56	.000 525	1.57	55
366	25.9	1.71	.000 515	1.75	59

| XLIX. |
| Δ = earth ; no vegetation. Form, trapezoidal. Bottom width = 6.5 |

No.	a	r	s	v	n
367	10.9	.96	.000 250	.89	57
368	17.0	1.32	.000 275	1.34	70
369	24.2	1.57	.000 246	1.36	69
370	30.8	1.78	.000 275	1.47	66

| L. |
| Δ = as in XLVII. Form, trapezoidal. Bottom width = 6.3 |

No.	a	r	s	v	n
371	11.8	1.05	.000 310	.82	45
372	18.0	1.42	.000 290	1.26	62
373	25.4	1.65	.000 330	1.30	56
374	32.0	1.85	do.	1.41	57

When the two foregoing canals were in earth, there was sometimes more vegetation on the upper part of the slopes than on the lower portion ; hence Δ was not constant for different values of r in the same section.

Rectangular Pipes.

In order to determine whether or not the resistance of the air upon the surface of an open conduit appreciably retarded the flow, M. Bazin made the following experiments with two rectangular wooden pipes, one 2.625 wide × 1.64 deep, and the other 1.575 wide × .984 deep ; the widths and conditions of surface being identical with the open conduits used in Series XIX. and XX. The first was laid with an inclination of .0049, and the second with an inclination of .0059. Both ends of each pipe were in all cases submerged, thus allowing the various changes in the hydraulic grade. The heads lost by "friction" were measured in both pipes by 5 piezometric tubes attached to the pipe ; No. 3 was near the centre of each pipe, and midway between Nos. 1 and 5, and Nos. 2 and 4 ; the piezometric heads were read by five open glass tubes placed side by side ; h was assumed to be the difference in elevation in tubes 2 and 4. Q was measured by flow through the same gates used for Series II. to XXVII.

TABLE LXIII.
Darcy and Bazin.—Rectangular Unplaned Wooden Pipes.

$$v = n\,(r\,s)^{½}$$

	LI. Section 2.625 × 1.64; r = .505 l = (distance piezometer No. 2 to No. 4) = 103.68						LII. Section 1.575 × .984; r = .303 l = (distance piezometer No. 2 to No. 4) = 49.21				
No.	Q	h	s	v	n	No.	Q	h	s	v	n
375	7.17	.023+.026	.00047	1.67	107.6	383	1.91	.020+.007	.00053	1.23	96.8
376	10.84	.056+.056	.00108	2.52	108.1	384	2.75	.026+.026	.00107	1.78	98.9
377	14.51	.092+.105	.00190	3.37	108.9	385	3.53	.046+.039	.00173	2.28	99.4
378	18.19	.148+.154	.00291	4.22	110.2	386	4.56	.069+.066	.00273	2.94	102.2
379	21.83	.210+.233	.00427	5.07	109.2	387	5.47	.102+.080	.00387	3.53	103.2
380	23.80	.256+.269	.00506	5.53	109.4	388	6.74	.164+.144	.00627	4.35	99.9
381	25.46	.282+.315	.00576	5.91	109.7	389	7.17	.180+.177	.00727	4.62	98.6
382	27.44	.328+.358	.00661	6.37	110.3	390	8.23	.220+.213	.00880	5.31	102.8

These experiments indicate a value of about 110 for n, with r = .505, and 100 for n with r = .303.[*] Referring to Series XIX. for r = .50, n is 110 ; and with Series XX. for r = .30, n is 98.

M. Bazin concludes from these experiments that the surface air, when there is no wind, does not appreciably retard the flow in open conduits.

They also indicate for the larger pipe that n increases only slightly with a change of v from 1.7 to 6.4.

---

[*] In these experiments readings for each pipe were taken from 5 piezometers, the extreme distances apart being 195 feet for conduit LL., and 98 feet for conduit LII. M. Bazin, however, deduces s from the 3 central piezometers, as given in Table LXIII. If s be deduced from the readings of the extreme piezometers, n for Series LI. ranges from 112 to 115.5, and for Series LII. from 98 to 103, as follows ;

	LI.			LII.	
No.	s	n	No.	s	n
375	.00044	112.0	383	.00050	100.0
376	.00096	114.4	384	.00103	100.5
377	.00173	113.9	385	.00173	99.4
378	.00266	115.3	386	.00287	99.8
379	.00389	114.4	387	.00390	102.8
380	.00468	114.1	388	.00620	100.4
381	.00525	114.8	389	.00733	98.1
382	.00603	115.6	390	.00920	100.6

Judging from the data given by M. Bazin, one would suppose that these latter values of s and n should be more reliable than the first.

Fteley and Stearns.

During the years 1878-1880 Messrs. Fteley and Stearns made a large number of experiments of the flow through various portions of the new Sudbury conduit, which forms a late addition to the water supply of the City of Boston, U.S.A. These experiments have been alluded to in the paper of these gentlemen, published by the Am. Soc. of C.E.,* but professional duties have thus far prevented them from giving to the world a full account of their investigations with conduits. They have most kindly placed their full results at our disposition, and we here desire to express our great obligations to them.

The experiments to be given are the most accurate ever made with a conduit of considerable size, and will constitute a most valuable addition to our store of knowledge in this important branch of the science of Hydraulics.

This conduit, except through hard rock tunnels, is built of well-formed hard brick, with good mortar joints; its general section is shown by the following sketch, the bottom invert having a radius of 13.22 feet, and a chord of 8.29 feet; its original section has been changed only very slightly by settlement, and there are very slight gains or losses by infiltration or leakage along its course; the curves of the sections experimented upon have a minimum radius of 1433 feet; its general inclination is 1 foot per mile.

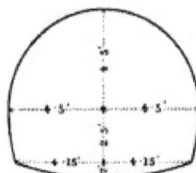

The discharge, or Q, was for Experiments Nos. 391 to 446 (except for No. 443) measured by the flow over a weir, 19 feet long, whose co-efficients of discharge with various heads, had been determined with great exactness by direct measurement;† this weir was placed at the upper end of the conduit. Most of the experiments were made with sections of the conduit only a short distance below the weir, so that only very slight corrections were required in any experiment for gains by infiltration or losses by leakage. These corrections were determined by actual measurement, when no water was flowing through the conduit.

The head h, or surface inclination, for all the experiments was ascertained by

---

* Transactions of Am. Soc. C.E. ; January, February and March, 1883.
† The experiments with this weir, where Q was directly measured, are given in the Chapter on Weirs ; our Weir Nos. 9 to 16 inclusive.

establishing two bench-marks, one at the upper and the other at the lower end of the longitudinal section ; a dam was placed below the lower bench-mark, the flow of water suspended in the conduit, and the difference in elevation between the two bench-marks determined by repeated readings with gauges of the surface of the stagnant water. As the conduit was arched over, this surface was very still, being thus protected from the disturbing effects of the wind ; this method hence insured great accuracy in the determination of h, which can be considered as being given within a limit of possible error of .0025. The value of s can, therefore, be considered as having been determined with great accuracy, except for such experiments with $l = 600$ (Nos. 447-459), where s is less than .0001, or h less than .06.

In Table LXIV. are given the experiments made in October and November, 1879, with the normal inclination of the conduit, and the general condition of its interior surface.

TABLE LXIV.

Fteley and Stearns.—Sudbury Conduit.

Δ = hard brick, fairly clean and smooth, with mortar joints well made.

Bottom inclination about .000 189. $v = n\,(r\,s)^{\frac{1}{2}}$

| | Sta. 17-59 ; hence $l = 4200$ feet. | | | | | | | Sta. 59-111.94.5 feet ; hence $l = 5294.5$ feet. | | | | |
No.	d	Q	v	r	s	n	No.	d	Q	v	r	s	n
391	1.518	20.086	1.827	1.0779	.000 192 77	126.7	400	1.505	20.086	1.844	1.0709	.000 189 29	129.5
392	2.014	32.755	2.131	1.3723	.000 190 93	131.6	401	2.003	32.755	2.113	1.3669	.000 190 08	133.0
393	2.037	33.314	2.139	1.3848	.000 192 24	131.1	402	2.023	33.314	2.155	1.3779	.000 190 05	133.2
394	2.513	46.638	2.351	1.6253	.000 192 31	133.0	403	2.499	46.648	2.366	1.6193	.000 189 88	134.9
395	2.519	47.179	2.372	1.6282	.000 192 36	134.0	404	2.499	47.224	2.395	1.6193	.000 192 10	135.8
396	3.010	62.372	2.564	1.8428	.000 188 84	137.5	405	3.002	62.372	2.572	1.8395	.000 190 29	137.5
397	3.561	79.696	2.720	2.0485	.000 191 86	137.2	406	3.548	79.696	2.731	2.0441	.000 190 05	138.6
398	4.012	94.414	2.831	2.1916	.000 189 48	138.9	407	4.008	94.394	2.834	2.1902	.000 188 55	139.5
399	4.552	111.434	2.926	2.3326	.000 192 15	138.2	408	4.541	111.548	2.937	2.3297	.000 188 86	140.0

In the foregoing series a practically constant head was maintained at the measuring weir, until the regimen of flow was fully established from Sta. 17 to Sta. 112 ; then readings of the surface heights were taken for the two adjoining longitudinal sections. Q is sometimes very slightly larger for the lower section than for the simultaneous experiment with the upper section, on account of minute accretions by infiltration.[*] Judging from the corresponding values of n for the two sections, it is probable that the wetted surface was slightly rougher for the upper section than it was for the lower one. The mean area from Sta. 17 to Sta. 112 was deduced from cross-sections of the conduit taken 25 feet apart ; these sections were very nearly uniform.

---

[*] In one case Q was smaller for the lower section ; this was due to a minute rise or fall of the surface during the experiment by which some of the water measured was stored in, or drawn from the upper portion of the conduit.

Table LXV. gives the results for the same longitudinal sections, with surface inclinations considerably differing from the normal or bottom inclination. The surface inclinations less than the normal slope, were caused by placing a dam of various heights at Sta. 112 ;* in order to obtain inclinations greater than the normal slope, two waste gates attached to the conduit at Sta. 112, were opened. The axial depths, from which the values of *r* have been deduced, are the means of the depths at the upper and lower ends of the longitudinal section, for each experiment. Hence in these experiments the various cross-sections were not uniform, and the flow was necessarily variable ; the deduced values of *n*, hence are approximative, as no attempt has been made to take into account this lack of uniformity. The last column in small type, *d'*, gives the mean axial depth in the conduit, which would have obtained with the same quantity of water flowing with the normal inclination of one foot per mile.

TABLE LXV.

Fteley and Stearns.—Sudbury Conduit.

Δ = hard brick fairly clean and smooth, with mortar joints well made.

Bottom inclination about .000 189. $v = n (r s)^{½}$

No.	*d*			*Q*	*v*	*r*	*s*	*n*	*d'*
	Sta. 17.	Sta. 59.	Mean.						
409	3.679	4.263	3.971	47.213	1.432	2.179	.000 0493	138.1	2.52
410	4.093	4.603	4.348	62.317	1.716	2.284	.000 0668	138.9	3.01
411	2.635	2.987	2.811	40.995	1.820	1.759	.000 1043	134.4	2.31
412	2.120	2.304	2.212	32.770	1.912	1.478	.000 1445	130.9	2.01
413	2.048	2.082	2.065	32.770	2.071	1.400	.000 1795	130.6	2.01
414	2.665	2.805	2.735	48.009	2.198	1.727	.000 1548	134.5	2.55
415	3.124	3.230	3.177	62.172	2.406	1.909	.000 1631	136.4	3.01
416	4.570	4.578	4.574	111.328	2.909	2.338	.000 1860	139.5	4.54

Surface slope less than bottom inclination.

Sta. 17-59 ; hence *l* = 4200 feet.

* The stations were 100 feet apart ; Sta. 0 being at the upper end of the conduit, and not far from the measuring weir.

TABLE LXV.—*continued.*

Sta. 59-111.94.5 ; hence *l* = 5 294.5 feet.

No.	d			Q	v	r	s	n	d'
	Sta. 59.	Sta. 111.94.5.	Mean.						
417	4.263	5.081	4.672	47.208	1.207	2.359	.000 0334	136.0	2.52
418	4.603	5.341	4.972	62.307	1.497	2.417	.000 0488	137.9	3.01
419	2.987	3.651	3.319	40.995	1.512	1.963	.000 0625	136.5	2.31
420	2.304	2.818	2.561	32.770	1.616	1.648	.000 0948	129.3	2.01
421	2.082	2.302	2.192	32.770	1.931	1.468	.000 1466	131.6	2.01
422	2.805	3.191	2.998	48.020	1.983	1.838	.000 1155	136.1	2.55
423	3.230	3.508	3.369	62.172	2.355	1.981	.000 1356	137.6	3.01
424	4.578	4.626	4.602	111.248	2.889	2.343	.000 1793	141.0	4.54

Surface slope greater than bottom inclination.

Sta. 17-59 ; hence *l* = 4200 feet.

No.	d			Q	c	r	s	n	d'
	Sta. 17.	Sta. 59.	Mean.						
425	2.027	1.977	2.002	32.988	2.164	1.366	.000 1998	130.8	2.02
426	2.496	2.430	2.463	46.843	2.416	1.602	.000 2041	133.6	2.51
427	2.982	2.898	2.940	62.297	2.630	1.814	.000 2082	135.3	3.01
428	3.479	3.401	3.440	78.766	2.792	2.006	.000 2070	137.0	3.53
429	3.990	3.936	3.963	95.032	2.888	2.177	.000 2006	138.2	4.03
430	3.824	3.602	3.713	95.013	3.098	2.099	.000 2111	137.7	4.05
431	4.541	4.239	4.390	124.190	3.386	2.294	.000 2600	138.6	4.96

Sta. 59-111.94.5 : hence *l* = 5294.5 feet.

No.	d			Q	c	r	s	n	d'
	Sta. 59.	Sta. 111.94.5.	Mean.						
432	1.977	1.621	1.799	32.988	2.448	1.251	.000 2553	137.0	2.02
433	2.430	2.060	2.245	46.850	2.687	1.495	.000 2580	136.8	2.51
434	2.898	2.516	2.707	62.297	2.886	1.714	.000 2602	136.6	3.01
435	3.401	3.131	3.266	78.766	2.957	1.943	.000 2389	137.3	3.53
436	3.936	3.820	3.878	95.004	2.955	2.151	.000 2102	139.0	4.03
437	3.602	2.160	2.881	95.013	4.103	1.789	.000 4604	142.9	4.03
438	4.339	2.635	3.437	124.190	4.407	2.005	.000 4913	140.4	4.96

In the following table are given experiments with various conditions of the wetted surface, at various longitudinal sections of the conduit. No. 443 was made on Nov. 12,

B B

1878, and for this experiment Q was determined by a current-meter, placed successively at various portions of the cross-section; from these measurements the mean velocity, v, was deduced. All the other experiments were made on April 1-2, 1880. In the table, l signifies, as usual, the total length of the section; l' the length with the particular condition of Δ; $l - l'$ is the length with normal condition of surface, being hard brick with good mortar joints.

TABLE LXVI.

Fteley and Stearns.—Sudbury Conduit.

Various values of Δ. $v = n (r \, s)^{\frac{1}{2}}$

Δ for l' = wash of pure Portland cement, over brick-work.

No.	Stations.	l	l'	d	Q	v	r	s	n
439	590.00.8 – 604.07.5	1406.7	1260	3.529	78.42	2.715	2.034	.000 1779	142.7
440	do.	do.	do.	3.038	62.75	2.574	1.847	.000 1810	140.8
441	604.07.5 – 619.29.5	1522	1153	3.569	78.42	2.675	2.053	.000 1790	139.5
442	do.	do.	do.	3.066	62.75	2.532	1.867	.000 1793	138.4

Δ for l = plaster of pure Portland cement, over brick-work.

443	Charles River Bridge	490		3.768	87.17	2.805	2.111	.000 1580	153.6
444	do.	do.		3.375	78.42	2.672	2.048	.000 1596	147.9
445	do.	do.		3.071	62.75	2.529	1.863	.000 1606	146.2

Δ for l' = concrete bottom; sides, rough rock.

Δ for $l - l'$ = concrete bottom and brick sides.

446	Beacon St. Tunnel	4613.7	4302.34	3.440	78.42	1.975	2.211	.000 2813	79.2

For No. 446, with Beacon Street tunnel—irregular sides of jagged rock—no experimental cross-sections were taken; the value of r, however, is not very far from the truth. In this tunnel there were 10 head walls; the tunnel being lined with brick at four places where the rock was soft, and also at its ends. The invert, or bottom of concrete, had a chord of about 10 feet.

Experiments Nos. 447 to 459 inclusive, were made with the section of the conduit between Stations 6 and 12, having a length of 600 feet; the mean area was very carefully determined by cross-sections taken $12\frac{1}{2}$ feet apart; the surface of the bricks had been well scraped, so that for this series Δ had a somewhat larger value than for the sections between Stations 17 and 112. These experiments were made chiefly to determine the values of n for small values of r, but some were made to test the effect of very small inclinations; these inclinations were obtained by placing a dam in the conduit at Station 13—100 feet below lower end of experimental section. As before remarked, for inclinations less than .0001, possible errors in the determination of h

would appreciably affect the deduced values of *n*. The amount of flow, *Q*, was determined by a weir, 5 feet long, placed near Station 0, whose co-efficients of discharge had been accurately determined.[*] The regimen of flow was almost perfectly established in these experiments.

TABLE LXVII.

Fteley and Stearns.—Sudbury Conduit.

Flow with various surface inclinations. Bottom slope about .000 16. Δ = hard brick, clean and smooth, with mortar joints well made, and surface carefully scraped clean from foreign substances.

Sta. 6 to 12 ; *l* = 600 feet. $v = n \, (r \, s)^{\frac{1}{2}}$

No.	*a*	*d*	*p*	*Q*	*v*	*r*	*h*	*s*	*n*
447	10.176	1.416	10.018	4.505	.4427	1.016	.008 42	.000 014 03	117.3
448	8.199	1.187	9.552	4.513	.5504	.8584	.014 75	.000 024 58	119.8
449	10.076	1.404	9.994	7.948	.7888	1.006	.023 00	.000 038 33	126.9
450	9.416	1.328	9.839	10.015	1.064	.9570	.044 75	.000 074 58	125.9
451	4.228	.719	8.585	4.560	1.079	.4925	.098 42	.000 164 0	120.0
452	7.253	1.076	9.335	7.965	1.098	.7778	.059 00	.000 098 33	125.6
453	5.077	.820	8.795	5.835	1.149	.5773	.095 75	.000 159 6	119.7
454	8.098	1.175	9.598	10.047	1.241	.8499	.066 92	.000 111 5	127.5
455	6.079	.939	9.040	7.888	1.298	.6725	.098 00	.000 163 3	123.9
456	7.068	1.055	9.280	10.060	1.423	.7616	.104 50	.000 174 2	123.6
457	6.951	1.041	9.262	10.001	1.439	.7513	.108 17	.000 180 3	123.6
458	8.598	1.233	9.646	13.488	1.569	.8913	.102 08	.000 170 1	127.4
459	8.523	1.224	9.628	13.443	1.577	.8852	.102 92	.000 171 5	128.0

CANALS.

Captain Allan Cunningham has lately published an account of many experiments,[†] made under his direction, upon the flow through the Ganges Canal. He deduces the following values for *n*, in $v = n \, (r \, s)^{\frac{1}{2}}$, *v* having been ascertained by extended float measurements.

* The experiments with this weir are given in our Chapter on Weirs, Nos. 17-34.
† Roorkee Hydraulic Experiments, Roorkee, 1881.

Δ	Site.	v	r	s	n
Masonry; apparently in pretty good order. (l = 87.2)	Solani (left).	3.5	6.4	.000 225	91
		4.1	7.9	.000 189	105
	Solani (right).	2.7	4.1	.000 205	95
		3.4	5.0	.000 240	101
		3.7	6.1	.000 220	100
		4.1	8.0	.000 190	104
	Solani (right), left branch closed.	1.24	4.2	.000 025	121
		2.5	2.6	.000 145	130
		4.8	3.7	.000 473	116
Sides, masonry; bottom, clay and boulders. Bed very irregular.	Solani (main).	.87	2.25	.000 148	48
		1.35	3.9	.000 088	73
		1.9	4.8	.000 200	61
		2.5	5.7	.000 161	82
		3.0	6.2	.000 171	94
		3.3	7.7	.000 214	81
		3.7	8.9	.000 227	82
		4.0	9.3	.000 227	87
Sides, masonry; bottom, earth. Beds very rough.	Belra.	3.1	8.0	.000 208	75
		3.2	9.0	.000 191	76
	Jaoli.	2.6	6.3	.000 140	89
		2.9	7.5	.000 160	85
	15 mile, old site.	4.0	8.6	.000 231	89
Earth. Beds very rough.	Kamehera.	2.7	4.1	.000 306	76
		2.8	4.5	.000 291	78
		2.9	4.8	.000 293	76

The above values of n appear to be too irregular to justify any general conclusion, except that n is higher for the Solani masonry aqueducts, than for other sections with higher values of Δ.

This work of Captain Cunningham contains much valuable information in regard to velocities in various parts of the section, effect of wind, et cet., et cet. .

The author has had a large experience in building small canals or ditches in California, for the supply of water for hydraulic gold mining. These ditches generally have inclinations of from 10 to 20 feet per mile, with r from 1 to 2 ; they frequently have sides of jagged rock, with pretty sharp curves. With $r = 1.4$, $r = 2.4$, and Δ high, n has a value of about 33 ; with $r = 2$, and Δ earth fairly smooth, n has a value of say 50 to 55. Flumes of unplaned plank, with sharp angular bends, with an inclination of

20 feet per mile, and $r = 1.5$, show a value of n of about 60; similar flumes with angular bends more carefully softened, with an inclination of 32 feet per mile, and $r = 1.2$, show a value of n of about 80; a flume, with an inclination of 32 feet per mile, with pretty sharp bends, $r = .90$, and $v = 4.3$, gave a value of n of about 59. With these flumes, however, it was impossible to measure r with any reasonable degree of accuracy, owing to the boiling action of the water as it passed by the angular curves.* With straight flumes, and $r = 1.3$, n had a value of 120 or thereabouts.

These experiences are not sufficiently exact to be of any especial service in this work. They simply fully confirm, if, indeed, any such confirmation be necessary, the general truth of the propositions established by MM. Darcy and Bazin, and Messrs. Fteley and Stearns, that in $v = n \, (r \, s)^{\frac{1}{2}}$, n increases rapidly with r, for values of r less than 2, and still more rapidly with decreasing values of Δ.

Rivers.

Some authors have in late years formulated expressions for the flow in large streams, giving much smaller exponents for s than $\frac{1}{2}$; Messrs. Humphreys and Abbot made this exponent $\frac{1}{3}$, and Hagen has even placed it at $\frac{1}{5}$. These expressions rest chiefly upon the experiments made by the Mississippi River Survey, and published in "Report upon the Physics and Hydraulics of the Mississippi River, by Captain A. A. Humphreys, and Lieut. H. L. Abbot, 1861."

The following table contains the 17 leading experiments with mean velocities made by Messrs. Humphreys and Abbot, with two previously made by Mr. C. Ellet; d is the maximum depth, and w the surface width.

---

* These ditches and flumes are built along the sides of precipitous mountain cañons, and it is not practicable to give the curves large radii. As plenty of fall is generally available they are given these exceedingly steep inclinations, so that a comparatively small section will carry a large quantity of water; the resulting high velocities are of much advantage in preventing the clogging of the canals by snow, during the very severe snow-storms which are phenomenal in their severity in certain portions of the Sierra Nevada range of mountains. Some of these ditches carry as much as 90 cubic feet per second; there is considerable abrasion on the bottoms and sides when the ditch is in earth, but not enough to be particularly objectionable.

TABLE LXVIII.

Humphreys and Abbot.[*]—*Mississippi and other Streams.*

$$v = n\,(r\,s)^{\frac{1}{2}}$$

No.	Stream.	w	d	a	p	r	v	s	n
460	Mississippi at Carrollton	2655	136	193 868	2693	72.03	5.929	.000 020 51	154.3
461	„ „ „	2656	136	195 349	2696	72.46	5.887	.000 017 13	167.1
462	„ „ „	2431	131	180 968	2461	73.53	4.034	.000 003 42	254.4
463	„ „ „	2429	132	183 663	2469	74.39	3.977	.000 003 84	235.3
464	„ „ Columbus	2214	88	148 042	2247	65.86	6.957	.000 068 00	103.9
465	„ „ Vicksburg	2729	100	178 137	2779	64.10	6.950	.000 063 79	108.7
466	„ „ „	2732	101	179 502	2783	64.52	6.824	.000 043 65	128.6
467	„ „ „	2507	63	78 828	2530	31.16	3.523	.000 022 27	133.8
468	„ „ „	2556	83	134 942	2589	52.12	5.558	.000 030 29	139.9
469	„ „ „	2580	90	150 354	2621	57.37	6.319	.000 048 11	120.3
470	Bayou Plaquemine	292	28	5 360	303	18.350	5.198	.000 206 44	84.5
471	„ „	268	24	4 259	278	15.320	3.959	.000 143 72	84.4
472	„ La Fourche	223	27	3 738	238	15.706	3.076	.000 044 68	116.1
473	„ „	223	24	3 025	232	13.039	2.843	.000 037 31	128.9
474	„ „	223	24	2 957	231	12.801	2.897	.000 036 55	129.8
475	„ „	223	23	2 868	230	12.470	2.789	.000 043 84	119.3
476	C. and O. Canal Feeder	23	7.6	121	32.7	3.700	3.032	.000 698 51	59.6
477	„ „ „	23	7.5	119	32.5	3.662	2.723	.000 698 51	53.8
478	Ohio at Pt. Pleasant	1073	8	7 218	1074	6.721	2.515	.000 093 34	100.4

It will be observed that the very high values of *n*, of 254.4 for No. 462, and 235.3 for No. 463, are with the respective inclinations of .000 003 42 and .000 003 84 ; such inclinations represent a surface fall of about .019 per mile. For any large river, especially a great stream like the Mississippi, any deductions based upon such infinitesimal slopes are absolutely worthless ; an instrumental error of .01 in levelling one mile is not at all improbable ; a comparatively small eddy at either end of the experimental section would produce abnormal elevations much more than these given slopes,[†] and doubtless extended measurements with the same experimental length would not unfrequently show the Mississippi to be running up hill ; these two experiments can hence be dismissed, as having not the slightest weight one way or the other.

We may perhaps with safety deduce from the foregoing table, that in $v = n\,(r\,s)^{\frac{1}{2}}$ with *r* = 4, *v* = 3, and Δ in earth fairly regular, *n* has a value of about 55 ; also that with

---

[*] Nos. 470 and 478 are given upon the authority of Mr. C. Ellet ; all the others were made under the direction of Messrs. Humphreys and Abbot.

[†] The length of the experimental section for Nos. 462 and 463, was a little less than 2 miles ; hence the total fall, or λ, was less than .04— about one half an inch.

r from 50 to 70, v from 3.5 to 7, with Δ very high, n will range from 100 to 150 ; also possibly, that with r from 3.5 to 7, the exponent of s should be somewhat less than $\frac{1}{2}$.

All determinations of the flow in large streams, depending upon the factors s, r and r, are necessarily very unreliable. The problem is almost infinitely complicated by variations in the cross-section, producing irregular currents, and often strong eddies not only on the sides but also near the bottom ; by the *vis-viva* of the stream as it enters the upper end of the experimental section ; by the condition of the stream below the section ; add to these the mechanical difficulties of obtaining correct values of h for small slopes even in calm weather, and the probable errors in the determination of v by floats or current-meters. We hence feel justified in assuming that the details of the problem presented in discussing the flow of a large stream are so very complicated as to be practically insoluble, and that experimental results thus far obtained must be regarded only as very rough approximations.

The views above expressed will doubtless seem heterodox to those hydraulicians who show with large streams results conforming, within two or three per cent., with some favorite formula, or to those mathematicians who think, by expressions more or less based upon very rough and insufficient experimental data, they can accurately state the laws governing the flow in a river, whose varying cross-sections are imperfectly known. In this connection the curious are referred to " Hydrologische Untersuchungen an den öffentlichen Flüssen im Königreiche Bayern, von J. Schmid," Munich, 1884, and to M. Boileau's last work " Notions Nouvelles d'Hydraulique, deuxième édition, 1881."

Note the judicious remarks of Captain Cunningham in regard to measurements of small surface inclinations, Roorkee Hydraulic Experiments, Vol. I., p. 326, et cet..

M. Boussinesq, in his *Essai sur la théorie des eaux courantes, Savants étrangers, Tomes XXIII. et XXIV.,* has discussed the flow of water in channels, through orifices, et cet., from an almost purely mathematical point of view. His reasoning is doubtless very ingenious and able, but his final conclusions do not always agree with the experimental data which we consider authentic.

Conclusions.

Effect of Changes in Δ.

The Darcy and Bazin series thoroughly illustrate the effect of changes in Δ, with r from .3 to 1. Series XXXVI., with a canal in earth and some vegetation (grass and weeds), with r from 1.1 to 1.7, shows values of n from 32.7 to 50.3 ; n increasing with r, and necessarily with v ; Series XXIV., in a semicircular conduit of pure cement plaster, with r from .37 to 1.03, shows values of n from 128.9 to 155.1, n increasing, as in the other case, with r and v.

The most striking illustration of the effect of a slight change in Δ for values of r less than 1, is given by Series XXIV. and XXV. ; the form in both cases was a semicircle with the same radius ($R = 2.05$), and with the same inclination of .0015 ; Δ for XXIV. was a plaster of pure cement, while for XXV. it was a plaster of mortar composed of $\frac{2}{3}$rds cement and $\frac{1}{3}$rd fine sand ; comparing the respective values of n for the two series, it will be seen that for equal values of r, n is from 8 to 10 per cent. higher with the smoother surface. Or, comparing Nos. 234 and 246, where a was very nearly the same, in the pure cement conduit Q was 41.07, while with the lining partly composed of sand Q was 38.35.

The best graphical comparison of these experiments, showing the effect of Δ, can be obtained by plotting the experimental curves for Series II., VII., III., IV. and V., with v and n for co-ordinates. Each series consists of 12 experiments, Q being the same for the respective experiment in each series, and its increment being nearly regular. That is to say ;

$$Q = 3.532 \text{ for Nos. } 8, 68, 20, 32 \text{ and } 44.$$
$$Q = 7.169 \text{ ,, ,, } 9, 69, 21, 33 \text{ ,, } 45.$$
$$Q = 10.842 \text{ ,, ,, } 10, 70, 22, 34 \text{ ,, } 46.$$
$$Q = 14.515 \text{ ,, ,, } 11, 71, 23, 35 \text{ ,, } 47.$$
$$Q = 18.188 \text{ ,, ,, } 12, 72, 24, 36 \text{ ,, } 48.$$
$$Q = 21.826 \text{ ,, ,, } 13, 73, 25, 37 \text{ ,, } 49.$$
$$Q = 25.463 \text{ ,, ,, } 14, 74, 26, 38 \text{ ,, } 50.$$
$$Q = 29.101 \text{ ,, ,, } 15, 75, 27, 39 \text{ ,, } 51.$$
$$Q = 32.738 \text{ ,, ,, } 16, 76, 28, 40 \text{ ,, } 52.$$
$$Q = 36.376 \text{ ,, ,, } 17, 77, 29, 41 \text{ ,, } 53.$$
$$Q = 40.014 \text{ ,, ,, } 18, 78, 30, 42 \text{ ,, } 54.$$
$$Q = 43.651 \text{ ,, ,, } 19, 79, 31, 43 \text{ ,, } 55.$$

The inclination was identical for the 5 series, being .0049. The form of section was rectangular, with widths varying from 5.94 to 6.53. Δ varied from a plaster of pure cement to pebbles from .10 to .13 in diameter embedded in cement ; r had a range from .17 to .99.

These conditions afford an excellent opportunity to determine roughly the effect of Δ upon n. The 5 experimental curves are plotted on Plate X., the value of r being written opposite each of the 60 experimental points.

We will use the following symbols for expressing the varying values of Δ ;

$Δ^0$ = plaster of pure cement.
$Δ^1$ = ,, ,, mortar, $\frac{2}{3}$ cement and $\frac{1}{3}$ fine sand.
$Δ^2$ = unplaned wood.
$Δ^3$ = brick, with joints fairly made.
$Δ^4$ = roughness as in Series XII.-XIV. (grating of wood).
$Δ^5$ = pebbles from .03 to .07 di., embedded in cement.
$Δ^6$ = ,, ,, .10 ,, .13 ,, ,, ,,

It will be seen by examining the 5 curves on Plate X. that there is a very marked lowering in the values of n, as the roughness of the surface increases ; for instance, with $r = .29$ for Series II., n is about 126.2, while with the same value of r for Series V., n is 47.5 ; we therefore see with same values of r and s, and a nearly identical water cross-section, that the discharge through the smooth conduit is 2.66 times greater, than the discharge through the rough conduit. It is certain that this great difference is almost altogether due to Δ, for, as will hereafter be shown, retaining r as a constant of .29, and increasing s for the rough surface, $Δ^6$, so that v will be the same (4.8) for $Δ^2$ as for $Δ^6$, the value of n for $Δ^6$ will be but little changed.

Comparing the experimental curves, it will be noticed that as Δ decreases, the curves become flatter; this warrants the belief that if the curves could be continued with increasing values of r, they would gradually approach each other, until with very high values of r they would be very nearly identical. It seems to us apparent that this must be true, for with r having a great value of say 100, the roughness indicated by $Δ^6$ would have hardly any appreciable effect upon the discharge, its retarding influences being practically confined to a comparatively narrow margin around the wetted border.

The effect of increasing r with different values of Δ, can be best appreciated by a comparison of the two extreme experimental curves. It will be seen, that for Series II.($Δ^2$), n is 135.1 for $r=.47$, and 138.2 for $r=.70$, showing an increase in n of 2.3 per cent. for this increase in r; for Series V. ($Δ^6$), the respective values of n for foregoing values of r, are about 56.5 and 64.5, showing an increment in n of 14 per cent. .

Nos. XXXIV. and XXXV. of the Darcy-Bazin series also forcibly illustrate the great effect upon the discharge caused by variations in Δ. For the first series the bottom and one side of the experimental canal were more or less covered with moss, grass and mud; the canal was then cleaned, and the experiments forming Series XXXV. made. Comparing Nos. 309 and 314, when Q and s were nearly the same;

XXXIV., No. 309 $Q=247$ $s=.0146$ $r=1.69$ $v=9.0$ $n=27.5$ $u=57$
XXXV., No. 314 $Q=244$ $s=.0142$ $r=1.49$ $v=11.3$ $n=21.6$ $u=77$

Filling the cleaned canal to the same height as obtained in No. 309, the discharge would have been one-half greater.

The experiments of Messrs. Fteley and Stearns also show the great importance of Δ, as will be seen by reference to Table LXVI.

Effects of Changes in v and r.

The effects upon n of changes in v and r are so closely related in conduits, that it is difficult to disintegrate them. The Darcy-Bazin and Fteley-Stearns experiments, however, afford sufficient data to enable us to point out roughly some general propositions, showing the general effect of changes in either v or r, with various values of Δ.

We will use for comparison three sets of the Darcy-Bazin series, namely VI.-VII.-VIII., XII.-XIII.-XIV., and XV.-XVI.-XVII. The form of cross-section was rectangular for the 9 series, with nearly even widths. There were three different inclinations for each set, and Δ was probably very nearly identical for each of the three series forming one set. For the first set Δ was unplaned plank; for the second a grillage of wooden laths; for the third a similar grillage, but with the interstices much wider, than for the second set.

These 9 series, forming the three sets for comparison, are plotted on Plate XI.,

c c

Fig. 1, with n and v for co-ordinates, the heavy lines representing the experimental curves, with the value of r written opposite each experimental point.   For each set we can now draw curves, representing within the limits of the experiments the values of n for constant values of r, with v a variable; these curves will therefore indicate the effect of changes in v, with r constant, for the three different conditions of wetted surface.

Examining the curve for $r=.60$, Set 1 with Δ^2, we see it is quite flat with v from 6 to 9, and gradually becomes sharper with velocities below 6.   It gives a value of n of about 106 with $r=3$, and about 117 with $r=9$.   The curves for this set indicate that the increase in n for each of the three experimental curves, as v and r increase, is due almost equally to v and r.

The curve for $r=.70$, Set 2, is nearly flat from $r=3$ to $r=6$, and then drops slightly to $r=7$.   This indicates for the roughness Δ^4 of these three experimental curves, that the increase in n, as v and r increase, is due almost entirely to v.

The curve for $r=.90$, Set 3, drops perceptibly as the velocity increases.   Hence with this roughness (slightly over Δ^6), the increase in n does not fully represent the influence of r, the effect of v being negative.

Some of the experiments of Fteley and Stearns given in Tables LXIV. and LXVII. are plotted on Plate XI., Fig. 2.   Those from Table LXVII., with considerably varying surface inclinations, enable us to draw similar curves showing the effect of r from 0.5 to 1.5 ; it will be observed that these curves are quite steeply inclined, showing the very marked influence of r upon n with small velocities.*   In the same figure are plotted Experiments Nos. 444 and 445 ; for these Δ had a considerably lower value than for the other experiments.

We can now with some safety generalize as follows in regard to our expression $v = n\,(r\,s)^{\mathbb{x}}$:

With very small velocities n increases rapidly with r, no matter whether Δ be large or small.

With increasing velocities, with low values of Δ, the effect of r constantly diminishes, until above some value its increase probably ceases to have any effect upon n.   Whether or not above this point n begins to diminish with v cannot be said.

With high values of Δ, n probably diminishes with an increase in r above 2.

The lower the value of Δ, and the higher the value of r, the greater will be the effect of v in increasing n.

An increase in r will always increase n, no matter what values Δ and v may have ; the higher the value of Δ, the greater will be the effect of r upon n.

Δ should be considered a function of r ; with Δ constant, as r increases the less will the effect of Δ be felt.

* It is to be remembered, however, that for the least velocities the values of v are the least trustworthy. The flow, also, was not absolutely uniform for these experiments (Table LXVII.).

It is proper to state that some of the Darcy-Bazin series present anomalous results, compared with those we have used for illustration.

Series LI. and LII., made with rectangular wooden pipes having the respective sections of 2.6 × 1.6 and 1.6 × 1.0, show a considerably less effect upon n by changes in r, than the curves on Plate XI. indicate. The heads for those pipes were measured by piezometric columns, which for such small inclinations and lengths, we regard as very unreliable ; by reference to the foot-note on page 181 it will be noticed that the readings from the different piezometers often gave discordant results. Comparing the two sets, consisting of Series XXVIII.-XXIX. and XXX.-XXXI., with a small conduit, first smooth, and then covered with cloth, it will be seen that the effect of r for the low value of Δ is quite large, while for the high value of Δ, n very rapidly decreases as r increases. In fact, by plotting Series XXX. and XXXI., with v and n as co-ordinates, it will be seen that with $r = .40$. and $r = .8$ for both series, n is in one case 43 and in the other 33.5. Such a condition would only be possible, by assuming that s has *directly* a considerable influence upon n. This assumption we do not consider to be a tenable one. The depths in the small trough used for Series XXVIII. to XXXI. were as low as .036 ; the surface of the water, when the cloth lining was used, appears to have been quite rough compared with such minute depths. The very exact measurement of n would hence be very difficult, and a slight probable error in its measurement would account for these apparent anomalies.

The data given in this chapter are hardly complete enough to fully warrant all the general principles which we have stated. They are, however, quite fully confirmed by the experiments with pipes, given in the next chapter.

Effect of Form.

Comparing Series XVIII. to XXIII. inclusive with Series VI. to VIII inclusive, all of Darcy and Bazin, it will be seen that the values of n correspond quite closely, taking into account the effects of changes in r and c. In all these series Δ was of unplaned plank, probably having nearly equal values. These experiments prove that neither d nor w are factors of considerable importance, with a comparatively low value of Δ. For instance ; for No. 185, form rectangular, $d = .87$, $w = 1.57$, with $r = .41$ and $v = 5.2$, n has a value of 105.7 ; while for No. 71, form also rectangular, $d = .46$, $w = 6.53$, with $r = .40$ and $v = 4.9$, n has a value of 109.4 ; also, for No. 216, form triangular, $d = 1.82$, $w = 3.64$, with $r = .64$ and $r = 6.6$, n has a value of 117.3 ; while for No. 76, form rectangular, $d = .78$, $w = 6.53$, with $r = .63$ and $r = 6.5$, n has a value of 116.4. These experiments show that triangular and trapezoidal forms have nearly the same values of n, as rectangular forms.

Contrasting circular with rectangular forms, Δ being the same (pure cement), it will be seen by reference to Series II. and XXIV., that with the circular form n is about 10 per cent. larger than for the rectangular form, making proper allowances for effects of r and v. Comparing Series VI. and XXVI., Δ being in both cases wood, and allowing for effects of r and v, it will be seen that n for the circular form is about 7 per cent. higher than for the rectangular one.

The high values of n for the Sudbury conduit, Fig. II., Plate XI., indicate that its form was nearly as favorable for discharge, as the circular.

Formulæ.

The following are the formulæ which have been most generally used, for the flow through conduits or canals, and rivers.

(1) Chezy ; $v = u\,(r\,s)^{\frac{1}{2}}$.

(2) Dubuat ; $v = \dfrac{88.51\,(r^{\frac{1}{2}} - .03)}{\left(\frac{1}{s}\right)^{\frac{1}{2}} - \text{Hyper. log.}\left(\frac{1}{s} + 1.6\right)^{\frac{1}{2}}} - .0894\,(r^{\frac{1}{2}} - .03).$

(3) Prony ;* $r\,s = a\,v + b\,v^2$, and $v = (10\,618\,r\,s + .055)^{\frac{1}{2}} - .234.$

(4) Humphreys and Abbot, for large streams ; $v = \left(a^{\frac{1}{4}}\left(\dfrac{225\,s^{\frac{1}{2}}}{r + \frac{1}{w}}\right)^{\frac{1}{2}} - .0388\right)^{2}.$

(5) Bazin ; $r\,s = A\,v^2$, A having following values depending upon Δ and r, viz ;

　　　　Very smooth walls (Δ^9), $A' = .000\,046\left(1 + \dfrac{10}{r}\right)$.

　　　　Fairly　..　　..　(Δ^9), $A'' = .000\,058\left(1 + \dfrac{23}{r}\right)$.

　　　　Rough masonry (about)　$A''' = .000\,073\left(1 + \dfrac{N.2}{r}\right)$.

　　　　Earth　　　　　　$A'''' = .000\,085\left(1 + \dfrac{4.1}{r}\right)$.

(6) Kutter ; $v = u'\,(r\,s)^{\frac{1}{2}}$, n' having following values depending upon Δ, s, and r, viz. ;

Call u co-efficient of rugosity, with following values ;

　　　　.009 well-planed timber.
　　　　.010 plaster in pure cement.
　　　　.011　　,,　of cement with one-third sand.
　　　　.012 unplaned timber.
　　　　.013 ashlar and brick-work.
　　　　.015 canvas lining.
　　　　.017 rubble masonry.
　　　　.020 canals in very firm gravel.
　　　　.025 rivers and canals, free from stones and weeds.
　　　　.030　,,　　　,,　with occasional stones and weeds.
　　　　.035　,,　　　,,　in bad order.

Then,

$$n' = \dfrac{41.66 + \dfrac{1.811}{u} + \dfrac{.002\,81}{s}}{1 + \dfrac{u}{r^{\frac{1}{2}}}\left(41.66 + \dfrac{.002\,81}{s}\right)}.$$

(7) Fteley and Stearns ; $v = 1.27\,r^{.69}\,s^{\frac{1}{2}}$.

Eytelwein and others have given u constant values, ranging from 92 to 100. The experiments heretofore given, which can be considered as entirely reliable, show that u may have a value anywhere from 30 to 155, with r not over 1.03, and with every

---

* In metrical measures $a = .000\,044$ and $b = .000\,309$.

probability that with very large values of r and low values of Δ, n will have very much larger values than 155.

The Dubuat formula is based upon the proposition that the velocity will be *nil* when $r = .0009$ and less. Dubuat thought that with $s = \frac{1}{1\,000\,000}$, v would also be *nil*; he assumed for canals that s must be about $\frac{1}{500\,000}$ in order to produce sensible motion. He also very erroneously assumed that Δ was unimportant.

The Prony formula is based upon the proposition of Coulomb, that the resistance opposed to a disc revolving in water is expressed by $a\,r + b\,v^2$. Prony assumes that changes in either r or Δ do not affect his co-efficients a and b.

The Humphreys and Abbot formula assumes that equal resistance is produced by the air as by the solid perimeter, and that the one-fourth power of s should be adopted, instead of the half power. The Darcy-Bazin experiments conclusively demonstrate that this first supposition is erroneous. With very large values of Δ, such as that with the ordinary section of a great river, it is probable that a somewhat lower power than $\frac{1}{2}$ would properly apply to s; we do not think it possible, however, that it can be as low as $\frac{1}{4}$, or .25; probably about .40 is the lowest value which can properly be assigned to this exponent.

The Bazin formulæ make v vary with r, and not with s; in them Δ is the controlling factor.

The Kutter formula endeavors to reconcile the Darcy-Bazin and some Swiss experiments with Nos. 462 and 463 of the Mississippi River Survey. These two experiments are the ones showing great values of n for very low values of s, and are considered by us to be absolutely unreliable. The salient features of this formula are; first, n' increases always with r; second, great importance is attached to Δ; third, with $r = 3.281$ (1 metre), with any inclination and any constant value of Δ, n' will always be constant, being $\frac{1}{n}$. As r increases above 1 metre, increasing values of s diminish n', until with very high values of r and very low values of s, n' becomes very large. As r diminishes below 1 metre, n' increases with s.

We differ entirely with Herr Kutter in his assumptions concerning the relative effects of r and s.

The Fteley and Stearns formula attributes the increased value of n with increasing values of r and v in the Sudbury conduit, entirely to r. This formula was proposed chiefly for this particular conduit, for which it gives satisfactory practical results. In our judgment, for this conduit the increase in n in Table LXIV. was due about equally to r and v.

We propose hereafter to attempt to show the values of n for smooth pipes; these values of n will also apply to open conduits, of circular, or nearly circular sections, with the same low value of Δ.

For conduits or canals with high values of Δ, we regard it almost useless to propose any formula, other than $v = n\ (r\ s)^{\frac{1}{2}}$; it is impossible to assign with any reasonable degree of accuracy, any numerical values to Δ for canals in earth or rock, or for rivers, and as Δ is such a controlling quantity in any equation of which it forms a part, such uncertainty must always make the equation simply approximative.

In practice, the constructing engineer, who wishes to know what value to give to n for a canal or conduit with rough surfaces, should look carefully over the Darcy-Bazin series, and keeping in mind our general propositions, he will then be able to make a pretty fair guess as to its value.

The reader is referred to the last portion of the succeeding chapter, for a more detailed analysis of the laws governing the flow through uniform channels.

CHAPTER VIII.

FLOW THROUGH PIPES.

THE total head, H, causing the flow of water through long cylindrical pipes is absorbed as follows :

FIRST.—By contraction as the water enters the mouth of the pipe.

SECOND.—In imparting velocity to the water. When there is no initial velocity of approach in the reservoir feeding the pipe, this is approximately represented by the *vis viva* of the escaping jet from the lower end of the pipe.

THIRD.—In overcoming various forms of resistance as the water passes through the pipe. This embraces friction and adhesion on the inner surface of the pipe, and also the retarding effects of minute cross-currents and eddies due to irregularities of the wetted surface, and varying velocities in the section.

The first and second absorptions of head may be termed *losses of head*, and are theoretically represented by h', in $v = o\,(2\,g\,h')^{\frac{1}{2}}$, or $h' = \dfrac{v^2}{2\,g\,o^2}$; o being the co-efficient of contraction at the entrance, and being practically identical with the co-efficient of discharge, c, for the pipe cut off at a short distance from the entrance. The effective head, h, which is absorbed in overcoming the total resistance after the regimen of flow has been established, is hence $H - h' = h$.

The flow through pipes is roughly represented by the well-known Chezy formula of $v = n\,(r\,s)^{\frac{1}{2}}$: v being the mean velocity of the water passing through the pipe ; n a co-efficient assumed as constant by some authorities, and as variable by many others ; r the "hydraulic mean radius," being the area of the cross-section divided by the length of the wetted perimeter of the section, or $\dfrac{D^2 \frac{\pi}{4}}{D \pi} = \dfrac{D}{4}$; s being the sin of the inclination, or $\dfrac{h}{l}$; l being the total length of the pipe, no matter whether it be horizontal, inclined, straight or curved.

With $v = n\,(r\,s)^{\frac{1}{2}}$

we have, $v = n\left(\dfrac{\frac{a}{p}\,h}{l}\right)^{\frac{1}{2}}$, or, $h = \dfrac{1}{n^2}\left(v^2\,p\,l\,\dfrac{1}{a}\right)$.

Therefore the head h, which is required in order to obtain a certain velocity, v, depends directly, upon the co-efficient $\frac{1}{n^2}$, upon the square of the velocity, upon the length of the wetted perimeter of the cross-section, upon the length of the pipe, and inversely upon the area of the cross-section.

As the effect of the head, h, depends upon the relation of gravity, which is slightly variable, the expression $v = \left(2 g h \frac{a}{p}\right)^{\frac{1}{2}}$, or $h = t \left(v^2 p l \frac{1}{a} \frac{1}{2g}\right)$, will be more accurate than $h = \frac{1}{n^2}\left(v^2 p l \frac{1}{a}\right)$. Comparing these two equations, we have $t = \frac{2 g}{n^2}$, or $n = (2 g)^{\frac{1}{2}}\left(\frac{1}{t}\right)^{\frac{1}{2}}$; hence, with $f = \left(\frac{1}{t}\right)^{\frac{1}{2}}$, $v = f (2 g)^{\frac{1}{2}} (r s)^{\frac{1}{2}}$ is the Chezy formula taking into account the variations in the value of $2 g$.*

The probable errors in the experimental data about to be discussed are so very much larger than any variations in $(2 g)^{\frac{1}{2}}$, that we will adhere in all the following reductions to the simpler form $v = n (r s)^{\frac{1}{2}}$. Another valid reason for preferring the co-efficient n rather than t is, that n can be shown with sufficient accuracy for all practical purposes by a whole number of two or three figures, while t is fractional; for instance, n is often assumed as a constant with a value of about $96 - v = 48 \left(\frac{D h}{l}\right)^{\frac{1}{2}} = 96 (r s)^{\frac{1}{2}}$ — while t is $\frac{2 g}{n^2} = .006\ 98$. It is apparent that 96 can much more readily be kept in remembrance, than a decimal of 4 or 5 figures.

The co-efficient n in the equation we have adopted for the reduction of our experimental data is a variable of wide range, and consequently the equation may be regarded as very imperfectly representing even the general laws governing the flow through pipes. Dubuat, Prony, and others have proposed formulæ with the endeavor by the use of two or more constant co-efficients, to obtain equations which will more accurately agree with experimental results. When, however, these more complicated expressions are applied to our experimental data, as will be shown at the close of this chapter, their results are often very far from the truth, so that but little, if indeed any advantage would result from employing them. In fact, it is very doubtful whether one general expression with constant co-efficients can be framed, which will with a fair degree of accuracy give the value of $v - r$, s and the condition of the wetted surface, or Δ, being known. Even were this possible, the expression would be so complicated as to be too cumbersome for practical use.

Admitting the approximate truth of the propositions upon which the Chezy

---

* t is generally termed the co-efficient of friction.

formula is based, it seems to us preferable to use it, rather than any more complicated formula, and by graphical comparisons of the values of n, deduced from experiments with widely varying values of r and v, and with very different conditions of the wetted or interior surface, endeavor to trace the causes producing the variations in n.

As will be seen hereafter, with pipes of very small diameters, or with exceedingly low velocities for larger diameters, the general laws indicated by the Chezy formula do not apply. In such tubes the temperature of the water is a very important element, the discharge under identical circumstances except as to temperature, being greatly increased by an increase in the temperature of the liquid from 45° to 70° F., a variation often met with in practice with conduit pipes.

Hagen and Grashof in their formulæ for pipes make T a factor of consequence, reasoning upon the general supposition that if the value of T is so important in considering the flow through small diameters, it must have some effect with larger diameters. This may perhaps be true, but existing experimental data with large diameters are not sufficiently precise or complete to indicate whether or not changes in T appreciably affect the discharge. With such variations in temperature as are ordinarily met with in practice with conduit pipes, it is probable that the value of T can be neglected without danger of serious error.

It seems probable that temperature has a very important effect upon the viscosity or adhesion of the water to the wetted surface of the tube; hence with very low velocities, when this adhesion is a very notable resistance to the flow, changes in T will very likely sensibly affect the discharge.

In experimentation with pipes it is necessary to determine:

First.—The total head, H, which is for a submerged discharge, the difference in elevation between the surface of the water in the feeding reservoir, and the surface in the outlet reservoir; and for the discharge free into the air, the difference in elevation between the surface in the feeding reservoir, and the centre of the escaping jet in the plane of the lower end of the pipe.* These differences in elevation can in general be readily and accurately obtained by two distinct determinations by an engineer's spirit level.

Second.—The total length, l, which should be measured following all the sinuosities of the pipe when it is curved. This dimension is more easily obtained than any of the others.

Third.—The mean diameter, D, from which in general the mean area, a, is deduced. Erroneous assumptions of the value of D are very frequent; even the direct measurement of the end diameters of each joint forming a line of pipe is not

---

* This last supposition is not absolutely correct, as has been shown in discussing the discharge through circular orifices. However, with the experiments given hereafter, errors from this incorrect assumption are so small that they can be neglected.

sufficiently accurate, as the interior forms of pipes made either of metal or glass are rarely cylindrical.

Darcy obtained the mean area of some of his experimental pipes, and hence D, by the measurement of the quantity of water contained in his pipe lines; this method may be slightly in error if intervals separate the several joints, as is often the case, for these spaces should not be considered as forming part of the mean area. In the New Almaden experiments we determined the mean area by filling each joint forming a line of pipe separately with water, and then weighing on accurate scales the total amount of water contained in the several joints; this method for small pipes is probably preferable to any other.

In examining the recorded experiments upon long pipes, we often find D given as 2, 3, 4, or some even number of feet or inches. These without doubt are simply the rough figures commercially given by the makers of the pipe, which may be in error 4 per cent. or more; such probable errors will affect the deduced values of n as much as 10 per cent.

FOURTH.—The amount of discharge in a known period of time, or q. This with small diameters, and consequent low values of Q, can be obtained with comparative ease by direct measurement in a vessel of known size. For large values of Q direct measurement is generally impracticable owing, either to lack of room, or to the cost of constructing vessels of proper size; it hence becomes necessary to measure the water over weirs or through orifices, either of which methods is more or less unsatisfactory. Such indirect measurements, when made by persons not conversant with the minor laws governing the discharge through openings, may very often be 5 per cent. or even more in error.

FIFTH.—The segregation of h' and h from H. This can be done by the theoretical assumption that $h' = \frac{v^2}{2\,g\,c^2}$, c being the co-efficient of contraction at the entrance, and having a minimum value of about .71 when the pipe extends into the feeding reservoir with its inner section unchanged, and a maximum value of about .99 when a bell or trumpet-shaped mouth-piece is attached to the entrance. Hence in practice by the use of an entrance of proper form c can be nearly eliminated from the foregoing equation, leaving only the loss of head in imparting velocity to the water to be taken into account.

Another method frequently used is by attaching piezometric tubes to the pipe, one being placed near the entrance and the other near the lower end; the difference in elevation between the surface of the water in these two tubes will represent with more or less accuracy the head, h, absorbed in overcoming the resistances in the pipe between the tubes.

Darcy adopted this latter method, using a column of mercury to determine high pressures in his piezometers, which he termed manometers. The results obtained by Darcy were so unsatisfactory, considering the time devoted to the execution of his experiments and the cost of his apparatus, that in all our experiments we preferred to use the theoretic computation of h.

The experimental proofs in regard to the accuracy of the expression $h' = \frac{v^2}{2\,g\,o^2}$, will be discussed in a following section of this chapter.

It has been our effort to critically examine every experiment with pipes, thus far given to the world by French, German, English, or American hydraulicians. Very many of them are palpably worthless, and their insertion here would only confuse the mind of the student in puzzling over the causes of their errors. We will, however, be compelled to use several experiments of very doubtful accuracy, in order to fill gaps where we have no more authentic data.

Quite a number of experiments, which we deem worthless, will also be given, some of which appear to contradict our final conclusions. It is too often the case that authors neglect to give data opposing their particular views, and we do not wish to lay ourselves open to the serious charge of suppression.

Short Pipes.

Experiments with short pipes, when the length was only a few diameters, have been made by Bossut, Dubuat, Michelotti and other hydraulicians. Venturi, Eytelwein and Francis have made interesting experiments with such pipes, having divergent adjutages (divergent conical discharge ends).

The only value the determinations of the flow through short pipes will have for us, will be in fixing the values of the co-efficients of contraction at the entrance of long pipes, in order to obtain h' in $h' = \frac{v^2}{2\,g\,o^2}$, the co-efficients of contraction and discharge being assumed as identical. We will hence devote but little space to the discussion of short pipes, the investigation being only of theoretical interest, as the discharge of water, Q, can be obtained with much more ease and far greater accuracy through orifices or over weirs, than through short pipes.

Bossut* with vertical cylindrical pipes of various diameters and lengths, the plane of the upper end of the pipes being flush with the bottom of the feeding reservoirs, obtained the following results.

---

* Traité théorique et expérimental d'Hydrodynamique, par M. l'Abbé Bossut. Tome Deuxième, Paris, 1786.

TABLE LXIX.

Bossut.—Flow through Vertical Cylindrical Pipes.

$(2g)^{1/2} = 8.02$ $Q = c \cdot a \cdot (2gh)^{1/2}$ $c_1 = \dfrac{v}{(2gh_1)^{1/2}}$

Bossut's No.	Head from Surface to lower End of Pipe h	Length of Pipe l	D	a	Time of Experiment t	Q	v	c	Head from Surface to upper End of Pipe h_1	c_1	Remarks
1	12.582	.355	.0888	.006 195	66	.143 3	23.13	.804	12.507	.816	Pipes of well-polished copper, of true cylindrical form.
2	12.584	.177	.0888	.006 195	65	.142 3	22.97	.804	12.507	.810	Water fills lower end of pipe.
3	12.640	.133	.0888	.006 195	65	.143 1	22.94	.804	12.507	.809	Water fills lower end of pipe.
4	12.640	.133	.0888	.006 195	91	.108 4	17.50	.614	12.507	.617	Escaping jet only touches upper edge of pipe.
5	4.085	.177	.0444	.001 549	60	.019 72	12.73	.785	3.908	.803	Water fills lower end of pipe.
6	4.085	.177	.0444	.001 549	80	.015 10	9.75	.601	3.908	.615	Escaping jet only touches upper edge of pipe.
7	4.085	.177	.0740	.004 302	24	.054 90	12.76	.787	3.908	.806	Water fills lower end of pipe.
8	4.085	.177	.0740	.004 302	30	.042 00	9.76	.602	3.908	.616	Escaping jet only touches upper edge of pipe.
9	2.131	.177	.0444	.001 549	85	.014 26	9.21	.786	1.954	.821	Water fills lower end of pipe.
10	2.131	.177	.0444	.001 549	110	.010 91	7.04	.602	1.954	.628	Escaping jet only touches upper edge of pipe.
11	2.131	.177	.0740	.004 302	30	.039 72	9.23	.789	1.954	.824	Water fills lower end of pipe.
12	2.131	.177	.0740	.004 302	40	.030 38	7.06	.603	1.954	.630	Escaping jet only touches upper edge of pipe.

Bossut assumes that the effective head, h, should be measured from the lower end of the pipe when the escaping jet clings to the sides of the pipe ; when the escaping jet only touches the upper edge or corner of the pipe, he considers that h should be measured from the point where the proper section of the contracted vein is smallest. As these experiments have been frequently commented upon as showing unusually small values of c, we have given them in detail. It is possible that the discharge would have been greater had the adjutages been horizontal instead of vertical, h being the same ; we have therefore calculated c, from h_0, the head h_0 being taken to the bottom of the feeding reservoir at the upper plane of the pipes, for comparison with c.

From these experiments Bossut deduces a table showing the discharge through a cylindrical pipe, with $D = .089$ and $l = .178$, with h from 1.07 to 16.0 ; calculating c from this table, it has a value of .8055 for the least head, gradually decreasing to .8005 for the largest head.[*]

D'Aubuisson[†] gives the following values of c, as deduced by the authorities named.

Authority.	D	l	h	c
Castel	.051	.131	.66	.827
"	"	"	1.58	.829
"	"	"	3.3	.829
"	"	"	6.6	.829
"	"	"	9.9	.830
Eytelwein	.085	.256	2.4	.821
Venturi	.131	.420	2.9	.822
Michelotti	.266	.709	7.1	.815
"	square .266	.709	12.5	.803
"	"	"	22.0	.803

Weisbach[‡] found a mean value of .815, c increasing as the size of the tube diminished. Buff found c from .825 to .855 as the head decreased from 2.8 to 0.1, with a tube .02 in diameter.

When the pipe projects into the feeding reservoir, c will have a much smaller value. Bidone,[§] with $D = .13$ and $l = .22$ found c to be .767. Weisbach found in a similar case a value of .71.

---

[*] Bossut places roughly $\frac{g}{2}$ at 15 French feet, instead of 15.1 which is nearly its correct value ; in our reductions we have made $\frac{g}{2} = 16.08$ English feet.

[†] Traité d'Hydraulique, 2nd ed., page 49.

[‡] Lehrbuch der Ingenieur, 4th revised ed.

[§] D'Aubuisson. Traité d'Hydraulique, 2nd ed., p. 38.

When a bell-shaped mouth-piece is used, as has been shown in Chapter III., c will have a value from .94 to very nearly 1..

We can hence assume that o, in $h' = \dfrac{v^2}{2g} \cdot \dfrac{1}{o^2}$, will have about the following values;

Cylindrical pipe, with mouth flush with side of reservoir	.825
,, ,, ,, ,, projecting into reservoir, and with sharp edges	.715
Bell-shaped mouth-piece, for small velocities...	.950
,, ,, ,, very high velocities	.995

When the end of the pipe is rough o will have a slightly smaller value. The above values will doubtless be somewhat modified by considerable variation in n.

Venturi attached conical divergent adjutages, and found a velocity at the smallest section of the pipe much larger than that apparently due to gravity. Eytelwein found the same results. The most interesting experiments with divergent adjutages are those made by Francis in 1854,[*] the pipes in all cases being submerged, and the head, h, hence being the difference in elevation of the surface in the two reservoirs.

The pipes used by Mr. Francis were of cast iron, first turned, and then ground with emery until the interior surfaces were quite smooth, but not having a bright polish. During the experiments the pipes were practically kept free from rust.

The several pipes could be screwed together so that one or more could be experimented upon. Their dimensions were as follows ;

Mouth-piece A {	Cycloidal curve, with inner diameter 1.4, and minor							
	diameter .1018					...	1.00 long	
{	Cylindrical section, diameter		.1018			...	.10 ,,	
Conical tube B, truncated cone,[†] larger diameter .1454						...	1.00 ,,	
,, ,, C,	,,	,,	,,	,,	.2339	...	1.00 ,,	
,, ,, D,	,,	,,	,,	,,	.3200	...	1.00 ,,	
,, ,, E,	,,	,,	,,	,,	.4085	...	1.00 ,,	
				Total length A B C D E		=	5.10	

The angle of divergence of each side, or β, was hence about 2° 30', or an angle of 5° for two opposite sides.

The amount of discharge was computed, by the flow over a sharp-crested weir, .658 long, with full contraction, the co-efficients of discharge being deduced from experiments of Poncelet and Lesbros—our Weir Nos. 35-40. The head, h, on this weir varied from .03 to .20 ; hence the discharge, or Q, cannot be considered as having been determined with great exactness, as the value of c for such small depths is more or less uncertain. Especially for the smaller heads on the measuring weir, we think that Mr. Francis has

---

[*] Lowell Hydraulic Experiments.

[†] B was slightly curved at its junction with the cylindrical part of A in order to avoid an abrupt angle.

placed Q at too low a value. In the following table we have only selected experiments where the depth on the weir was over .07, so that any errors in Q will not very largely affect the general accuracy of the deduced co-efficients.

TABLE LXX.

Francis.—Flow through Submerged Tubes.

$c' = \dfrac{Q}{a'(2gh)^{1/2}}$; $c'' = \dfrac{Q}{a''(2gh)^{1/2}}$; a' = area of smallest section = .008 061; a'' = area of discharge end of tube.

Figure.	h	c'	c''	Figure.	h	c'	c''	Figure.	h	c'	c''
A	.53	.927	.937	A B C	.14	1.98	.376	A B C D	1.18	2.43	.245
"	.78	.935	.935	"	.21	2.03	.384	"	1.36	2.43	.244
"	.96	.928	.928	"	.31	2.07	.392	"	1.39	2.26	.228
"	1.23	.933	.933	"	.50	2.12	.401				
"	1.40	.937	.937	"	.71	2.15	.408	A B C D E	.11	2.05	.128
"	1.52	.944	.944	"	1.10	2.16	.410	"	.15	2.11	.131
				"	1.31	2.12	.402	"	.21	2.17	.135
A B	.20	1.48	.726					"	.31	2.26	.140
"	.30	1.51	.742	A B C D	.13	2.08	.209	"	.42	2.31	.143
"	.40	1.54	.754	"	.27	2.23	.227	"	.64	2.30	.143
"	.85	1.59	.780	"	.44	2.26	.227	"	.96	2.31	.144
"	1.45	1.60	.782	"	.67	2.30	.232	"	1.29	2.39	.149
"	1.47	1.57	.772	"	.92	2.38	.239	"	1.36	2.33	.144
								"	1.11	2.25	.140

It will be observed that with the divergent tubes A B, et cet., the co-efficients c' and c'' increase with the head, h, until $h = 1.3$ or thereabouts, and then decrease. This is especially noticeable with the Figure A B C D E, where, with $h = 1.29$, $c' = 2.39$, and with $h = 1.41$, $c' = 2.25$, showing a diminution in the value of c' of 6 per cent. with a slight increase in head. This indicates that with very large heads these large values of c' would not be maintained, that is to say with the same form of adjutage.

The experiments made by us, and described in Chapter IX., with divergent adjutages showed that there was no sensible increase in discharge caused by the adjutages. In these experiments, however, the velocity of the escaping jets was so great—about 144 feet per second—that the jets doubtless did not come in contact with the divergent sides, so these experiments cannot be regarded as conclusive.

Daniel Bernoulli investigated this subject and concludes by the law of living forces that with such pipes, the discharge end of the pipe being full, the discharge should be in proportion to the area of the end at the point of discharge. Venturi attributes the increased discharge, caused by the addition of such adjutages, to atmospheric pressure, combined with what he terms *the lateral communication of motion in fluids*. D'Aubuisson inclines to the view that it is caused by molecular attraction.

When the discharge is free into the air, at the moment the jet escapes from the narrow section at **a** in the following sketch, the jet has its normal form as shown by

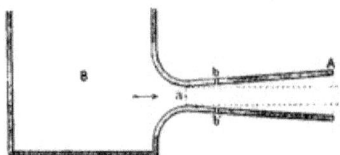

the dotted lines ; the jet at once begins to abstract the surrounding air, and, if the angle β be not too great, will soon drive out all the air ; as the air is exhausted, a partial vacuum is formed, and atmospheric pressure gradually forces the jet to fill the tube from **a** to **A**. This is shown to be true by boring holes through the tube in the plane **b b'** ; when this is done, air rushes in through these holes, and the jet nearly maintains its normal position. If atmospheric pressure be removed, the jet will also retain its normal position. Now, if the pressure in the feeding tank **B** be very great, the pressure of the atmosphere will not be sufficient to compel the jet to fill the tube, the jet will retain its normal shape, and a vacuum will be formed and maintained between the jet and the sides of the tube ; this was doubtless the case with our experiments with great heads.

In the author's engineering experience he has frequently seen the effect of this last phenomenon. The most notable case was with a dam built in California in 1857, for the purpose of impounding water for the supply of canals belonging to hydraulic gravel mines. This dam was a structure formed by filling a wooden crib with loose stone ; it was made water-tight by a lining of boards on its upper slope ; the water was drawn from the reservoir by a long wooden box passing through the dam, the discharge being regulated by gates placed at the upper end of the box, which were moved by wooden rods coming to the surface in the reservoir ; these gates were about 70 feet below high-water mark. The water escaping from the gates under great pressure, soon abstracted the air from the upper half of the box or conduit, a vacuum was thus formed, and atmospheric pressure on the outside finally crushed, in several places, the top planking of the box, allowing the entrance of the air. After these openings were made, the crushing or collapsing action of course ceased.

If the discharge be submerged, and the pressure H, or $H_s - H_i$, still be very large, the jet will retain its normal shape, and a vacuum will be formed between the jet and the sides of the tubes, as was the case with free discharge.

The phenomenon of increased discharge, caused by a divergent adjutage, is precisely analogous to the movement of water through a long pipe, having a stricture, or narrowing of section similar to the point **a** in the sketch, situate some distance from the outlet end ; the resistances encountered by the water from the point of stricture to the outlet end force the water to fill the pipe from **a** to **A**.

If the water moved in a solid mass, with each fillet in the same section having a uniform velocity with parallel movement, Bernoulli's proposition would be applicable ; but this condition is very far from being the case. Hence any such stricture in a pipe must result in a loss of force ; the greater the velocity the larger will be the proportionate loss of head.

PIPES OF VERY SMALL DIAMETERS.

The flow of liquids through pipes of very small diameters, forming capillary tubes, has been investigated by Poiseuille, and by Hagen and Jacobson. These researches have shown that the laws governing the flow through such tubes are not the same as those in general governing the flow through pipes of larger diameter.

Dr. Poiseuille's investigations appear to have been conducted with great patience and care, and we will here give an abstract of his paper, published in Vol. IX., Savants étrangers, of the Academy of Sciences, Paris, 1846. In this abstract all linear measures will be given in millimetres, all measures of capacities in cubic millimetres, and temperatures by the Centigrade scale.

A closed glass vessel, or "ampoule," of known size, was filled with the liquid to be experimented upon ; on the upper end of the ampoule a constant pressure was maintained by a pipe leading to a large air chamber or receiver ; the pressure was directly measured by means of a vertical column either of distilled water or mercury ; the pressure was varied by pumping air into the receiver until the desired amount of compression was obtained. To the lower end of the ampoule the glass tube to be experimented upon was attached in a horizontal position. The temperature of the liquid was kept almost absolutely constant during each experiment. The volume of discharge was determined by noting the time required to empty the ampoule, the pressure remaining constant ; index points, one on the upper and the other on the lower side of the ampoule, watched by reading microscopes, determined the beginning and the ending of the times.

Let P = pressure on the mouth of the tube, measured by millimetres of mercury, having a temperature of $10°$ C.

„ P' = pressure on the mouth of the tube, measured by millimetres of distilled water.

„ D = mean diameter of the tube in millimetres.

„ l = length of the tube in millimetres.

„ t = time of emptying the ampoule in seconds.

„ T = temperature by Centigrade scale.

„ v = mean velocity per second through tube.

„ q = absolute volume contained in measuring ampoule.

„ Q = volume of distilled water in cubic millimetres, discharged in one second.

E E

Let Q = weight in milligrammes of distilled water discharged in one second, no matter of what temperature (1 cubic millimetre of distilled water at 4° weighing 1 milligramme).

,, K, K', K'' and K''' = constant co-efficients.

Experiments were made with T from 0.5° to 45.1°; with 10 nearly cylindrical glass tubes, D having a value from .652 to .0139. The original length of the tubes was 962 for the largest diameter, and generally 100 for the smaller diameters; by successively cutting off portions of these tubes the length was reduced, being in one case only 1.. The liquid used for determining the laws governing the discharge was distilled water.

FIRST. *Effect of the Pressure.*—When l, D, q and T were constant it was found that t was in the inverse ratio of P, except when l was not largely in excess of D. This law for tubes, with $D = .142$ and less, was shown by experiment to be almost mathematically exact. For instance, in the series with $l = 75.05$, $D = .113$, $T = 10°$, and $q = 6448.2$:

No. Experiment.	Pressure of Mercury.	Observed Time. t	Calculated Time. t'
1	55.286 mms.	21 430″	21 399″
2	97.923 ,,	12 079″	12 083″
3	148.275 ,,	7 981.5″	7 978.9″
4	193.247 ,,	6 100″	modulus
5	387.695 ,,	3 052″	3 051.6″
6	739.467 ,,	1 600″	1 599.9″
7	774.891 ,,	1 526.5″	1 526.8″

With the largest tube, $D = .66$, the agreement between t and t' was not so close.

The experiments showed that this law as to pressure had the following limits, depending upon some ratio between l and D;

D	The Law applies when ;	The Law does not apply when ;
.66	l was from 962 to 384.	l was from 200 to 10.8.
.142	l ,, ,, 100 ,, 15.7.	l ,, ,, 9.5 ,, 1..
.132	l was 364.	
.113	l was from 100 to 23.6.	l ,, ,, 9. ,, 3.9.
.09	l was 258.	
.085	l was from 100 to 10.1.	l was 6.0.
.045	l ,, ,, 100 ,, 10..	l ,, 3.35.
.029	l ,, ,, 23.1 ,, 2.1.	
.0139	l was 66.	

In all cases when l was less than the required multiple of D, with increasing pressures t diminished less rapidly than $\frac{1}{P}$, and in general as l was shortened more and more, this divergence from the law became greater. For instance, with $D = .142$, q being also constant ;

l	P	t
9.55 {	23.64	5570°
	774.6	207
6.775 {	24.75	3829
	773.8	165.75
1. {	4.783	3937
	148.47	267
	773.7	95

SECOND. *Effect of Length.*—When P, q and T were constant, and D nearly constant, t was in direct proportion to l. This law was not demonstrated by experiment with the same exactness as was done in proving the law of pressure, probably owing to small experimental errors in measuring the varying mean sections of the tube as its length was reduced.

THIRD. *Effect of Diameter.*—When l, P and T were constant, the discharge per second, Q, varied quite closely with the direct ratio of the fourth power of D.

In all the experiments just summarized, T was constant at 10°. We hence have for this value of T, the following general law for the flow of water through glass pipes of very small diameters, where l is very considerable in proportion to D, and where $T = 10°$ C ;

The discharge, Q, varies in direct proportion with the pressure and with the fourth power of the diameter, and in the inverse ratio of the length.

This law is expressed by the formula $Q = K \frac{P D^4}{l}$, K being a constant, and having a value of very nearly 2495. The equation becomes, when the pressure is measured by distilled water, $Q = K' \frac{P' D^4}{l}$, K' having a constant value of about 183.78.

We therefore have $v = \frac{4 K}{\pi} \times \frac{P D^2}{l}$, or $v = 3177 \frac{P D^2}{l}$. Hence in such tubes the velocity is in direct proportion to the pressure and to the square of the diameter, and in the inverse ratio of the length.

FOURTH. *Effect of Temperature.*—With variations in T from 0.6° to 45.1°, K in the foregoing equation had the following values, P being nearly 776 ;

D = .141 l = 100.5		D = .085 l = 100.3		D = .044 l = 50.2		D = .029 l = 23.1	
T	K	T	K	T	K	T	K
0.6°	1875	0.5°	1856			0.5°	1851
5°	2158	5°	2158	5°	2157	5°	2159
10°	2497	10°	2497	10°	2495	10°	2495
20°	3237	20°	3235	20°	3240	20°	3232
30.1°	4067	30.1°	4062	30.07°	4064	30.05°	4054
40.1°	4969	40.1°	4971	40°	4978	40.1°	4970
45°	5444	45.1°	5449	45.1°	5445	45°	5449

It will be seen by an examination of the foregoing table that T had a constant effect upon the discharge, with widely varying values of D and l. It is hence apparent that the general law before given applies irrespective of temperature; that is to say, with varying values of P, D and l, so long as T is constant the co-efficient K will be constant.

The influence of changes of temperature as here indicated is quite remarkable; the discharge is trebled when T is increased from 0° to 45°, and a change in T from 5° to 20° increases the discharge one-half.

Dr. Poiseuille proposes the following final formula, l being sufficiently great in relation to D; $Q = K''' (1 + .033\,68\,T + .000\,221\,T^2)\dfrac{P\,D^4}{l}$, K''' being a constant with a value 1836.7. In this expression Poiseuille thinks that $2g$ should be introduced as a variable, g having a value of 9.808 metres at Paris. In this formula Q' is necessarily the weight, in milligrammes of water discharged in one second, and not its volume in cubic millimetres which varies with T.

With P' = pressure in water, we have

$$Q = K''' (1 + .033\,68\,T + .000\,221\,T^2)\frac{P'\,D^4}{l},$$

K''' having a constant value of 135.282.

This final formula gave closely agreeing results when applied to experiments where l varied from 100 to 23, D from .141 to .029, and P was 776. . With $l = 400$, $D = .652$, and P' from 269 to 2011, the agreement was not quite so close.

A commission of the Academy, of which Arago and Regnault were members, personally verified the accuracy of Dr. Poiseuille's experiments. The report of this commission is to be found in the Comptes Rendus of the Academy, Vol. XV., p. 1167, 1842.

FIFTH. *Flow of other Liquids.*—With q, D, P, l and T constant, q being the capacity of the ampoule, distilled water discharged in $t = 568''$; when 250 parts of ioduret of potassium was added to 500 parts of water, t was $475''$.[*]

With q, D, P, l and T constant (D being about .35 mm.), t had the following values :[†]

Distilled water	$t = 527''$
Pure alcohol	$t = 683''$
86.5 parts water and 73.5 parts alcohol	$t = 1732''$
1276 ,, ,, 73.5 ,, ,,	$t = 694''$

The final Poiseuille formula, disregarding the change in volume produced by variation in temperature, becomes, in English feet, $c = 52\,500\,(1 + .033\,68\,T, + .000\,221\,T,') \, s\, D^2,\, T,$ being the temperature in degrees of the Centigrade scale.

Dr. Poiseuille, in making his reductions, did not take into consideration either the head lost in contraction at the entrance of his tubes, or the head absorbed in imparting velocity. This head, h', for the experiments with some of his shorter tubes, was quite notable.

The experiments of Poiseuille, Hagen, and Jacobson have been quite ably discussed by Dr. Lampe in Vol. XIX. of Der Civilingenieur, to which paper the reader is referred.

Other Experiments.

Hagen's experiments, with D from 2.8 to 6 mm., to some extent showed that the Poiseuille law would apply to these larger diameters.[‡]

Jacobson, with $D = 5.108$ mm. and maximum value of $l = 2518.9$ mm., the tube being of polished brass, believed that the flow followed the Poiseuille law.[§]

Professor Osborne Reynolds[ǁ] has lately investigated the flow through glass and lead pipes, and has found, even with diameters as large as 1 inch, the flow follows the Poiseuille law up to a certain limit of velocity. At the end of this chapter the experiments of Professor Reynolds will be again alluded to.

EXPERIMENTAL DATA.

Couplet.

The very early experiments of Couplet made at Versailles, an account of which he published in the Memoirs of the Academy of Sciences, 1732, "*Recherches sur le Mouvement des Eaux*," give us the most reliable data extant for the flow through long pipes with small velocities. M. Couplet made 15 experiments, from which we will select 6 with a pipe 5 French inches in diameter, and about 1170 toises in length : the pipe for the first 320 feet in length was of stoneware, and for the remaining 7163 feet of lead. There was one rather abrupt bend and several easy bends. The comparative elevations of the inlet tank and the vertical outlet end of the pipe, were determined by attaching

---

[*] L'Académie des Sciences, Comptes Rendus, vol. xxiv., 1847.
[†] L'Académie des Sciences, Savants étrangers, vol. ix., p. 538, 1846.
[‡] Abhandlungen über den Einfluss der Temperatur auf die Bewegung des Wassers in Röhren. Berlin, 1854.
[§] *Vide* Der Civilingenieur, vol. xix., pp. 28-30.
[ǁ] Transactions Royal Society, 1883.

to the outlet an additional vertical pipe, and allowing the stagnant water to take its hydrostatic level; H can therefore be considered as having been very closely determined. The discharge, or q, was ascertained by dividing the escaping jet into two sections, which were measured one after the other, by a measuring vessel having a capacity of .6277 cubic foot. The times appear to have been measured by a one-half second pendulum; they were too short for very exact determinations of t and Q. As the pipe was of lead—except 320 feet of stoneware—its inner surface was probably in fairly good condition.

There was some danger of accumulation of air at one anticlinal point in the vertical plane of this pipe.

As the velocities were very low, h will be assumed as identical with the measured head, H; corrections for losses of head due to contraction at the entrance, and to imparting velocity to the water, will not sensibly affect the deduced values of n.

TABLE LXXI.

Couplet.—Earthen and Leaden Pipe at Versailles.

$$v = n (r s)^{\frac{1}{2}}$$

No.	l	D	a	t	Q	v	$H = h$	s	n
1	7483.4	.4441	.1549	32.5 / 75	.027 68	.1787	.4939	.000 066	65.9
2	"	"	"	23 / 39	.043 39	.2801	1.0066	.000 135	72.5
3	"	"	"	18.5* / 27.5	.056 76	.3664	1.4876	.000 199	78.0
4	"	"	"	15.5 / 24.5	.066 12	.4269	1.8725	.000 250	81.0
5	"	"	"	15 / 21	.071 74	.4632	2.1315	.000 285	82.4
6	"	"	"	13½ / 20	.073 23	.4728	2.2203	.000 297	82.4

It seems probable that the foregoing values of n are approximately correct. The length l, the head H, and the discharge Q, appear to have been measured with care; admitting that D may be somewhat in error, as it was a constant for the 6 experiments, we still have fairly reliable data for the construction of the form of curve for n, indicating the effect of variations in the value of v for low velocities.

Couplet's other experiments deserve a passing notice, and we will give 7 of them; the other 2 were with a compound conduit. How H was measured for them does not appear. Q for Nos. —— was obtained

---

* Couplet gives t for No. 3* as 32 half-seconds, which is presumably an error in typography, as his reductions are based upon 37 half-seconds.

† Experiment No. 6 was repeated with similar results.

by the use of the measuring vessel of .6277 capacity (896 French cubic inches). For Nos. ι and η, q was obtained by using the outlet tank, having an area of about 1983 French square feet; the respective depths in this tank for the two experiments were 10 and 9 French inches.

The pipe for Nos. α–λ was of iron, except for 31 feet at the two ends; its diameter was 4 French inches; it had two quite abrupt bends, and several easy bends; it was doubtless an old pipe, more or less covered with incrustations.

The pipe for Nos. δ and ϵ was of iron, except for 35 feet at the ends, which were of lead; its diameter was 6 French inches; the curves were easy; it apparently had been laid only a short time before these experiments, replacing the 4-inch pipe just described, and hence may have been nearly new; ϵ and H both have small values; the deduced values of n are hence rather unreliable.

Experiments Nos. ζ and η were with pipes belonging to a conduit system of 5 pipes laid side by side, two having a diameter of 18 French inches, and the other three a diameter of 12 French inches; the total length of each pipe was about 600 toises, of which 58 feet was of lead, and the remainder of iron; there were several bends, but no abrupt ones. Q for No. ζ was obtained by closing the inlet valves of 4 pipes, and opening the valve of one of the 18-inch pipes. For No. η, the inlet valves of the 3 12-inch pipes, and the valve of the 18-inch pipe (before experimented upon) were opened simultaneously, and the discharge measured; the proper discharge, as before determined, for the 18-inch pipe was deducted, and the result divided by 3;[*] hence No. η represents the mean value of the discharge from 3 12-inch pipes. Apparently these two experiments are the most reliable of the 13 given, as in them ι had quite a large value, H a considerable value, and also the given diameters for such large pipes are presumably more accurate than the diameters given for the smaller pipes.[†]

For the first 5 experiments H is assumed as identical with λ; the last 2 are corrected for loss of head in imparting velocity, but with no correction for contraction at entrance.

Couplet.—Pipes at Versailles. $v = n (r s)^{.5}$

No.	l	D	t	Q	v	H	h'	h	s	n
α			34.5	.01819	.1806	.798	...	.798	.000421	30.0
β	1898.5	.3553	20.0	.03138	.3106	1.865	...	1.865	.000982	33.9
λ			15.5	.04050	.4086	2.753	...	2.753	.001450	36.0
δ	1825.4	.5329	11.5	.05468	.2447	.266	...	.266	.000146	55.5
ϵ			8.0	.07846	.3518	.406	...	.406	.000225	60.3
ζ	3836.8	1.5087	720	6.961	3.478	12.900	.188	12.712	.003313	95.6
η		1.0638	360	1.862	2.087		.068	12.832	.003345	69.9

Dubuat, and after him Prony, selected Nos. 1-5 inclusive and No. ζ, and disregarded the other 6 experiments just given. This was disingenuous, as No. η is on its face entitled to equal weight with No. ζ. The reason of this omission was doubtless the fact that these experiments contradicted Dubuat's proposition that Δ was a factor of no importance, and Prony's formula, in which the co-efficients do not vary with changes in r.

To us these experiments—Nos. α · η—appear entirely reasonable. The 4-inch pipe was in all probability old and rough, with its original section, a, considerably diminished by incrustations; the low values of n are hence not at all improbable. The values of n for the 6-inch pipe are nearly double those for the smaller pipe; this was due to better condition of surface and larger value of r. The value of n of 69.9 for the 12-inch pipe, while it is 95.6 for the 18-inch, is fully accounted for by smaller values of v and r, and also by the

---

[*] For No. ζ, q was 4152.04 French cubic feet, with $t = 720''$; hence $Q = 5.7667$. For No. η, q was 3737.12 French cubic feet, with $t = 360''$; hence Q' for 4 pipes was 10.3809; (10.3809 − 5.7667) ÷ 3 = 1.5381 (French) = Q for No. η.

[†] That is to say, an error in D of $\frac{1}{8}$ inch in a 5-inch pipe will make a much greater error in the deduced value of n than a similar error in the diameter of a 12-inch pipe.

assumption that incrustations had proportionately more largely decreased a for the small pipe than for the large pipe.

The omission of these experiments well illustrates the danger of neglecting data which contradict a favorite theory, no matter how logically the theory appears to have been deduced. Had Dubuat given the proper weight to these experiments, he would not have enunciated a fallacy, which, under the influence of his great name, deceived hydraulicians for the better part of a century.

Bossut.

The Abbe Bossut made quite careful experiments with pipes at Mésières, the results of which were first published in 1771. His second or new edition was published in Paris in 1786, "*Traité théorique et expérimental d'Hydrodynamique,*" from which we have compiled the following data.

The diameters of the pipes experimented upon were respectively 1, 1½ and 2 French inches; the largest pipe was thought by Bossut to be about $\frac{1}{50}$th of an inch above the stated diameter; we will hence assume the diameter of this pipe to have been 2.01 French inches. The pipes were straight, and the discharge in all cases was free into the air. The pipes for experiments 7 to 30 inclusive obtained their supply from a closed rectangular tin box, 1 foot square; this box was connected, by a pipe about 8 or 9 inches in diameter, with the feeding reservoir in which the surface of the water was kept at a constant height; presumably the ends of the pipes were just flush with the side of the box. For the two series embraced by Nos. 7 to 30, the diameters were 1½ and 2.01 French inches, with heads respectively 1 and 2 French feet, measured from the axis of the pipes to the surface of the water in the reservoir; the pipes being horizontal. The co-efficient of contraction at entrance, n, will be assumed at .8125, that being about its value as shown by Bossut.

In all Bossut's experiments q was directly measured, apparently with much exactitude.

TABLE LXXII.

Bossut.—Straight Horizontal Tin Pipes.

$D = 1\frac{1}{2}$ and 2.01 French inches; l from 30 to 180 French feet.

$$o = .8125 \qquad 2g = 64.3 \qquad k' = \frac{v^2}{2\,y\,o^2} \qquad v = u\,(r\,s)^{\frac{1}{2}}$$

Series I. and II.

No.	l	D	a	t	Q	c	H	k'	h	s	u
7	191.84	.1184	.011 01	50	.012 29	1.116	1.0658	.0293	1.0365	.005 40	88.7
8	159.87	"	"	50	.013 76	1.249	1.0658	.0368	1.0290	.006 44	90.5
9	127.89	"	"	50	.015 78	1.432	1.0658	.0484	1.0174	.007 96	93.3
10	191.84	"	"	60	.018 48	1.678	2.1315	.0664	2.0651	.010 76	94.0
11	95.92	"	"	45	.018 53	1.682	1.0658	.0667	.9991	.010 42	95.8
12	159.87	"	"	60	.020 57	1.868	2.1315	.0822	2.0493	.012 82	95.9
13	63.95	"	"	45	.022 85	2.075	1.0658	.1014	.9644	.015 08	98.2
14	127.89	"	"	60	.023 48	2.132	2.1315	.1071	2.0244	.015 83	98.5
15	95.92	"	"	50	.027 46	2.493	2.1315	.1465	1.9850	.020 69	100.7
16	31.97	"	"	45	.032 44	2.946	1.0658	.2044	.8614	.026 94	104.3
17	63.95	"	"	50	.033 72	3.062	2.1315	.2209	1.9106	.029 88	103.0
18	31.97	"	"	50	.047 47	4.310	2.1315	.4376	1.0939	.052 98	108.8
19	191.84	.1785	.025 03	75	.036 42	1.455	1.0658	.0499	1.0159	.005 30	94.6
20	159.87	"	"	75	.040 71	1.626	1.0658	.0623	1.0035	.006 28	97.2
21	127.89	"	"	75	.046 05	1.840	1.0658	.0798	.9860	.007 71	99.2
22	95.92	"	"	70	.052 94	2.115	1.0658	.1054	.9604	.010 01	100.1
23	191.84	"	"	65	.055 00	2.197	2.1315	.1138	2.0177	.010 52	101.4
24	159.87	"	"	65	.061 09	2.441	2.1315	.1403	1.9912	.013 45	103.5
25	63.95	"	"	70	.064 97	2.596	1.0658	.1587	.9071	.014 19	103.2
26	127.89	"	"	65	.068 71	2.745	2.1315	.1775	1.9540	.015 28	105.1
27	95.92	"	"	60	.079 54	3.178	2.1315	.2379	1.8936	.019 74	107.1
28	31.97	"	"	70	.089 67	3.583	1.0658	.3024	.7634	.023 88	109.7
29	63.95	"	"	60	.093 62	3.830	2.1315	.3439	1.7876	.027 96	108.2
30	31.97	"	"	60	.130 99	5.233	2.1315	.6453	1.4862	.046 48	114.9

The pipe of $1\frac{1}{2}$ French inches diameter was then placed on an incline having an angle from the horizon of 6° 31'. The pipe at its upper end was connected with a feeding reservoir, where a constant head of 10 French inches was maintained above the centre of the inlet end of the pipe; this head being that which was required to impart velocity and to overcome loss by contraction at the entrance. The discharge was measured for three separate lengths, the incline being the same for each length experimented upon.

TABLE LXXIII.
Bossut.—Straight Inclined Tin Pipe.
$D = 1\frac{1}{2}$ French inches. $l = 59$, 118 and 177 French feet.   Series III.

$$o = .8125 \qquad 2\,g = 64.3 \qquad h' = \frac{v^2}{2\,g\,o^2} \qquad v = n\,(r\,s)^{\frac{1}{2}}$$

No.	l	D	a	t	Q	v	H	h'	h	s	n
31	62.88	.1184	.011 01	60	.067 66	6.143	8.02	.89	7.13	.1134	106.0
32	125.76	„	„	60	.067 73	6.150	15.16	.89	14.27	.1135	106.1
33	188.64	„	„	60	.067 81	6.157	22.29	.89	21.40	.1134	106.2

A lead pipe, whose diameter was 1 French inch and very uniform, was laid in a horizontal position, and supplied with the respective heads of 4 and 12 French inches above the axis of the pipe.   The inlet end of the pipe was connected by a conical mouthpiece with a supply pipe of 2 French inches diameter ; in the supply pipe was a cock having an opening of about $1\frac{1}{2}$ inches; we will therefore assume $o = .900$.

TABLE LXXIV.
Bossut.—Straight Horizontal Lead Pipe.   Series IV.

$$o = .900 \qquad 2\,g = 64.3 \qquad h' = \frac{v^2}{2\,g\,o^2} \qquad v = n\,(r\,s)^{\frac{1}{2}}$$

No.	l	D	a	t	q	v	H	h'	h	s	n
34	53.29	.088 81	.006 195	120	.006 73	1.086	.3353	.0236	.3327	.006 24	92.2
35	„	„	„	60	.012 26	1.979	1.0658	.0752	.9906	.018 59	97.4

Dubuat.

The Chevalier Dubuat made a number of experiments with pipes having diameters from $\frac{1}{12}$th to 2 French inches, which are described in the second volume of his "Principes d'Hydraulique," Paris, 1786.[*]   The experiments with pipes of less diameter than 1 inch are of very doubtful value, and we will only select those where the diameters were 1 and 2 inches.

Dubuat apparently possessed much less skill as an experimenter than Bossut. The following determinations hence are not entitled to as much weight as those just given from Bossut.

Nos. 53 to 59 inclusive are evidently anomalous, and were not used by Dubuat for purposes of comparison.   He states that the regimen of flow was not well established for Nos. 53 to 56 inclusive.

---

[*] Dubuat's first work, in which he announced his great discovery of the general laws governing the flow in uniform channels, was published in 1779.

TABLE LXXV.
Dubuat.—Straight Tin Pipes.

$D = 1$ and 2 French inches. $a = .8125$ $2y = 64.3$ $k' = \dfrac{v^2}{2g \cdot a}$ $v = u \, (rs)^{\frac{1}{2}}$

No.	Discharge	l	D	a	Q	c	H	k'	h	s	u
36	In water	65.46	.088 81	.006 195	.000 875	.141	.013 32	.000 47	.012 85	.000 196	67.6
37	„ „	„	„	„	.001 994	.322	.044 41	.002 44	.041 97	.000 641	85.3
38	„ „	12.301	„	„	.004 78	.772	.062 17	.014 03	.048 14	.003 91	82.8
39	„ „	65.46	„	„	.005 74	.927	.373 0	.020 3	.352 7	.005 39	84.8
40	„ „	„	„	„	.007 33	1.183	.526 7	.032 9	.493 8	.007 54	91.4
41	„ „	„	„	„	.008 31	1.342	.691 0	.042 4	.648 6	.009 91	90.5
42	„ air	„	„	„	.008 96	1.446	.795 8	.049 3	.746 5	.011 41	90.9
43	„ water	„	„	„	.009 15	1.476	.795 8	.051 4	.744 4	.011 37	92.9
44	„ air	„	„	„	.011 00	1.775	1.094 2	.074 3	1.019 9	.015 58	95.5
45	„ „	„	„	„	.011 54	1.862	1.216 7	.081 7	1.135 0	.017 34	94.9
46	„ „	„	„	„	.012 02	1.941	1.296 7	.088 8	1.207 9	.018 43	95.9
47	„ water	„	„	„	.015 77	2.546	2.104 9	.152 7	1.952 2	.029 82	98.9
48	„ „	12.301	„	„	.016 14	2.606	.532 9	.160 0	.372 9	.030 31	100.4
49	„ air	10.391	„	„	.032 08	3.179	1.598 6	.631 8	.966 8	.093 03	113.9
50	„ water	12.301	„	„	.032 36	5.223	1.860 6	.642 6	1.218 0	.099 01	111.4
51*	„ air	10.391	„	„	.039 23	6.333	2.368 4	.944 6	1.423 8	.137 0	114.8
52	„ „	„	„	„	.046 73	7.544	3.197 3	1.340 7	1.856 6	.178 7	119.8
53†	„ „	2.131	„	„	.033 8	5.453	.799	.700	.099		170.
54	„ „	„	„	„	.046 7	7.539	1.599	1.339	.260		145.
55	„ „	„	„	„	.058 6	9.456	2.398	2.106	.292	...	171.
56	„ „	„	„	„	.067 5	10.892	3.220	2.795	.425	...	164.
57	„ „	.355	„	„	.063 3	10.544	2.105	2.618	−.213	...	(†)
58	„ „	22.670	.1776	.024 78	.129 3	5.217	1.450	.641	.809	.035 7	131.
59	„ „	„	„	„	.190 0	7.668	3.228	1.385	1.843	.081 3	128.

English Experiments.

W. A. Provis‡ published, in 1838, an account of 208 experiments with a lead pipe of various lengths, with a diameter of $1\frac{1}{2}$ inches. Provis was evidently ignorant of the laws, as enunciated by Dubuat, governing the flow of water in pipes, and in his account of his experiments does not give sufficient data to enable one to compute the final results properly.

---

* No. 51 was thought to be doubtful, owing to accidental crushing of pipe.
† Dubuat gives value of *v* for No. 53 as 5.202, showing error either in computation or typography.
‡ Transactions Institution of Civil Engineers, vol. ii., p. 201, 1838.

James Leslie[*] in 1855 gives the results of 51 experiments with a lead pipe having a diameter of $2\frac{1}{2}$ inches, with lengths from 10 to 1086 feet, and with heads from $\frac{3}{16}$th of an inch to 10 ft. $1\frac{3}{8}$ inches. His two series with the larger values of l, appear to be somewhere near the truth. For smaller values of l his results are often palpably ridiculous; for instance, with $l = 25$, and $H = .083$, his determined velocity is 1.37; this gives h a value of .026, and n the absurd value of 186. His experiments evidently were conducted with but little skill.

Thomas Duncan[†] gives a number of experiments with lead pipes, the largest diameter being 1 inch; these appear to have but little merit.

The following table gives the results of a number of experiments with English water mains, doubtless all of cast-iron. Nos. A to R inclusive are given by Mr. James Simpson,[‡] and appear to have been made under his direction; Nos. S to U inclusive are given by Mr. Thomas Duncan.[§] In all these experiments H is assumed as being h.

Main from Brixton to Streatham. $D = 1.$; $l = 5200.$					Main from Belvedere Road to Brixton. $D = 1.583$; $l = 22440.$					Main at Liverpool. $D = 1.$; $l = 8140.$			
No.	H	Q	v	n	No.	H	Q	v	n	No.	H	Q	v	n
A	40.	3.417	4.351	99	F	43.5	5.517	2.803	101	K	27.	2.075	2.642	92
B	38.	3.417	4.351	102	G	41.	5.383	2.735	102	L	24.75	1.983	2.525	92
C	16.75	2.283	2.907	102	H	34.	4.967	2.524	103	M	18.	1.733	2.207	94
D	19.	2.283	2.907	96	I	27.5	4.450	2.261	103	N	12.	1.483	1.888	98
E	4.	1.142	1.454	105	J	24.	4.050	2.058	100	O	5.5	1.117	1.422	109

Main at Carlisle. $D = 1.$; $l = 6600.$					Main from Ditton to Brixton. $D = 2.5$; $l = 54120.$					Main at Liverpool. Green Lane to Kensington. Six sharp bends in pipe. $D = 1.$; $l = 8160.$			
No.	H	Q	v	n	No.	H	Q	v	n	No.	H	v	Q	n
P	46.	3.150	4.011	96	R	25	8.683	1.769	104	S	28.88	2.52	1.98	85
Q	34.5	2.800	3.565	99						T	11.02	1.64	1.29	89
										U	1.235	1.09	.86	96

Mr. Bidder[‖] states that an engineer had determined the flow through an earthen-

---

* Proceedings Institution of Civil Engineers, vol. xiv., p. 51, 1855.
† Proceedings Institution of Civil Engineers, vol. xii., p. 501, 1853.
‡ Proceedings Institution of Civil Engineers, vol. xiv., p. 316, 1855.
§ Proceedings Institution of Civil Engineers, vol. xii., p. 499, 1853.
‖ Proceedings Institution of Civil Engineers, vol. xiii., p. 115, 1853.

ware pipe at Alnwick, with $D = 1.5$, $l = 2310$., and $s = \dfrac{1}{400}$. The first experiment was with the pipe just full at its upper end, the depth of water at the lower end being 1.04. The second experiment was with a head at the inlet of 1.083 above top of pipe, the water having at the lower end a depth of 1.25 ; hence total head $= \dfrac{2310}{400} + 1.083 + \dfrac{D}{2} =$ $7.608 = H$. The observed velocities are given respectively as 3.581 and 3.915. Hence,

| No. V | $v = 3.581$ | $s = .0025$ | | $r = .375$ | $n = 117$ |
| ,, W | $v = 3.915$ | $H = h = 7.608$ | $l = 2310$ | $r = .375$ | $n = 111$ |

Possibly the hypothesis upon which s is based in No. V may be somewhat in error, and the given value of n, 117, may hence be somewhat too high.

In none of the foregoing experiments—Nos. A to W—is given any account of the methods by which Q and v were determined, nor is any mention made of the condition of the wetted surface. The Simpson series show a value of n of about 100 ; in those of Duncan and Bidder n ranges from 85 to 117. They all generally indicate that as r diminishes, n increases. It seems to us that they are not sufficiently authentic to warrant the determination of any curve or curves for values of n, with varying values of r, v and Δ, based upon their results.

Mr. Hawksley states that a large sum of money, 7189l., was expended upon experiments with pipes, in connection with the Metropolitan Drainage question. The experimentation, however, was so badly conducted that the final results were not even collated.[*] We have before alluded to the very astonishing results found with weirs, presumably by the same investigators. It is much to be regretted that in the only instance where liberal governmental or municipal aid has been extended in England for the purpose of increasing our knowledge of the science of Hydraulics, the work was placed in incompetent hands. In this regard the judicious liberality of the French Government affords a pleasing contrast ; this liberality has given the world the benefit of the labors of Bossut, Dubuat, Poncelet, Lesbros, Darcy and Bazin.

Scotch Authorities.

Jardine[†] states the flow through a lead pipe connected with the Edinburgh water supply to have been 11½ cubic feet per minute for the years 1738-1742, when the discharge was at its maximum ; for this pipe $l = 14\,930$, $D = 4\frac{1}{2}$ inches, and $H = 51$. ; hence $r = 1.71$, and $n = 96$. .

Nos. 60 to 63 inclusive are given upon the authority of Mr. James Leslie.[‡] Both the Crawley and the Colinton pipes were mains of the Edinburgh Water Company. The Crawley pipe was 30 years old, with many incrustations on its inner surface.

---

* Proceedings Institution of Civil Engineers, vol. xiv., p. 293.
† Edinburgh or Brewster's Encyclopædia, vol. xi., p. 526, article " Hydrodynamics."
‡ Proceedings Institution of Civil Engineers, vol. xiv., 1855.

The Colinton pipe was 8 or 9 years old ; the condition of its inner surface is not stated ; it had a total length of 29 580 feet, and experiments were made as to its discharge, first for its entire length, and afterwards by dividing it into sections having lengths respectively of 25 765 and 3815 feet ; the shorter section was at the upper end of the pipe proper ; Q for No. 61 is stated to have been the mean of 15 observations, the maximum value being 10.00 and the minimum 9.375 ; Q for No. 62 is stated to have been the mean of 26 observations, the maximum value being 7.50 and the minimum 7.142. In none of these 4 experiments given by Leslie, is the manner of obtaining Q given. The two pipes were doubtless of cast iron, and the given diameters the approximate or commercial size of the pipes when laid.

No. 64 is given upon the authority of Mr. James M. Gale.* This pipe conducted the Loch Katrine supply of water, or a portion of this supply, to the City of Glasgow. Its total length appears to have been $3\frac{3}{10}$ miles. It was coated with the preparation of Dr. Angus Smith (probably asphaltum and coal-tar), and its inner surface was in fairly good condition ; from the statements made during the discussion of Mr. Gale's paper, it appears that the pipe had rusted a little, especially on the bottom, which was thought to have been caused by the boots of the workmen, who walked through the pipe, removing or breaking the asphaltum coating. It was supplied with water at its inlet by an aqueduct. The normal flow was 23 430 000 gallons per diem, or, as stated by Professor Macquorn Rankine, 2607.5 cubic feet per minute ; when, however, the feeding aqueduct was quite full, thus giving a somewhat larger head, the flow increased to 25 millions of gallons per diem. The diameter was 4 feet, being doubtless the founder's measurement. The inclination, or slope of hydraulic grade-line, was 5 feet per mile, or $s = \frac{1}{1056}$. In what way the value of Q was obtained, does not appear.

TABLE LXXVI.

Pipes belonging to the Edinburgh and Glasgow Water Supply, of Cast Iron.

$$2g = 64.3 \qquad h' = \frac{v}{2g} \qquad v = n (r s)^{\frac{1}{2}}$$

No.	Name.	l	D	a	Q	v	H	h'	h	s	n
60	Crawley.	44 400.	1.25	1.227	4.350	3.463	226.	.19	225.81	.005 086	86.9
61		29 580.	1.333	1.396	9.517	6.816	420.	.72	419.28	.014 174	90.2
62	Colinton.	25 765.	„	„	7.333	5.252	230.	.43	229.57	.008 910	96.4
63		3 815.	„	„	20.250	14.503	184.	3.27	180.73	.047 374	115.1
64	Loch Katrine.		4.	12.566	43.158	3.438				.000 947	112.4

* Transactions Institution of Engineers in Scotland, vol. xii., 1860.

Darcy.

The most elaborate and costly series of experiments upon the flow of water through pipes thus far executed, are those which were made under the direction of M. H. Darcy, while he was in charge of the water service of the City of Paris, during the years 1849-1851.[*]

Experiments were made with 22 pipes of wrought and cast-iron, sheet-iron covered with bitumen, lead, and glass. The pipes were straight, with a slight upward inclination towards the outlet. The glass pipe was 147 feet long; the lead pipes were 172 feet long; the other pipes were about 367 feet long. The diameters varied from .04 to 1.64 feet. The character of the inner surfaces varied from that of glass to that of old cast-iron.

The upper or inlet end of each pipe was successively attached to the end of a closed cylinder, 3.28 feet in diameter, by a square joint. A constant head for each experiment was obtained by a feeding pipe connecting with this cylinder, and by which the heads or pressure at the cylinder could be varied at will.

To determine the losses in head from the cylinder to nearly the lower or outlet end of the pipe, four tubes were attached to the pipe and one to the cylinder, numbered from 1 to 5; at their junctions the axes of these tubes were at right angles to the axis of the pipe; where these tubes were screwed into the cast pipes, their lower ends were made as nearly as was possible flush with the interior wall of the conduit pipe; with thin pipes these tubes were soldered to the outside over small holes pierced through the pipe. These pipes were connected with the experimental conduit as follows; No. 1 near the outlet end of the pipe; No. 3 was about $16\frac{1}{2}$ feet (5 metres) from the inlet end at the cylinder, and No. 2 midway[†] between Nos. 1 and 3; No. 4 about 1 foot from the inlet; No. 5 was attached to the large feeding cylinder. For instance, for Series No. 22, with $D = 1.64$, these piezometric tubes were connected as follows; No. 1, 20.84 feet from outlet end; No. 2, 164.045 feet (50 metres) from No. 1; No. 3, 164.045 feet from No. 2; No. 4, 15.42 feet from No. 3, and .98 foot from the inlet at the cylinder; the total length of the pipe hence being 365.33 feet. Small pipes supplied by the flow through these tubes were brought together near the centre of the experimental conduit, and connected with vertical column or reading-pipes of glass, placed side by side; when the pressures were large, the indicated pressures were measured at the same point by vertical columns of mercury in glass tubes, called quicksilver manometers.

By this arrangement it was thought that the losses in head from the feeding cylinder could be directly and accurately measured by comparisons of the several heights indicated in the reading-tubes. Supposing that the regimen of flow in the

---

[*] Recherches expérimentales relatives au mouvement de l'eau dans les toyaux. Paris, 1857.

[†] The tube No. 2 was not exactly midway between Nos. 1 and 3 in the glass pipe, but nearly so.

experimental pipe had been fully established by the time the water reached piezometer
No. 3, the difference in elevation between reading-tubes Nos. 3 and 5 would indicate
the losses in head due to contraction at the inlet, and to imparting velocity to the water,
plus the loss by friction, et cet., in the known length from the inlet to No. 3; as
the distances between Nos. 3 and 2, and Nos. 2 and 1 were equal, it is evident that
the difference of height in the reading-tubes 3 and 2, should be the same as for that
between reading-tubes 2 and 1, provided the character of the wetted surface and the
cross-section of the pipe were uniform for its entire length between piezometers 1 and
3, and also provided that the retardation by friction, et cet., was independent of the
amount of pressure, which was necessarily greater between 3 and 2, than between 2
and 1. It may be here incidentally remarked that these experiments proved that such
changes in pressure did not sensibly affect the flow, proving the correctness of Dubuat's
demonstration of this fact. The results obtained by the comparison of the heights in
the various reading-tubes, will be discussed in the following section of this chapter
treating of the determination of h'.

Although some of the experimental results tend to show that the regimen of flow
had not always been fully established at piezometer No. 3, thus making the difference
between the readings at Nos. 3 and 2 larger than was properly due to the retarding
influences between piezometers Nos. 3 and 2, still we will follow M. Darcy, and estimate the inclination, or s, by the difference of elevation in tubes 3 and 1, divided by
the length of the pipe between those piezometers.

The absolute volume of discharge, q, was directly measured in tanks of considerable capacity, so that the length of time, t, of each experiment was amply great :[*] Q
was therefore apparently determined with great accuracy. The mean sections and
diameters were obtained by either measuring the volume of water contained in the
experimental pipe *in situ*, or by measuring the volume of water contained in the several
joints, or by two direct linear measurements at each of the two ends of each joint.
These measurements appear to have been carefully made; hence it can be assumed that
a and D were determined with sufficient accuracy, thus giving data for the determination of the correct values of the mean velocity, $\dfrac{Q}{a} = v$.

It is proper to state that there are very many errors in the data given by Darcy
in his published volume; many of these errors arise from careless proof reading, but
some of them appear to be blunders of the author in transferring from one set of tabulated results to another; these latter errors necessarily affect the accuracy of his final
reductions and comparisons. So far as discovered these errors have been corrected in
the following table. Such carelessness appears strange, when one considers the time
and money which were devoted to the execution of the experiments.

The given length, l', is the total length of each pipe from the feeding cylinder to

---

[*] In these experiments t ranged from 135 to 3240″; it was, however, rarely less than 240″.

the outlet end. The lengths between piezometers 3 and 1, from which s has been calculated, were as follows ; for the glass pipe, 147.18 ; for the lead pipes, 164.04 (50 metres); for all the other pipes, 328.09 (100 metres).

TABLE LXXVII.

Henry Darcy.—Experimental Conduit Pipes. $v = u (r s)^{\frac{1}{2}}$

New Drawn Iron (ordinary Water or Gas Wrought-Iron Pipes)

I. l = 374.6 D = .0400 T = 66°–80°				II. l = 372.2 D = .0873 T = 55°–60°				II.* l = 372.2 D = .0873 T = 66°–72°				III. l = 371.9 D = .1296 T = 53°–58°			
No.	s	v	n	No.	s	v	n	No.	s	v	n	No.	s	v	n
65	.000 85	.113	38.7	78	.000 33	.190	70.7	91	.017 58	1.604	81.9	98	.000 22	.205	76.9
66	.001 84	.236	54.9	79	.001 52	.430	74.6	92	.009 82	1.181	80.7	99	.000 78	.365	73.6
67	.003 04	.384	69.6	80	.004 87	.814	78.9	93	.001 77	.459	73.9	100	.001 82	.606	79.0
68	.005 33	.482	66.0	81	.010 15	1.207	81.1					101	.003 36	.858	82.3
69	.007 54	.554	63.8	82	.019 37	1.713	83.3	94	.000 2	.116	55.4	102	.006 30	1.252	86.3
70	.016 59	.755	58.6	83	.031 26	2.188	83.8	95	.000 235	.126	55.5	103	.012 86	1.835	89.9
71	.025 80	.942	58.6	84	.043 48	2.612	84.8	96	.000 25	.133	56.8	104	.023 89	2.585	92.9
72	.034 72	1.125	60.4	85	.063 16	3.153	84.9	97	.000 55	.282	81.5	105	.031 23	3.002	94.4
73	.043 99	1.286	61.3	86	.100 22	4.052	86.7					106	.043 48	3.503	95.7
74	.063 64	1.568	62.6	87	.105 71	4.203	87.5					107	.123 15	6.301	99.8
75	.085 54	1.880	64.3	88	.178 26	5.518	88.5					108	.175 55	7.564	100.3
76	.178 62	2.776	65.7	89	.256 01	6.555	87.7					109	.221 08	8.521	100.0
77	.344 26	3.921	66.8	90	.309 52	7.166	87.2								

New Lead Pipes.

IV. l = 172.0 D = .0459 T not given.				V. l = 172.4 D = .0886 T not given.				VI. l = 172.4 D = .134? T not given.			
No.	s	v	n	No.	s	v	n	No.	s	v	n
110	.000 64	.131	48.4	117	.009 44	.213	68.3	124	.000 82	.394	75.0
111	.003 36	.541	87.2	118	.003 00	.617	75.7	125	.003 62	.906	82.1
112	.008 62	.807	81.1	119	.008 14	1.089	81.1	126	.007 78	1.404	86.8
113	.025 26	1.463	85.9	120	.022 68	1.959	87.4	127	.023 10	2.398	93.2
114	.061 46	2.402	90.4	121	.054 36	3.350	96.5	128	.056 00	4.318	99.5
115	.114 38	3.438	94.9	122	.105 00	4.718	97.8	129	.110 74	6.316	103.5
116	.161 48	4.232	98.3	123	.146 32	5.309	96.8	130	.158 82	7.562	103.5

* For series II*, a copper diaphragm, pierced with a hole .0016 in diameter, was placed at the head of the pipe II.

TABLE LXXVII.—*continued.*

Sheet-Iron, covered with Bitumen. New Pipes.

VII. $l=371.8$ $D=.0879$ $T=51°-60°$			VIII. $l=365.1$ $D=.2710$ $T=70°-72°$			IX. $l=365.3$ $D=.6430$ T not given.			X. $l=365.5$ $D=.935$ $T=70°$		
No.	*s*	*v*	*n*	No.	*s*	*v*	*n*	No.	*s*	*v*	*n*	No.	*s*	*v*	*n*
131	.000 22	.098	44.8	143	.000 27	.328	76.7	155	.000 2	.591	104.1	166	.000 7	1.296	101.3
132	.000 67	.302	78.7	144	.000 66	.577	86.4	156	.000 48	.912	103.8	167	.002 55	2.782	114.0
133	.002 26	.509	72.4	145	.002 03	1.171	99.9	157	.001 29	1.529	106.2	168	.004 33	3.868	121.6
134	.006 09	.889	76.8	146	.006 29	2.182	105.7	158	.003 3	2.589	111.1	169	.006 85	4.902	122.5
135	.011 33	1.260	79.8	147	.012 20	3.117	108.4	159	.005 80	3.530	115.6	170	.011 9	6.673	126.5
136	.022 21	1.860	84.2	148	.022 85	4.442	112.9	160	.011 9	5.436	124.3	171	.020 44	8.852	128.1
137	.030 35	2.224	86.1	149	.031 07	5.292	115.3	161	.012 0	5.509	125.4	172	.028 07	10.522	129.9
138	.045 40	2.798	88.6	150	.040 70	6.148	117.1	162	.021 0	7.411	127.6				
139	.118 46	4.813	94.3	151	.071 70	8.438	121.1	163	.029 7	9.000	130.2				
140	.179 85	6.099	97.0	152	.106 54	10.535	124.0	164	.036 4	10.013	130.9				
141	.244 19	7.228	98.7	153	.138 80	12.034	124.1	165	.121 56	19.72	141.0				
142	.307 14	8.225	100.1	154	.156 05	12.786	124.3								

New Glass Pipe.		
XI. $l=147.2$ $D=.1630$ T not given.		
No.	*s*	*v*	*n*
173	.000 96	.502	80.3
174	.003 45	1.024	86.3
175	.007 71	1.591	89.8
176	.023 18	2.930	95.3
177	.057 62	4.849	100.1
178	.111 91	6.916	102.4

TABLE LXXVII.—*continued.*

Old Cast-Iron Pipes, and also the same cleaned.

XII. $l=374.9$ $D=.1178$ $T=45°$				XIII. No. XII. cleaned. $l=374.9$ $D=.1194$ $T=39°-43°$				XIV. $l=366.3$ $D=.2606$ T not given.				XV. No. XIV. cleaned. $l=366.3$ $D=.2628$ T not given.			
No.	s	v	n	No.	s	v	n	No.	s	v	n	No.	s	v	n
179	.000 25	.167	61.7	186	.000 71	.371	80.5	193	.000 65	.403	62.0	199	.000 84	.633	85.2
180	.000 71	.266	58.1	187	.001 80	.617	84.1	194	.002 50	.823	64.5	200	.002 94	1.263	90.9
181	.001 83	.426	58.1	188	.006 51	1.270	91.1	195	.007 25	1.463	67.3	201	.007 23	2.014	92.4
182	.006 7	.830	59.1	189	.014 41	1.972	95.1	196	.016 10	2.224	68.7	202	.007 37	2.047	93.0
183	.015 25	1.250	59.0	190	.030 18	2.927	97.5	197	.031 00	3.054	67.9	203	.015 57	2.835	88.6
184	.032 40	1.808	58.5	191	.039 66	3.392	98.6	198	.045 35	3.747	68.9	204	.029 38	4.095	93.2
185	.041 55	2.077	59.4	192	.046 50	3.694	99.1					205	.044 73	5.007	92.4

New Cast-Iron Pipes.

XVI. $l=366.1$ $D=.2687$ $T=57°-65°$				XVII. $l=365.7$ $D=.4495$ $T=59°-61°$				XVIII. $l=365.4$ $D=.6168$ T not given.			
No.	s	v	n	No.	s	r	n	No.	s	v	n
206	.000 2	.289	78.8	219	.000 24	.489	94.1	229	.000 27	.673	104.2
207	.000 83	.561	75.1	220	.000 87	.978	98.9	230	.001 75	1.631	99.3
208	.002 32	1.175	94.1	221	.002 09	1.601	104.5	231	.003 68	2.487	104.4
209	.003 31	1.841	97.5	222	.004 75	2.503	108.4	232	.008 05	3.701	105.0
210	.010 2	2.595	99.1	223	.012 60	4.196	111.5	233	.013 40	4.882	107.4
211	.022 55	3.868	99.9	224	.022 25	5.623	112.5	234	.023 5	6.342	107.7
212	.032 08	4.652	100.2	225	.033 18	6.883	112.7	235	.038 1	8.222	107.3
213	.040 41	5.154	98.9	226	.039 05	7.484	113.0	236	.109 80	14.183	109.0
214	.095 47	8.048	100.5	227	.098 52	11.942	113.5	237	.145 91	16.168	107.8
215	.099 04	8.160	100.0	228	.167 56	15.397	112.2				
216	.119 78	8.924	99.5								
217	.168 07	10.623	100.0								
218	.170 72	10.712	100.0								

TABLE LXXVII.—*continued.*

Old, Cleaned, and New Cast-Iron Pipes.

XIX. An old pipe. $l=365.3$ $D=.7979$ T not given.				XX. No. XIX. cleaned. $l=365.3$ $D=.8028$ T not given.				XXI. An old pipe very well cleaned. $l=365.3$ $D=.9744$ $T=69°-72°$				XXII. A new pipe. $l=365.3$ $D=1.6404$ T not given.			
No.	z	v	n	No.	z	v	n	No.	z	v	n	No.	z	v	n
238	.000 94	1.007	73.6	246	.000 52	.912	89.3	254	.000 28	.800	96.9	262	.000 45	1.380	101.6
239	.002 02	1.483	73.9	247	.001 65	1.762	96.8	255	.001 19	1.765	103.7	263	.000 45	1.472	108.4
240	.004 73	2.320	73.5	248	.004 98	3.113	98.5	256	.003 68	2.700	105.5	264	.000 6	1.559	99.4
241	.011 50	3.629	75.8	249	.011 55	4.659	96.8	257	.005 37	3.789	104.8	265	.001 2	2.602	117.3
242	.022 90	5.075	75.1	250	.020 35	6.247	97.7	258	.011 05	5.420	104.5	266	.001 25	2.609	115.2
243	.032 00	6.014	75.3	251	.027 35	7.238	97.7	259	.023 05	7.841	104.6	267	.002 1	3.416	116.4
244	.047 05	6.801	75.2	252	.037 30	8.438	97.5	260	.032 05	9.183	103.9	268	.002 3	3.633	119.0
245	.139 81	12.576	73.3	253	.113 43	14.754	97.8	261	.0407	10.368	104.1	269	.002 6	3.674	112.5
												270	.002 5	3.700	115.6

Iben.

Herr Otto Iben has collated in his "*Druckhöhen-Verlust in geschlossenen eisernen Rohrleitungen, Hamburg,* 1880," a large number of experiments of the flow through cast-iron pipes, made during the years 1874-1879 by engineers in charge of the water works of Hamburg, Stuttgart, and other German cities. We understand that Herr Iben simply acted as editor in compiling the data given to him, and hence is not responsible for the accuracy of the experimental work.

The heads lost by the retardation of flow through the various pipes experimented upon, were in all cases determined by piezometers, either directly by vertical columns of water or quicksilver, or by Bourdon gauges.

In the second general series of experiments made at Hamburg, the results of which will be summarized in the following table, the several elements were determined as follows :

Q by absolute measurement of q ; D, for the new pipes, by carefully measuring a few joints of each commercial size in use, by end diameters and also by quantity of water contained in the several joints ; for the old pipes, although doubtless their diameters were considerably diminished by incrustations, D was assumed to be the same as for new pipes of the same commercial or founder's size ; the loss of head, h, was generally determined by Bourdon gauges, the gauges being carefully compared for correction by attaching them to a vertical stand-pipe before and after each series of experiments; for Series X. open glass tubes were used for determining the pressures or piezometric heads, which permitted very exact readings. The new pipes* had all been coated with tar, and hence probably had quite smooth interior surfaces. Old pipes, Series XI. and XIII., had also been coated with tar before they were laid in place. The following table gives a summary of the results deduced from these experiments.

TABLE LXXVIII.

Hamburg Water Mains, 1877-1879. $v = n \, (r \, s)^{\frac{1}{2}}$

New Cast-Iron Pipes, coated with Tar. (Nos. V. and V*. had been 3 months in use.)

D = .335				D = .499						D = 1.001				D = 1.667			
l = 415. I.		l = 397. II.		l = 1073. IV.		l = 930. V.		l = 930. V*.		l = 1795. VIII.		l = 1125. IX.		l = 580. X.		l = 3514. XIX.	
c	n	c	n	v	n	c	n	c	n	c	n	c	n	v	n	c	n
1.0	79	1.0	132	2.0	82	2.3	111	3.2	89	1.6	85	1.7	200	0.4	123	0.7	105
1.7	92	2.1	97	3.3	87	3.3	85	4.4	92	2.1	97	1.8	149	0.9	106	1.6	110
2.1	90	2.8	97	3.9	88	4.3	101	5.3	91	2.3	102	2.2	131	1.3	109	1.9	109
2.3	90	4.3	105	4.8	92	4.7	91			2.6	112	2.6	122	1.7	110	2.5	109
2.8	91	5.5	104	5.3	87	5.1	90			3.2	108	2.9	129	2.2	112		
						5.3	90			3.8	121	3.2	132	2.5	113		
						5.6	91			4.1	114	3.5	125	3.0	114		
						6.0	92			4.8	125	3.8	123	3.6	115		
						6.5	92					4.5	122	4.8	116		
						7.1	90					6.0	149	6.1	114		
						8.0	92					7.0	148				
						8.7	95					7.2	129				
												7.9	129				
												8.5	129				
												8.7	122				
												10.5	132				

---

* Series V. and V*. were with a pipe, which had been three months in use.

TABLE LXXVIII.—*continued.*

Old Cast-Iron Pipes. (Nos. XI. and XIII. had been coated with Tar.)

$D=.335$		$D=.499$				$D=1.001$										$D=1.667$	
$l=373.$		$l=900.$		$l=916.$		$l=2149.$		$l=7179.$		$l=1808.$		$l=1736.$		$l=781.$		$l=4403.$	
19 years old.		13 years old.		19 years old.		2 years old.		14 years old.		15 years old.		22 years old.		22 years old.		25 years old.	
III.		VII.		VII*.		XI.		XIII.		XIV.		XV.		XV*.		XXII.	
v	n	v	n	v	n	v	n	v	n	v	n	v	n	v	n	v	n
0.3	25	0.8	26	0.8	23	0.6	74	0.7	71	0.9	67	0.8	50	0.4	42	1.6	64
0.7	20	1.3	31	1.3	26	0.7	77	1.2	75	1.2	61	1.1	47	0.6	43	2.2	60
1.0	20	1.5	34	1.5	27	0.8	81	1.6	78	1.8	58	1.5	47	1.5	44	2.7	60
1.6	21	1.7	34	1.7	27	1.2	85	2.3	79	2.3	59	1.9	45	2.4	45	3.1	59
2.1	21	1.8	34	1.8	27	1.2	85	2.7	80	2.5	59	2.4	45	2.9	45	3.6	59
2.4	21	1.9	34	1.9	27	1.3	89	3.2	82	2.6	58	2.6	47	3.5	46	4.0	58
						1.6	92	3.9	80			3.1	46	4.4	45	4.3	58
						2.4	86					3.5	46			4.5	58

A glance at the foregoing results, especially for the new pipes, will show that the deduced values of *n* jump about in the most irregular manner, in one case *n* having the absurd value of 200, with $D=1$, and $s=.000\ 29$. The results obtained from the old pipes possess some value in indicating a very greatly reduced discharge from pipes unprotected by asphaltum coating, and which have been several years in use.

Herr Iben gives 47 experiments made at Stuttgart, which have no more value than the Hamburg ones just summarized.

He also quotes 8 experiments made at Bonn, which possibly are entitled to some weight. The Bonn pipe was a long force-main, supplied by pumps at the lower end; it was new, with no obstructions of consequence such as valves or sharp bends, and had been coated with a preparation of asphaltum. *Q* was determined by absolute measurements of *q* in a reservoir, and also by capacity of the plungers of the pumps; the two measurements agreed closely. Owing to the disturbed condition of the surface of the water in the reservoir, *q* was probably not very accurately determined. The head, *h*, was determined by manometers at the pumps (probably Bourdon gauges attached to the air chambers of the two pumps). There was a small amount of air in the pipe, which may have affected accuracy of results.

TABLE LXXIX.

Bonn Water Works.

New Cast-Iron Pipe, coated with Asphaltum.

$$D = 1.004 \qquad l = 17\,684 \qquad T = 41° \qquad v = n\,(rs\,)^{\frac{1}{2}}$$

No.	Q	v	h	s	n	No.	Q	v	h	s	n
271	1.341	1.598	21.3	.001 21	90.1	271′	1.230	1.553	21.3	.001 21	89.3
272	1.675	2.116	32.8	.001 86	96.1	272′	1.665	2.104	34.4	.001 95	95.1
373	2.054	2.595	47.6	.002 69	99.9	273′	2.074	2.620	45.9	.002 60	102.6
274	2.474	3.125	60.7	.003 43	106.5	274′	2.451	3.096	64.0	.003 62	102.7

The experiments given by Herr Iben have been discussed at length by Herr Albert Frank, in Der Civilingenieur, Vol. XXVII, 1881, "*Die Formeln über die Bewegung des Wassers in Röhren.*"

Lampe.

In all probability the most accurate experiments of the flow through a long pipe, where the loss of head from friction, et cet., has been measured by piezometers, are those made in 1869-1871 by Prof. Dr. C. J. H. Lampe, and described by him in Der Civilingenieur, Vol. XIX., 1873, " *Untersuchungen über die Bewegung des Wassers in Röhren.*" We will therefore describe these experiments at some length.

The pipe experimented upon was a cast-iron conduit, which conveyed by gravity a supply of spring water to the town of Danzig. It had a total length of 46 352 feet ;* the lower 9040 feet in length had a considerably steeper inclination than the upper portion, so there were two hydraulic-grade lines ; the 4 experiments, however, were made upon the upper portion of the pipe. The pipe was laid in 1869 ; it was covered with 5 feet of earth ; it had 3 curves, each of 10.3 feet radius, and a number of very easy curves as it followed the general contour of the ground. Its section was very uniform, and its diameter was almost exactly 16 Rhenish inches, or 1.373 English feet ; the constructing engineer states that the maximum deviation from this diameter was only .008 foot. The pipe was coated with a patent varnish, which did not appreciably diminish its section. Examination showed that from 1869 to 1871 the character of the inner surface had very slightly changed ; the only material adhering to the surface in 1871 could be readily removed by rubbing with the finger, and there being no signs of rust. The joints were 12 feet in length, united by lead and hemp packings. There were 26 air-cocks attached to the pipe along its course, at anticlinal points, the respective distances and elevations of which had been carefully determined.

---

* Dr. Lampe gives most of his measurements in Rhenish feet ; he assumes 1½ Rh. ft. = .418 47 metre ; hence 1 Rh. ft. = 1.029 72 English feet, which is the ratio we have used in reducing his data to English measurements.

The mean velocity was ascertained by measuring the discharge, or q, in a masonry reservoir, situate at the outlet end of the pipe, which had an area of about 15 160 Rhenish square feet, and a depth of several feet, thus affording an excellent opportunity for the accurate determination of Q and v.

The pressures were determined by connecting a quicksilver manometer, first with one of the air-cocks and then with another; these pressure determinations were not synchronous with the measurement of q, but in some instances several days apart. As in none of the 4 experiments was the pipe filled at its inlet, this lack of synchronism appears to form a dangerous source of error, the pipe being fed by the flow from springs, whose discharge must necessarily have been more or less irregular. Dr. Lampe, however, states to us, in an explanatory letter, that he is satisfied no serious error could have arisen from this cause, there having been directly preceding or during these intervals no rains of consequence to notably affect the flow from the feeding springs; subsequent readings of the manometer, shortly after these intervals, also verified the constancy of flow. He considers that errors from this source will not change the given results more than $\frac{1}{2}$ of one per cent. .*

Another source of error in these experiments, arises from the fact that the pipe had two hydraulic-grade lines. Although the several pressures were all determined above the anticlinal point at which these lines united, still this condition of the pipe was more or less unfavourable to extreme accuracy of observation, as the sucking or siphon action as the water passed over this summit could not have been perfectly regular, as doubtless the amount of air accumulating at this summit varied slightly from time to time; the consequent intermittent sucking action of the water as it flowed below this summit, hence may have appreciably affected the nearest piezometers. The very great length of that portion of the conduit experimented upon, however, makes it improbable that this last source of error could considerably affect the accuracy of the determination of the general value of s.

In the following table we give the details of these piezometric measurements for Dr. Lampe's first and third experiments, our Nos. 275 and 277, expressed in Rhenish feet.

---

* There are several typographical errors in Dr. Lampe's paper in Der Civilingenieur, especially as to dates, which he has been kind enough to correct for us.

TABLE LXXX.

Dr. Lampe.—Piezometric Determinations. Danzig Conduit Pipe.
(Pipe Experiments Nos. 275 and 277.) Elevations and distances given in Rhenish feet.

Data for No. 275. October, 1869.

No. of Air-cock.	Elevation of Zero at each Air-cock.	Measured Head or Pressure at each Air-cock, in Feet of Water.	Piezometric Elevation at each Air-cock.	Calculated Elevation at each Piezometer.	Error.	Loss of Head between the several Air-cocks.	Distance apart of several Air-cocks	Between the several Air-cocks.
22	− 69.38	29.33	− 40.05	− 39.76	−.29	3.01	1 406.7	.002 14
21	− 92.97	49.91	− 43.06	− 42.50	−.56	11.48	6 309.6	.001 82
17	− 105.20	50.66	− 54.54	− 54.81	+.27	5.51	3 107.4	.001 77
13	− 110.38	50.33	− 60.05	− 60.87	+.82	3.86	1 147.8	.003 36
12	− 88.75	24.84	− 63.91	− 63.11	−.80	4.08	2 525.1	.001 62
10	− 108.65	40.66	− 67.99	− 68.03	+.04	4.79	2 539.1	.001 89
9	− 99.78	27.00	− 72.78	− 72.99	+.21	4.17	2 346.4	.001 86
8	− 100.62	23.67	− 76.95	− 77.37	+.42	5.17	2 447.1	.002 11
7	− 111.70	29.58	− 82.12	− 82.14	+.02	5.48	2 555.1	.002 14
6	− 102.93	15.33	− 87.60	− 87.12	−.48	2.71	1 802.4	.001 50
5	− 95.64	5.33	− 90.31	− 90.64	+.33			
					+.02	50.26	26 086.7	.001 927

Data for No. 277. October, 1870.

No. of Air-cock.	Elevation of Zero at each Air-cock.	Measured Head or Pressure at each Air-cock, in Feet of Water.	Piezometric Elevation at each Air-cock.	Calculated Elevation at each Piezometer.	Error.	Loss of Head between the several Air-cocks.	Distance apart of several Air-cocks	Between the several Air-cocks.
25	− 60.46	11.32	− 40.14	− 48.75	−.39	1.04	990.5	.001 05
24	− 65.25	15.07	− 50.18	− 50.11	−.07	3.67	2 781.2	.001 32
23	− 68.53	14.68	− 53.85	− 53.94	+.09	1.06	945.0	.001 12
22	− 69.38	14.47	− 54.91	− 55.24	+.33	1.97	1 406.7	.001 40
21	− 92.97	36.09	− 56.88	− 57.18	+.30	6.41	4 555.1	.001 41
19	− 109.83	46.54	− 63.29	− 63.44	+.15	.76	605.6	.001 25
18	− 108.10	44.05	− 64.05	− 64.28	+.23	1.80	1 148.9	.001 57
17	− 105.20	39.35	− 65.85	− 65.96	+.01	3.08	2 095.5	.001 47
15	− 120.47	51.54	− 68.93	− 68.74	−.19	.71	580.2	.001 28
14	− 118.13	48.46	− 69.67	− 69.54	−.13	.73	431.7	.001 69
13	− 110.38	39.98	− 70.40	− 70.14	−.26	4.73	3 672.9	.001 29
10	− 108.65	33.52	− 75.13	− 75.19	+.06	6.82	4 785.5	.001 43
8	− 100.62	18.67	− 81.95	− 81.78	−.17	6.44	5 002.2	.001 29
6	− 102.93	14.54	− 88.39	− 88.66	+.27	2.79	1 802.4	.001 55
5	− 95.64	4.46	− 91.18	− 91.14	−.04			
					−.19	42.04	30 803.4	.001 365

The calculated piezometric elevations, given in smaller type in the preceding table, are for a straight line* uniting the assumed elevations in the extreme piezometers in each experiment, being the lines of most probable inclination, as calculated by Dr. Lampe; s for No. 275 is assumed to be .001 9504, and for No. 277 to be .001 3761.

From the table it will be observed that the values of s between the several piezometers range for No. 275 between .001 50 and .003 36, and for No. 277 between .001 05 and .001 69. These variations are altogether too great to arise from possible variations in a and Δ; in fact, a critical comparison of the piezometric errors in the four experiments shows that these errors must be attributed chiefly to incorrect readings, or rather to erroneous indicated heights in the manometers. These errors, however, are not cumulative, as the maximum deviation of the indicated piezometric heads from the straight line of most probable inclination is, for No. 275 .82 foot, and for No. 277 .39 foot. We can hence conclude that, while for comparatively short lengths such as 1000 and 2000 feet, the piezometric heights are sometimes very inaccurate and misleading, still for the total lengths of 25 000 to 30 000 feet they can be relied upon as not being far from the truth. As the readings at the various piezometers were not made at the same moment, that fact may in part account for the defective results.

In the following table the final results of these four experiments are given in English measures; for No. 278 s is assumed to be .000 5936 instead of .000 5915 as stated by Dr. Lampe. For No. 275 T was about 47°.

TABLE LXXXI.

Lampe.—Danzig Conduit Pipe. Cast-Iron, coated with Smooth Varnish, and free from Rust.

$$D = 1.373 \qquad v = n\,(rs)^{\frac{1}{2}}$$

No.	Date.	Q	v	l	h	s	n
275	October, 1869.	4.575	3.090	26.862	52.393	.001 950	119.4
276	March, 1871.	4.011	2.709	31 719	51.692	.001 630	114.6
277	October 8, 1870.	3.671	2.479	31 719	43.650	.001 376	114.1
278	„ 1, „	2.334	1.577	23 414	15.085	.000 5936	110.5

Kirkwood.

Mr. James P. Kirkwood[†] describes two experiments made at his instance, the first under the direction of Gen. George S. Greene, with a main in the City of New York and the second by a subordinate of Mr. Kirkwood, with a main belonging to the Water Works of Jersey City, U.S.A. Both pipes were cast-iron mains connecting reservoirs,

---

* Strictly speaking this "straight" line, is a curved "level" line, nearly following the general curvature of the earth's surface; the distances doubtless being measured along the inclined line of the pipe.

† The Brooklyn Water Works. Van Nostrand, N.Y. 1867.

and in both cases q was determined by the lowering of the surface of the water in the upper or feeding reservoir.

The Croton (New York) main is said to have been "heavily tuberculated"; its diameter is given at 3 feet, being doubtless its original founder's size; it had three quarter-circle curves, each of 90 feet radius; losses of head due to primary contraction and to imparting velocity will be disregarded.

The Jersey City main was laid in 1858; the date of the experiment is not given, but apparently was in 1860 or perhaps in 1859; its general diameter was 20 inches, but for 128 feet in length the diameter was 24 inches; this enlargement will be assumed to counterbalance losses of head due to contraction and to imparting velocity; the water passing through the pipe came from the Passaic River. Six determinations are given, which probably formed one series, as the respective heads consecutively decrease from 30.262 to 26.325; the maximum value of n is 74.1, with $h = 29.758$ and $v = 1.51$; the minimum value of n is 71.1, with $h = 27.302$ and $v = 1.39$. We will take the mean values of the six determinations, as given by Mr. Kirkwood.

TABLE LXXXII.

Kirkwood.—Croton and Jersey City Cast-Iron Mains. No. 279 much incrusted; Condition of No. 280 not stated.

$$r = v \, (c \, s)^{\frac{1}{2}}$$

No.	Name.	D	q	r	l	h	s	n
279	Croton.	3.	21.204	3.000	11.217	20.215	.001 802	81.6
280	Jersey City.	1.667	3.137	1.438	29.715	28.128	.000 947	72.4

Mr. Kirkwood does not sufficiently describe the details of these experiments to enable one to judge of the degree of accuracy obtained in making them. If the Jersey City pipe had only been 1 or 2 years in use, as seems to have been the case, the very low value of 72.4 deduced for n seems altogether improbable.

Rochester Main.

Mr. J. Nelson Tubbs, Chief Engineer of the Rochester Water Works, New York, gives the following data in regard to the Hemlock Lake conduit-pipe bringing water by gravity to the town of Rochester.[*]

The upper portion of the conduit is 3 feet in diameter, is 50 776 feet long, and has a fall in this distance of 27 feet below mean surface of lake; this section is of riveted wrought-iron, $\frac{3}{16}$ inch thick. The lower portion of the conduit is 2 feet in diameter, is 51 493 feet long, and has a fall of 116.72 feet to its outlet at the storage-reservoir; of this section, 35 772 feet is of cast-iron, and the remaining 15 723 feet of riveted wrought-iron from $\frac{3}{16}$ to $\frac{1}{4}$ inch in thickness. The flow through this pipe, presumably

[*] Annual Report of the Executive Board, City of Rochester, N.Y., for the year 1876. Rochester, 1877.

soon after it was laid, was measured by noting the rise of the water in the lower reservoir for a period of 8 hours. The banks of the reservoir were new, and it was hence thought that the measured volume, or q, was somewhat underestimated, owing to absorption in the fresh-made banks. The flow was found to be at the rate of 9 292 800 gallons in 24 hours, or, say $Q = 14.378$.

Disregarding the small losses of head due to bends, contraction at entrance, and imparting velocity, we have a total head $27.00 + 116.72 = 143.72$; assuming that n is constant in $v = n \left(\dfrac{r\,h}{l} \right)^N$ for both sections of the pipe, the frictional head, h, for the lower section will be 7.701 times that for the upper section; hence,

TABLE LXXXIII.

J. Nelson Tubbs.— Rochester Compound Pipe.

$v = n \left(\dfrac{r\,h}{l} \right)^N$; n being constant.

No.	D	Q	c	l	h	n
281	3.	14.378	2.934	50 776	16.517	130.2
	2.		4.577	51 495	127.203	

Rosemary Pipe.

In the Transactions of the Am. Soc. of C.E., January, 1885, Mr. F. P. Stearns has been kind enough to give at our request a description of some very carefully conducted experiments of the flow through a cast-iron conduit pipe, having a diameter of 4 feet and a length of about 1800 feet; this pipe forming a portion of the Sudbury conduit line for the water supply of the City of Boston, U.S.A., and being known as the Rosemary pipe. The pipe had been laid three years before these experiments; it had originally been coated with a preparation of coal-tar and asphaltum, and at the time of these experiments its interior surface appeared to be very nearly in as good order as when the pipe was first laid. Q was determined by the flow over a weir, ten miles distant from the pipe, and an additional allowance was made for the infiltration into the conduit between the measuring weir and the pipe; the co-efficients of discharge for this weir had been determined with great accuracy, so that for these experiments Q can be considered to be given with sufficient exactness.

The head lost by friction, et cet., was determined by two piezometric columns of water placed near the extremities of the pipe; the piezometric tubes were small smooth brass tubes, laid in the conduit pipe, and resting immediately upon its bottom; the ends of these tubes were plugged, and holes drilled through their upper sides, near their ends; these holes were true and cylindrical, and at right angles to the axis of the conduit. The relative elevations of the zeros on the two piezometric scales were very accurately determined. Much care was taken to prevent accumulations of air in the piezometric tubes. The pipe was of nearly uniform diameter, so that the given mean area was within a small fraction of the truth.

TABLE LXXXIV.

F. P. Stearns. —Rosemary Cast-Iron Pipe, coated with Asphalt and free from Rust.

D = 4.00 $v = n (r s)^{\frac{3}{5}}$ T = about 38°

No.	Q	r	l	h	s	n
282	32.867	2.6155		.5557	.000 3180	146.7
283	16.972	3.738	1747.2	1.243	.000 7115	140.1
284	62.391	4.965		2.133	.001 221	142.1
285	77.852	6.195		3.230	.001 849	144.1

It will be noticed that n for No. 282 has an abnormally large value compared with the three succeeding experiments. No reason is known for this palpable error, which apparently amounts to at least 5 per cent. in the value of n; it is probably due to the inherent defects of piezometric measurements in the determination of h, which in this case appears to be about 10 or 12 per cent. in error, if Nos. 283, 284 and 285 are assumed to be correct.

Three experiments were made with the same pipe, the heads being measured by the difference in elevation of water at the two ends of the pipe; in a and b the pipe was not full at the inlet end. Making no corrections for losses of head in imparting velocity, et cet., these experiments gave the following values for n'.

No. a $v = 2.497$ $n' = 130.2$
No. b $v = 3.121$ $n' = 130.6$
No. c $v = 4.437$ $n' = 136.1$

Correcting No. c for loss in imparting velocity, n would have a value of 148.5. In these three last experiments the attending circumstances were not favourable for great exactness.

Hamilton Smith, Jun.

The conduit pipe of the Spring Valley Mining Company, at Cherokee, Butte County, California, was laid in the year 1871, and at this time was the most daring work of the kind ever constructed. It is an inverted siphon, having a length of about 2½ miles, an approximate diameter of 30 inches, and has a maximum depression of 887 feet below the hydraulic-grade line. It is made of double-riveted sheet-iron, which is subjected to a constant maximum tensile strain of 17 549 lbs. per square inch.*

---

* This pipe is described at some length in Vol. VI. of the Trans. Am. Inst. of Mining Engineers, in a paper by Aug. J. Bowie, Jun., Esq., "*Hydraulic Mining in California.*" See also Vol. XIII., p. 55, Trans. Am. Soc. of C.E., February, 1884, paper by the author on "*Water Power with High Pressures.*"

The length and head are taken from surveys made by engineers of the Spring Valley Co.; the author cannot state whether or not they were determined with proper accuracy. The given diameter of 2.43 was directly measured by the author at the only uncovered part of the pipe; it is probably slightly less than the mean diameter for the entire length. Q was determined by the superintendent of the company by the flow through standard orifices; three tests were made, under different circumstances, giving following results;

1st	...	$Q = 51.54$
2nd	...	$Q = 48.52$
3rd	...	$Q = 52.85$
Mean of 1st and 2nd tests	...	$Q = 50.0$

It is most likely that the adopted value of Q is somewhat too low, as in calculating the discharge through the orifices, no correction was made for partial suppression of bottom contraction.

The pipe had been 5 years in use; its interior surface at the time of the experiment was very smooth, with the exception of the rivet-heads, which for over half its length formed noteworthy obstructions. The pipe had originally been coated by a bath in boiling asphaltum and coal tar.

TABLE LXXXV.

Cherokee Wrought-Iron Conduit Pipe; 5 Years in Use.

$$v = n \, (r \, s)^8$$

No.	D	Q	v	l	H	s	n	Velocity of Stones sent through Pipe.
286	2.43	50.0	10.78	12798	150.0	.011 72	127.8	9.0

The inlet end of the pipe is funnel-shaped for some little distance, so that no correction need be applied for contraction or imparting velocity.

Stones, weighing as much as 25 lbs. each, were sent through the pipe, with a velocity of 9.0; their velocity was doubtless much retarded by striking the rivet heads, which for over half the length of the pipe were of considerable size.

In Chapter X. will be found a detailed account of 53 experiments made by the author at New Almaden, California. It will only be necessary here to summarize their results.

TABLE LXXXVI.

Hamilton Smith, Jun., 1877.—Small Pipes of Various Kinds.

$$T = 57° \text{ to } 68° \qquad h = H - \frac{r^2}{2\,g\,\sigma^2} \qquad c = u\left(\frac{D}{4}\,\frac{h}{l}\right)^{\frac{1}{2}} = u\,(r\,s)^{\frac{1}{2}}$$

I. New wrought-iron, un-coated, no funnel. $l = 60.172 \quad D = .0878$				II. Same pipe as No. 1, with funnel-shaped mouth-piece. $l = 60.247 \quad D = .0878$				III. Same pipe as No. 1, coated with asphalt; with funnel. $l = 60.264 \quad D = .0873$				IV. First two joints pipe No. 1, coated with asphalt; with funnel. $l = 16.685 \quad D = .0876$			
No.	s	c	u	No.	s	c	u	No.	s	v	u	No.	r	c	u
287	.126 99	5.325	100.9	294	.130 25	5.387	100.8	301	.130 64	5.443	101.9	305	.230 92	6.882	96.8
288	.100 70	4.673	99.4	295	.102 03	4.706	99.5	302	.103 38	4.761	100.2	306	.159 84	5.621	95.0
289	.074 30	3.948	97.8	296	.076 15	3.999	97.8	303	.052 19	3.224	95.5	307	.076 90	3.775	92.9
290	.050 31	3.175	95.5	297	.051 57	3.236	95.9	304	.026 93	2.220	91.6	308	.042 06	2.333	88.0
291	.025 76	2.154	90.6	298	.026 41	2.199	91.3								
292	.012 27	1.421	86.6	299	.012 47	1.435	86.8								
293	.007 50	.958	74.7	300	.007 85	1.052	80.1								

V. Old wrought-iron, un-coated, 52.7 feet. New, coated, 7.4 " Funnel mouth-piece. $l = 60.217 \quad D = .0853$				VI. New wrought-iron, un-coated. No funnel. $l = 60.127 \quad D = .0523$				VII. New glass. Funnel mouth-piece. $l = 63.902 \quad D = .0764$				VIII. New glass. No funnel. $l = 34.941 \quad D = .0622$			
No.	s	c	u	No.	s	v	u	No.	s	c	u	No.	s	v	u
309	.133 12	4.266	80.1	316	.133 22	3.878	92.9	322	.129 18	5.009	100.6	327	.132 51	4.375	96.3
310	.105 10	3.773	79.7	317	.105 29	3.385	91.2	323	.102 06	4.383	99.3	328	.095 85	3.666	95.0
311	.077 66	3.211	78.9	318	.077 84	2.863	89.7	324	.075 30	3.685	97.2	329	.054 26	2.652	91.3
312	.052 37	2.619	78.2	319	.052 99	2.295	87.2	325	.050 77	2.945	94.6	330	.017 97	1.398	83.6
313	.027 06	1.829	76.1	320	.027 49	1.578	83.2	326	.025 01	1.955	89.5				
314	.012 56	1.195	73.0	321	.013 12	1.029	78.6								
315	.007 65	.910	71.2												

TABLE LXXXVI.—*continued.*

IX. New glass. No funnel. $l = 11.127$　$D = .0418$				X. New wood (bored). No funnel. $l = 62.05$　$D = .1052$			
No.	*s*	*v*	*n*	No.	*v*	*v*	*n*
331	.230 88	4.139	90.4	335	.131 15	3.986	67.9
332	.165 90	3.659	87.9	336	.103 06	3.519	67.6
333	.098 41	2.719	84.8	337	.076 10	3.008	67.2
334	.064 35	2.077	80.1	338	.050 94	2.469	67.5
				339	.024 19	1.653	65.5

　　Nos. 331 to 334, with the small glass pipe, Series IX., are the least reliable of the foregoing experiments, its mean section not having been as accurately determined as was the case with the other pipes. By plotting the several series, with *v* and *n* as co-ordinates, it will be found that very smooth curves are formed, No. 338 being the only experiment of the 53 which is notably irregular; *n* for No. 338 should probably be 67.0, instead of 67.5, *vide* note in Table XCVII. These experiments can be considered entirely reliable so far as the experimental data is concerned; the danger of error in them is the doubt as to whether or not $\dfrac{v^2}{2\,g\,d}$ correctly represents $H - h$.

　　The following experiments were all with riveted sheet-iron pipes, and will be described in detail in Chapter X. The condition of surface was about the same for Nos. 340 to 355 inclusive, and can be considered quite smooth; the rivet heads for half the length of No. 356, Texas Creek pipe, formed noteworthy obstructions; the curves in all these pipes somewhat retarded the flow, as did also irregularities at the joints. The danger of experimental error in these experiments, consists chiefly in the value of Q, which was measured by the flow over weirs or through orifices; in this regard No. 355, or the Humbug pipe, is the least trustworthy of the experiments. In all of them, except No. 356, h' is a considerable fraction of H, and the deduced values of h may therefore possibly be appreciably in error.

TABLE LXXXVII.

Hamilton Smith, Jr.—Riveted Sheet-Iron Pipes, Smooth Interior Surfaces, except Rivet Heads and Joints.

$$h = H - \frac{c^2}{2g\sigma^3} \qquad v = n\left(\frac{D}{4}\cdot\frac{h}{l}\right)^{\frac12} = n(rs)^{\frac12}$$

T for Nos. 340 to 354 about 55°; for No. 356 from 50° to 60°.

No.	D	Q	r	l	H	h'	h	s	n	Maximum Velocity of Stones or Wooden Blocks sent through Pipe.
340	.9105	6.525	10.021	684.8	24.220	1.562	22.658	.033 09	115.5	9.42
341	.911	5.644	8.659	697.0	19.005	1.166	17.839	.025 59	113.4	
342	.911	4.515	6.927	713.9	12.850	.746	12.104	.016 95	111.5	5.79
343	.911	3.972	6.094	721.3	10.200	.578	9.622	.013 34	110.6	
344	.911	3.071	4.712	730.6	6.555	.345	6.210	.008 50	107.1	
345	1.056	9.376	10.706	684.9	24.510	1.783	22.727	.033 18	114.4	
346	1.056	7.572	8.646	699.6	16.690	1.163	15.527	.022 19	113.0	7.4 (?)
347	1.056	6.097	6.962	709.2	10.885	.754	10.131	.014 28	113.4	
348	1.056	4.024	4.595	718.4	5.130	.328	4.802	.006 68	109.4	
349	1.230	14.365	12.090	684.4	24.390	2.274	23.116	.032 31	121.3	11.88
350	1.230	12.587	10.593	695.6	18.925	1.746	17.179	.024 70	121.6	9.58
351	1.230	10.054	8.462	705.0	12.715	1.114	11.601	.016 46	119.0	
352	1.230	8.690	7.314	710.7	9.550	.832	8.718	.012 37	119.1	
353	1.230	8.128	6.841	712.4	8.545	.728	7.817	.010 97	117.8	
354	1.229	5.199	4.383	719.9	3.915	.299	3.616	.005 02	111.6	
355	2.154	45.92	12.605	1193.8	22.067	2.471	19.596	016 41	134.1	11.24
356	1.416	31.721	20.143	4438.7	303.62	7.46	296.16	.066 72	131.1	20.94

It is probable that *n* for No. 349 is somewhat too low, and also that *n* for No. 350 is slightly too low; *n* for Nos. 345 and 346 is known to be too low, owing to the stoppage of a stone in the pipe. *Q*, and hence *v* and *n*, for No. 355 is probably too high, as will be seen by reference to Chapter X.

Clarke.

Mr. Eliot C. Clarke, in his account of the "Main Drainage Works of the City of Boston, Mass., Second Ed., 1886," describes a number of experiments made as to the flow through the main sewage tunnel, lately built by the City of Boston, from the main-land, under Dorchester Bay, to Squantum Neck; the tunnel is placed at an average elevation of 142 feet below low tide.   This tunnel has a circular section, with a diameter of 7.5 feet, the enclosing walls being of hard brick.   It is an "inverted siphon," with a total length of 7166 feet, through which the sewage flow from the city is all discharged ; the flow is caused by gravity ; the sewage is pumped from a lower level to a reservoir at the inlet of this tunnel-pipe ; the elevation of the surface in this reservoir or pen-stock is sufficient to produce the flow by gravity through the tunnel, and through an additional conduit, to the discharging reservoirs on Moon Island.

The pumps were started in January, 1884, and have been continuously in operation, except occasional stoppages of a few hours, up to the present time—February, 1886. The ordinary velocity through the tunnel is seldom greater than one foot, and often less than one-half foot.   The tunnel can at any time be flushed by running the pumping machinery to its full capacity; the first flushing was done on June 12, 1884, which removed a considerable deposit of soft mud, horse manure, et cet. ; since that date the tunnel has been flushed at regular intervals of about two weeks.   The extra supply for this flushing is obtained by pumping salt water, which is added to the ordinary sewage flow.

The following table shows the results of 7 experiments ; Q for the first five was ascertained by the registered strokes of the several pump-pistons, with an allowance for "slip," which was determined a few days before Experiment No. 361 by actual measurement in a reservoir ; for the last two experiments Q was obtained by measurement of q in a reservoir ; these determinations of q probably were within 1 per cent. of the truth : for the first 4 experiments a correction was made to the observed head, for loss of head at entrance; in the other 3 experiments the head was measured a short distance below the mouth of the tunnel, and no such correction was required.   The tunnel has one quarter-turn of 9.75 radius, and one angle of 23½°; the outlet end of the tunnel is divergent, which probably compensates for these two bends.

TABLE LXXXVIII.

Clarke.—Flow through Dorchester Bay Tunnel-Pipe.

$l = 7166.$; $D = 7.5$, and $r = 1.875$; (a for Nos. 357 and 358 diminished considerably by sewage deposits.) Δ = hard brick, modified more or less by sewage-slime.

No.	h	s	Q	v	u	Date.	Character of Flow.
357	.520	.000 0726	41.06	.929	80.	June 6, 1884	Sewage.
358	.566	.000 0790	44.08	.998	82.	„ 9, „	„
359	3.648	.000 5091	176.24	3.989	129.	„ 12, „	About ¼ sewage and ¾ salt water.
360	.297	.000 0414	42.64	.965	102.	„ 13, „	Sewage.
361	4.165	.000 5812	173.56	3.929	119.	Oct. 20, „	About ¼ sewage and ¾ salt water
362	3.975	.000 5547	167.78	3.798	118.	Aug. 28, 1885	„ „ „ „ „ „
363	3.680	.000 5135	166.51	3.769	121.	Sep. 25, „	„ „ „ „ „ „

The last three experiments—361, 362 and 363—are the most trustworthy of the series.

Some experiments were also made of the flow through the conduit, which receives its supply from the outlet end of the Dorchester Bay tunnel ; this conduit is a square tight wooden flume or pipe 6. × 6., about one mile in length, and is made of planed plank placed lengthwise ; this, of course, gives a lower value of Δ than would have been the case with plank placed crosswise. The experimental section of the flume was 2486.5 feet in length, and straight. When the sewage in the discharging basins or reservoirs on Moon Island has a low level, the conduit is about one-half filled, but when the basins are nearly full the conduit is completely filled, and becomes a pipe. The ordinary velocity in the conduit is about 3 feet. From its bottom to the ordinary flow line, the sides are covered with a slimy deposit from ⅛ to ½ of an inch in thickness; above this line, on the sides and top, there is some slime, but not so much as below. This deposit is not removed by flushing, although when flushing is being done, the velocity at the lower end of the conduit is 7 feet.

The following results were obtained in October, 1884, Q having been determined with approximate accuracy by the strokes of the pump-pistons;

r	v	n	Conduit or Pipe.	Character of Flow.
1.45	2.94	117.	Conduit half-filled.	Sewage.
1.41	2.87	117.	„ „ „	„
1.5	4.8	135.	„ full, and hence a pipe.	⅓ sewage, and ⅔ salt water.

It would probably be dangerous to draw any general conclusions from the foregoing experiments, as in some of them Δ and *n* were not constant, and in others the value of Q, and hence *v*, was only approximative.

Other Experiments.

Many other experiments given by French, German, English and American authorities have been carefully examined, but are regarded as too doubtful in point of accuracy to be of value. In many of them the inclination was measured by Bourdon gauges ; with ordinary values of *s* and *l*, such determinations of *h* are too inaccurate to give anything but very rough approximations of the true values of *s*.

h'.

Whether or not the expression $h' = \dfrac{v^2}{2\,g\,\phi^2}$ correctly represents the losses of head due to contraction at the entrance and to imparting velocity to the water, can be determined in two ways :

First.—By having pipes with a and Δ constant, and with l and s variable. We will assume, and we think with entire safety,[*] that s should not be regarded as a factor *directly* influencing the value of n, but that v—which is the result of s—is the factor to be taken into consideration. Now, with two pipes having a and Δ constant, l considerably different, and H for each pipe of a value to make v identical, it is apparent that if, by the use of the expression $v = n \left(\dfrac{H - \dfrac{v^2}{2\,g\,\phi^2}}{l} \, v \right)^{\frac{1}{2}}$, n has similar values, the expression is correct ; and conversely, if v differs beyond the limit of probable experimental errors the expression is inaccurate.

Second.—By the use of piezometric columns along the course of a pipe, whereby h can be directly determined for a given part of the length ; h being ascertained for a portion of the length, if the total head H and the total length l' be known, the value of h' can be easily calculated.

Direct Method of Proof.

The Bossut experiments, Table LXXII., afford fairly accurate and complete data in this regard. By reference to Plate XII., where they are plotted with v and n as co-ordinates, it will be seen that they present quite uniform experimental curves, the curve for the larger diameter of .178 being somewhat higher than for the one with $D = .118$. Now, in these experiments l varied from 32 to 192, and the several corrections of h' applied to H are proportionately widely apart. Slight probable variations in a and Δ, aside from probable errors in H and Q, will fully account for the irregularities in these curves.

Hence we may say that the Bossut experiments confirm the general accuracy of our expression $h' = \dfrac{v^2}{2\,g\,\phi^2}$. The three experiments of Bossut given in Table LXXIII., when the inclination of the pipe proper was constant, also warrant this deduction.

Plotting Experiments Nos. 36 to 52 inclusive of Dubuat on the same sheet, we see an experimental curve much more irregular than those of Bossut, No. 37 apparently differing 12 per cent. from the most probable curve for the series. So far as the effect of h' is concerned, it will be observed that Nos. 38 and 48, where h' is a large fraction of

---

[*] The only experimental data, possessing any moderate claims to accuracy, which contradict this assumption, are the Darcy-Bazin series, Nos. XXVIII. to XXXI. inclusive, with very small conduits. We have shown in the preceding chapter that these seeming contradictions can be fully accounted for by *probable* experimental errors.

H, agree quite closely with the curve formed by the other experimental points, when h' was a much smaller fraction of H.

Therefore it can be said that the Dubuat series proves that $h' = \dfrac{v^2}{2\,g\,o^2}$ is approximately correct.*

On the same sheet we will now plot Series I., II., III. and IV., of the New Almaden experiments made by the author, and given in Table LXXXVI.

For   I. ; $l = 60.2$; Δ = new wrought-iron ; end of pipe flush with inlet tank.

,,   II. ; $l = 60.2$; Δ = ,,   ,,   ; funnel-shaped mouth-piece.

,, III. ; $l = 60.3$; Δ = coating of asphalt ;   ,,   ,,   ,,

,, IV. ; $l = 16.7$, and Δ and mouth-piece as in Series III.

The diameters varied from .0873 to .0878.

The inlet end of the pipe for Series I. was a little rough, and o was assumed at .80 ; for the funnel-shaped mouth-piece used for the other series, o was placed at .98. The uncoated pipe, Series I. and II., was new and free from rust ; hence the varnish applied, in Series III. and IV., very slightly diminished the value of Δ.

By reference to Plate XII., it will be observed that each of these four experimental curves is almost perfectly symmetrical ; this proves that the experimental values of H and Q were accurately determined. Contrasting the curves for Series I. and II., where a was the same, they will be seen to be practically identical ; this shows that the corrections for contraction for Series I. are the proper ones, or more exactly, that in $h' = \dfrac{v^2}{2\,g} \times \dfrac{1}{o^2}$, the use of the factor $\dfrac{1}{o^2}$ gives correct results.

Comparing Series II. and III., it will be seen that for the higher velocities the curve for III. is slightly the higher ; this was to be expected, as for III. Δ had a slightly lower value.

Comparing Series III. and IV., a notable variation will be observed, increasing with v, the value of n being 7 per cent. lower for IV. than for III., with $v = 5.5$. In both series the effect of o was nearly eliminated, the value of h' being nearly altogether due to the head required to impart velocity, or $\dfrac{v^2}{2\,g}$; the relative correction of h', or $\dfrac{h'}{H}$, was nearly three times greater for Series IV. than for Series III. This seems to prove that $\dfrac{v^2}{2\,g}$ does not sufficiently represent the loss of head absorbed in imparting velocity.

There appears to be no good reason to doubt the accuracy of the experiments constituting Series IV.† Of the experimental elements, l, Q and H can be considered free from possible error of consequence. Owing to the smaller quantity of water contained in this pipe, a cannot be considered to have been quite as accurately ascertained

---

* Nos. 53 to 69 of Dubuat are not used, as they present anomalous and contradictory results.

† The accuracy of the experimental data for Series III. is absolutely proved, by the agreement of its curve with the curves for Series I. and II., and also by the curve for Series VII., with a glass pipe of nearly similar dimensions.

as in Series III.—the diameter and area of both pipes having been deduced from the weight of water in them. It is possible that Δ may have been higher for IV. than for III. Taking into consideration all these chances of variation or error, it is not probable that they could have caused the marked difference between the two curves.

The relative corrections of h', for the New Almaden Series VII., VIII. and IX., with glass pipes, were considerably different, as will be seen by reference to Table XCVII. Examining Plate XIII., where the curves for these series are plotted, it will be noticed that the curves indicate no serious error due to incorrect assumptions of the value of h'. The values of n are largest for the largest diameter, and least for the least diameter; the differences between the curves can be accounted for by the variation in v. The relative differences of the correction $\dfrac{v^2}{2g}$ were, however, for these three pipes considerably less than for Series III. and IV., and hence they are not entitled to as much weight, so far as the accurate determination of the true value of h' is concerned.

Taking all the foregoing facts into consideration, it can be assumed that the New Almaden experiments indicate that the loss of head absorbed in imparting velocity is somewhat greater than $\dfrac{v^2}{2g}$.

Unfortunately, these experiments were not reduced until a year or more after they were made, and it was then impracticable to repeat them, and to further investigate the proper value which should be given to h'.

Indirect Method of Proof.

Darcy's experiments should afford data from which the exact value of h' can be deduced.

It will be shown in the next section of this chapter, that a piezometric column should truly indicate the hydraulic head in a pipe or conduit, when the mouth of the orifice is exactly flush with the side of the conduit, and its axis normal with the axis of the conduit. M. Darcy states that his piezometric tubes were thus attached to his experimental conduit pipes, and hence the heights of his piezometric columns should represent the true head at each of his 5 piezometers.[*]

The head indicated by his piezometer No. 5 should represent the effective height of the water at the inlet. Supposing the regimen of flow to have been fully established at piezometer No. 3, the difference of height between columns Nos. 5 and 3 should indicate the losses of head due to contraction and impartation of velocity, and to

---

[*] An account of the arrangement of M. Darcy's piezometers, with a description of his methods of experimentation, will be found in the preceding section of this chapter.

" friction," between No. 3 and the inlet end of the pipe. The loss due to friction can be segregated, by assuming it to be in the direct proportion of this length—No. 3 to the inlet end—to the length between Nos. 1 and 3; the loss of head, due to friction alone, between Nos. 1 and 3 is given by the respective heights of these two piezometric columns. This supposition involves the assumption that a and Δ are the same for the length 1 to 3 as they are for the length 3 to the inlet; any variations in a and Δ are probably not large enough to produce serious error.

Unfortunately M. Darcy does not give the height of the surface of the water in his outlet tank. Were this height known, we would have a reliable check upon the accuracy of his piezometric readings.

We will select from the Darcy Series, Nos. I., II. and III., with small wrought-iron new pipes, and Nos. XVI., XVII., XVIII., XIX. and XX., with cast-iron pipes of larger size. Nos. VII., VIII., IX. and X., with sheet-iron pipes covered with bitumen, present very discordant results, and will not be used; in fact, M. Darcy rejects Series IX. and X., as he conjectures that small pieces of bitumen may have interfered with the flow in the piezometers nearest the inlet.

Piezometer No. 5 was attached to the inlet tank or cylinder. The positions of the other 4 piezometers, for the eight selected series, were as follows, the given distances being in metres;

	Distances in Metres.				
No. of Series.	From Outlet End to Piezometer No. 1.	From Piezometer No. 1 to No. 2.	From Piezometer No. 2 to No. 3.	From Piezometer No. 3 to No. 4.	From Piezometer No. 4 to Inlet End.
I.	8.13	50.	50.	5.883	.167
II.	8.35	50.	50.	4.805	.30
III.	8.40	50.	50.	4.67	.29
XVI.	6.59	50.	50.	4.716	.28
XVII.	6.297	50.	50.	4.73	.43
XVIII.	6.78	50.	50.	4.293	.30
XIX.	6.912	50.	50.	4.17	.276
XX.	6.912	50.	50.	4.17	.276

The pipes all appear to have had a square and even connection with the inlet tank; hence o will in all cases be assumed to have a value of .82.

In the following table:

The first column gives our number of the experiment.

The second column; the mean velocity.

The third column ; the difference of height of piezometric columns Nos. 1 and 2.

The fourth column ; the difference between piezometers Nos. 2 and 3.

The fifth column ; the difference between piezometers Nos. 1 and 3, which we assume to be the measure of the frictional loss of head.

The sixth column ; the difference between piczometers Nos. 3 and 4.

The seventh column ; the difference between piezometers Nos. 4 and 5.

The eighth column ; the difference between piezometers Nos. 3 and 5, being the total loss of head between No. 3 and the inlet end of the pipe.

The ninth column ; the frictional loss of head between No. 3 and the inlet, obtained by dividing the loss of head indicated in the fifth column, by the distance between piezometers Nos. 1 and 3—in all cases 100 metres—and multiplying the dividend by the distance from piezometer No. 3 to the inlet end of the particular pipe.

The tenth column ; the difference between the total loss of head between piezometer No. 3 and the inlet, and the frictional loss as given in the ninth column ; this difference should represent the *true* value of h'.

The eleventh column ; the theoretical value of h', obtained by the expression $h' = \dfrac{v^2}{2\,g\,d^3}$.

The twelfth column ; the difference, or lack of agreement, between the true and the theoretical values of h'.

All these measures are given in metres.

TABLE LXXXIX.

Values of h' deduced from the Darcy Experiments.

$2g = 19.62$ $o = .82$ All values given in metres.

Series I.—New Wrought-Iron. $D = .0122$ $l = 114.18$

Maximum oscillation .005 at piezometer No. 5.

1	2	3	4	5	6	7	8	9	10	11	12
									h'		
No.	c	1—2	2—3	1—3	3—4	4—5	3—5	h_r	True.	$\frac{v^2}{2g\,o^3}$	Error.
65	.034	.043	.042	.085	.004	.007	.011	.005	.006	0	+.006
66	.072	.092	.092	.184	.009	.004	.013	.011	.002	0	+.002
67	.117	.154	.150	.304	.018	.006	.024	.018	.006	.001	+.005
68	.147	.276	.257	.533	.027	.005	.032	.032	.000	.002	-.002
69	.169	.378	.376	.754	.038	.006	.044	.046	-.002	.002	-.004
70	.230	.875	.784	1.659	.070	.006	.076	.100	-.024	.004	-.028
71	.287	1.365	1.215	2.580	.120	-.006	.114	.156	-.042	.006	-.048
72	.343	1.793	1.679	3.472	.154	.009	.163	.210	-.047	.009	-.056
73	.392	2.253	2.146	4.399	.200	.010	.210	.266	-.056	.012	-.068
74	.478	3.207	3.057	6.264	.289	.010	.299	.379	-.080	.017	-.097
75	.573	4.328	4.226	8.554							
76	.846	8.900	8.962	17.862	.832	.024	.856	1.081	-.225	.054	-.279
77	1.193	17.146	17.280	34.126	1.591	.102	1.693	2.083	-.390	.108	-.498
		40.810	40.266	81.076							

Series II.—New Wrought-Iron Pipe. $D = .0266$ $l = 113.455$

Maximum oscillation .04 at piezometer No. 5.

No.	c	1—2	2—3	1—3	3—4	4—5	3—5	h_r	True.	$\frac{v^2}{2g\,o^3}$	Error.
78	.058	.018	.015	.033	0	.001	.001	.002	-.001	0	-.001
79	.131	.079	.073	.152	.0055	.0015	.007	.008	-.001	.001	-.002
80	.248	.2515	.2355	.487	.020	.008	.028	.025	.003	.005	-.002
81	.368	.520	.495	1.015	.046	.013	.059	.052	.007	.010	-.003
82	.522	.982	.955	1.937	.087	.027	.114	.099	.015	.021	-.006
83	.667	1.583	1.543	3.126	.140	.049	.189	.160	.029	.034	-.005
84	.796	2.211	2.137	4.348	.200	.067	.267	.222	.045	.048	-.003
85	.961	3.184	3.132	6.316	.299	.093	.392	.322	.070	.070	0
86	1.235	5.049	4.973	10.022	.451	.216	.667	.512	.155	.116	+.039
87	1.281	5.326	5.245	10.571	.419	.231	.650	.540	.110	.124	-.014
88	1.682	8.874	8.952	17.826	.768	.311	1.079	.910	.169	.214	-.045
89	1.998	12.790	12.811	25.601	1.155	.403	1.558	1.307	.251	.303	-.052
90	2.184	15.230	15.722	30.952	1.506	.534	2.040	1.580	.460	.362	+.098
		56.0975	56.2885	112.386							

TABLE LXXXIX.—continued.

Series III.—New Wrought-Iron Pipe.   D = .0395   l = 113.36
Maximum oscillations .020 at piezometers Nos. 3, 4 and 5.

1	2	3	4	5	6	7	8	9	10	11	12
									K'		
No.	v	1—2	2—3	1—3	3—4	4—5	3—5	k,	True.	$\frac{l^2}{2gv^2}$	Error.
98	.063	.012	.010	.022	0	.001	.001	.001	0	0	0
99	.111	.044	.034	.078	.003	.002	.005	.004	.001	.001	0
100	.185	.093	.089	.182	.009	.002	.011	.009	.002	.003	−.001
101	.262	.173	.163	.336	.014	.010	.024	.017	.007	.005	+.002
102	.382	.331	.319	.650	.029	.016	.045	.032	.013	.011	+.002
103	.559	.654	.632	1.286	.061	.032	.093	.064	.029	.024	+.005
104	.788	1.217	1.172	2.389	.117	.063	.180	.118	.062	.047	+.015
105	.915	1.59	1.533	3.123	.154	.083	.237	.156	.082	.063	+.019
106	1.095	2.213	2.135	4.348	.213	.120	.333	.216	.117	.091	+.026
107	1.920	6.277	6.038	12.315	.601	.384	.985	.611	.374	.279	+.095
108	2.305	8.980	8.573	17.553	.901	.512	1.413	.871	.543	.403	+.139
109	2.597	11.437	10.981	22.408	1.147	.657	1.804	1.111	.693	.511	+.182
		33.011	31.679	64.690							

Series XVI.—New Cast-Iron.   D = .0819   l = 111.586
Maximum oscillations .040 at piezometers Nos. 3, 4 and 5.

No.	v	1—2	2—3	1—3	3—4	4—5	3—5	k,	True.	$\frac{l^2}{2gv^2}$	Error.
206	.088	.010	.010	.020	.001	.001	.002	.001	.001	.001	0
207	.171	.045	.038	.083	.005	.005	.010	.004	.006	.002	+.004
208	.358	.117	.115	.232	.007	.014	.021	.012	.009	.010	−.001
209	.561	.273	.258	.531	.021	.035	.056	.027	.029	.024	+.005
210	.791	.52	.50	1.02	.04	.07	.11	.051	.059	.047	+.012
211	1.185	1.145	1.11	2.255	.105	.130	.235	.113	.122	.106	+.016
212	1.418	1.628	1.58	3.208	.115	.220	.335	.160	.175	.152	+.023
213	1.571	2.0515	1.99	4.0415	.145	.28	.425	.202	.223	.187	+.036
214	2.453	4.721	4.826	9.547	.288	.661	.949	.177	.472	.456	+.016
215	2.487	5.034	4.870	9.904	.446	.636	1.082	.495	.587	.469	+.118
216	2.720	6.106	5.872	11.978	.418	.848	1.266	.598	.668	.561	+.107
217	3.238	8.603	8.304	16.907	.665	1.076	1.741	.840	.901	.795	+.106
218	3.265	8.646	8.426	17.072	.641	1.106	1.747	.853	.894	.808	+.086
		38.7995	37.899	76.6985							

TABLE LXXXIX.—*continued.*

Series XVII.—New Cast-Iron. $D = .137$ $l' = 111.477$
Maximum oscillations 0.30 at piezometers Nos. 3 and 5.

1	2	3	4	5	6	7	8	9	10	11	12
										N	
No.	v	1—2	2—3	1—3	3—4	1—5	3—5	h_f	True	$\frac{c^2}{2\,g\,a^2}$	Error.
219	.149	.012	.012	.024	.001	.003	.004	.001	.003	.002	+ .001
220	.298	.049	.038	.087	.001	.009	.010	.004	.006	.007	− .001
221	.488	.112	.097	.209	.007	.020	.027	.011	.016	.018	− .002
222	.763	.251	.224	.475	.013	.053	.066	.025	.041	.044	− .003
223	1.279	.670	.590	1.260	.047	.133	.180	.065	.115	.124	− .009
224	1.714	1.18	1.045	2.225	.080	.240	.320	.115	.205	.223	− .018
225	2.098	1.758	1.560	3.318	.130	.345	.475	.172	.303	.334	− .031
226	2.281	2.065	1.840	3.905	.155	.405	.560	.202	.358	.394	− .046
227	3.640	5.162	4.690	9.852	.353	1.014	1.367	.510	.857	1.004	− .147
228	4.693	8.787	7.969	16.756	.486	1.811	2.297	.868	1.429	1.669	− .240
		20.046	18.065	38.111							

Series XVIII.—New Cast-Iron. $D = .188$ $l' = 111.373$
Maximum oscillation .060 at piezometer No. 4.

229	.205	.014	.013	.027	.003	.005	.010	.001	.009	.003	+ .006
230	.497	.085	.090	.175	.010	.01	.020	.008	.012	.019	− .007
231	.758	.188	.180	.368	.017	.048	.065	.017	.048	.043	+ .008
232	1.128	.420	.385	.805	.020	.115	.135	.037	.098	.096	+ .002
233	1.488	.700	.640	1.340	.07	.155	.225	.062	.163	.168	− .005
234	1.933	1.16	1.09	2.25	.10	.28	.38	.103	.277	.283	− .006
235	2.506	1.955	1.855	3.81	.19	.46	.65	.175	.475	.476	− .001
236	4.323	5.704	5.276	10.980	.675	1.338	2.013	.304	1.509	1.417	+ .092
237	4.928	7.523	7.068	14.591	.730	1.774	2.504	.670	1.834	1.841	− .007
		17.749	16.597	34.346							

TABLE LXXXIX.—*continued.*

Series XIX.—Old Cast-Iron. $D = .2432$ $l = 111.358$
Maximum oscillations .060 at piezometers Nos. 3, 4 and 5.

1	2	3	4	5	6	7	8	9	10	11	12
									h'		
No.	v	1—2	2—3	1—3	3—4	4—5	3—5	h_f	True.	$\frac{v^2}{2g\,c^2}$	Error.
238	.307	.019	.015	.094	.005	.010	.015	.004	.011	.007	+.004
239	.452	.104	.098	.202	.009	.018	.027	.009	.018	.015	+.003
240	.707	.238	.235	.473	.020	.040	.060	.021	.039	.038	+.001
241	1.106	.585	.565	1.150	.060	.085	.145	.031	.094	.093	+.001
242	1.547	1.160	1.130	2.29	.120	.170	.290	.102	.188	.181	+.007
243	1.833	1.620	1.580	3.20	.170	.230	.400	.142	.258	.255	+.003
244	2.073	2.085	2.020	4.105	.220	.300	.520	.182	.338	.325	+.013
245	3.833	7.154	6.827	13.981	.781	1.223	2.004	.622	1.382	1.114	+.268
		12.995	12.500	25.495							

Series XX.—No. XIX cleaned. $D = .2447$ $l = 111.358$
Maximum oscillations .12 at piezometers Nos. 3, 4 and 5.

1	2	3	4	5	6	7	8	9	10	11	12
246	.278	.027	.025	.052	.005	.005	.01	.002	.008	.006	+.002
247	.357	.085	.080	.165	.010	.020	.030	.007	.023	.022	+.001
248	.949	.253	.245	.498	.03	.065	.095	.022	.073	.068	+.005
249	1.420	.590	.565	1.155	.055	.145	.20	.051	.149	.153	-.004
250	1.904	1.035	1.00	2.035	.12	.255	.375	.090	.285	.275	+.010
251	2.206	1.385	1.35	2.735	.16	.32	.48	.122	.358	.369	-.011
252	2.572	1.890	1.84	3.73	.22	.46	.68	.166	.514	.501	+.013
253	4.497	5.838	5.505	11.343	.701	1.428	2.129	.504	1.625	1.533	+.092
		11.103	10.610	21.713							

An examination of the foregoing table shows:

The heights indicated by piezometers Nos. 1, 2 and 3 are often discordant; supposing α and Δ to have different values in the two sections of the pipe (1-2 and 2-3) such variations would be constant for all the experiments of the particular series, and affect the readings of the three piezometric columns for each experiment the same way. Instead of this being the case, sometimes the indicated loss of head between piezometers Nos. 1 and 2 is greater, and sometimes less, than the indicated loss of head between Nos. 2 and 3.

The differences between the piezometric columns 3 and 4, and 4 and 5, also appear to be very contradictory.

Of the selected experiments. 44 indicate that $\frac{v^2}{2\,g\,o^2}$ is too small, while 37 indicate that it is too great ; of the eight selected series, four generally indicate that $\frac{v^2}{2\,g\,o^2}$ is too small, and the other four that it is too great. Series I. is palpably in error, as the frictional loss of head between piezometer No. 3 and the inlet, is greater than the total indicated loss of head. It can be safely assumed that $\frac{v^2}{2\,g\,o^2}$ cannot be too large a correction ; therefore, examining the + errors of the table alone, we see that they rarely form a considerable fraction of the " true " value of h'.[*]

We can fairly deduce from the preceding observations, that the Darcy experiments indicate that $\frac{v^2}{2\,g\,o^2}$ quite closely represents the true value of h'. Also, that our primary assumption, that his piezometers indicated the true head, is correct ; for were this not the case, and the piezometric columns Nos. 1 to 4 inclusive more or less lowered by the " sucking " action of the flowing stream—as M. Darcy surmises—their true heights would have been higher than the observed heights ; No. 5, being attached to the large feeding tank or cylinder, would not be nearly as much depressed ; hence the difference between the observed heights of Nos. 4 and 5 would be less than that given in our table ; this would result in indicating that $\frac{v^2}{2\,g\,o^2}$ was in nearly all cases too large a correction. Such a conclusion, as before stated, we cannot admit.

Conclusion.

In the preceding data, the only proofs entitled to much weight, which show that $\frac{v^2}{2\,g\,o^2}$ gives too small values for h', are Series III. and IV. of the author's New Almaden experiments. These experiments were made with such care, that we feel disposed from them alone, to draw the final conclusion that, so far as loss of head absorbed in the impartation of velocity is concerned, $\frac{v^2}{2\,g}$ does not sufficiently represent this loss. How much the expression is in error, we are not prepared to say, but probably not very much.

The head absorbed in the impartation of velocity to the flowing stream is in part, or in whole, represented by the *vis-viva* of the escaping jet from the outlet end of the

---

[*] Not considering low velocities, where the probable errors of observation would fully account for any discrepancy given.

pipe. This *vis-viva* represents the force of each fillet of the jet, or, considering that the maximum velocity is in the axis of a circular pipe, and the minimum velocity by its walls, dividing the section into concentric rings, the mean of the sum of the squares of the velocities of the several rings will much more nearly indicate the true force, and will doubtless be slightly different from the square of the mean velocity.

It is, however, not worth while to discuss this question from a theoretical point of view, as it can only be accurately and satisfactorily determined by experiments made with the proper care. A pipe, consisting of joints of drawn brass tubing, where o and Δ are almost exactly constant, should be employed. The inlet tank should be of relatively large dimensions; a funnel-shaped mouth-piece should be attached, so that o will be nearly l.. The surface height in the inlet tank should be carefully compared with the height of the centre of the outlet end, when the discharge is free, and with the surface of the water in the outlet tank, when the discharge is submerged; a submerged discharge will be required for low velocities. By varying the length of the pipe, the exact value of h' can be readily deduced. To eliminate possible variations in o and Δ, for the short lengths the experiments should be repeated with different joints. Piezometers could also be attached to such a pipe, and the causes of their generally erroneous indications be ascertained. Several series, with different diameters, should be made to determine whether or not variation in D affects h'. Q must be absolutely determined by measurement of q in a vessel of ample size.

PIEZOMETERS.

By attaching vertical tubes to an experimental conduit pipe, at points along its course after the regimen of flow has been established, theoretically the frictional loss of head, h, between two such attachments can be determined, by measuring the difference in elevation of the heights of the columns of water in the two tubes. These heights can be ascertained directly, by the use of vertical glass tubes; and indirectly, either by columns of mercury in glass tubes, or by Bourdon pressure gauges.

This method, at first sight, appears to be greatly preferable to the direct measurement of the total head H, where considerable corrections are sometimes necessary for the primary losses of head, due to contraction at the entrance, and to the impartation of velocity to the flowing stream. Induced by these seeming advantages of piezometric measurement, in late years nearly all experimenters have determined h by piezometric columns; the author being almost the only modern experimenter, who has preferred to deduce h from the observed value of H.

Dubuat assumed that the pressure on the interior walls of a pipe was diminished by the head generating the velocity of flow. Navier disputed this proposition, but, in general, writers upon Hydraulics during the last century have followed the theory of Dubuat. If this supposition be true, either in whole or in part, the indicated piezometric heads, after the regimen of flow has been fully established, should indicate a less head than that due to the hydraulic head; in such case, however, the piezometric columns should indicate the hydraulic inclination for a pipe of uniform section, as the lowering effect upon each column would be identical.

Darcy thought that the stream flowing through a pipe would have a "sucking" action as it passed by the mouth of a piezometric tube, and hence the height of the column in the tube would be more or less lowered.[*] In the preceding section of this chapter, we have shown that such a deduction is hardly warranted by M. Darcy's own experiments.

Mr. Hiram F. Mills has lately very fully investigated the question of piezometric measurements applied to an open conduit, and has given a description of his experiments in the Proceedings of the American Academy of Arts and Sciences, Vol. VI., New Series, Boston, 1879.

His experimental conduit was a straight trough of wood, of uniform section, 30 feet long, 1 foot deep, and $\frac{8}{10}$ths of a foot wide. This trough could be inclined at pleasure, so that velocities up to nearly 9 feet per second were obtained. At distances of 2.5 feet apart, 9 openings, of various sizes and shapes, and in wood, brass and iron, were pierced in each side of the conduit; each of these 18 openings was connected by a pipe with a small open tin reservoir or box, having a horizontal section of .9 × .5. The height of the surface of the water in the centre of the conduit was observed, and simultaneous observations made of the height of the surface in the two reservoirs opposite the central station. In all, some 6000 observations of surface heights were made.

These experiments gave the following results:

First.—When the mouth of the opening was flush with the side of the conduit, and the axis of the opening perpendicular (normal) to the side, the surface of the water in the piezometric reservoir had very nearly the same elevation as the surface of the water in the centre of the conduit. The maximum deviation of any single experiment is not given; the maximum mean deviation of any one series appears to have been +.0176, when the velocity, v, in the conduit was 7.8. With high velocities, in general the water in the reservoir was slightly higher than in the conduit.

Second.—When the mouth of the opening was flush with the side of the conduit, and the axis of the opening horizontally inclined with the line of the conduit, the water in the reservoir was somewhat higher than in the conduit, when the inclination was down stream; on the other hand, when the inclination was up-stream, the water in the reservoir was somewhat lower than in the conduit.

Third.—When a pipe, projecting into the conduit, was attached to one of these inclined orifices, there was a much larger difference of elevation between the water in

---

[*] Recherches expérimentales, relatives au mouvement de l'eau, dans les tuyaux. Paris, 1807, p. 217.

the reservoir and in the conduit, than when the mouth of the orifice was flush with the side of the conduit.

Fourth.—When a projecting pipe with a square end was attached in line with an orifice, whose axis was perpendicular to the side, the water in the reservoir was considerably lower than in the conduit.

With a projecting pipe, the difference in elevation between the conduit and a reservoir, amounted to .44, with $c = 8.0$; this appears to have been the greatest difference observed.[*]

The following general conclusion can be drawn from Mr. Mills' experiments; When the axis of an orifice is normal to the side of a straight conduit, and the mouth of the orifice absolutely flush with the side of the conduit, the piezometric column will truly indicate the surface elevation of the water in the conduit; but, especially with high velocities, a very slight deviation from the given conditions will cause notable errors in the piezometric columns. If the mouth of an orifice is placed so that the current of the stream impinges upon the section of the mouth, the piezometric column will be higher than the surface in the conduit; on the other hand, if the line of the current forms an acute angle with the axis of the orifice, the piezometric column will be lower than the surface in the conduit.

There is every reason to believe that the foregoing propositions will equally apply to piezometers attached to a pipe under pressure.

The sucking or lowering effect of a current passing by a projecting pipe, has long been utilized by the steamboatmen of the Mississippi River, who, strange to relate, discharge the water from their coal barges by boring holes in the bottom! This rather surprising feat is accomplished by inserting a spoon-shaped iron, which projects below the bottom, with the outer side of the spoon pointing up stream ; when the empty coal barges—large flat-bottomed boats—are being towed up stream, the sucking action of the current keeps the barge free from water. Naturally, when the boat is not in motion, the holes are plugged.

These experiments of Mr. Mills simply prove that the summit of the liquid column n a piezometric tube, properly attached to a conduit, is in the true line of hydraulic inclination ; for the axial surface line in his open conduit was the true "hydraulic-grade line."

---

[*] In this experiment air was drawn in through the piezometric pipe, so this difference of .44 probably does not fully represent the full "sucking" effect of the current. Mr. Mills states that in other similar experiments, the details of which are not given, the difference between the surface of the piezometric column and the hydraulic-grade line was greater than $\frac{v^2}{2g}$.

The following sketch illustrates the effect of attaching piezometers to a pipe under pressure.

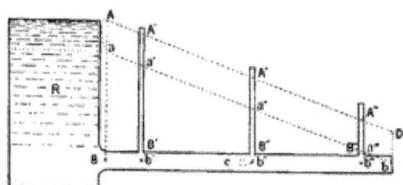

Let the horizontal pipe **B b**, of uniform section and uniform condition of interior surface, be attached to the large reservoir **R**, by a trumpet-shaped mouth-piece, so that the co-efficient of contraction will be nearly unity; we can hence neglect the primary loss of head due to contraction at the entrance. To the pipe attach the vertical tubes, **A′ B′**, **A″ B″**, and **A‴ B‴**, open at the top and connected with the pipe in the manner prescribed by Mr. Mills. Let the reservoir be kept constantly full with the liquid, and assume **A a** as the head required to impart velocity to the stream flowing through the pipe. Assume that uniform motion is fully acquired at the point **B′**. The liquid in the three tubes will then rise to the points a′, a″, and a‴, all in the straight line **a b**.* The line a a′a″a‴b will then represent the hydraulic pressure line, or the hydraulic-grade line, and the pressure against the interior walls of the pipe at the point **B′** will be represented by the line a′ b′, at **B″** by the line a″ b″, and at the end of the pipe by 0. This will be absolutely true in regard to pressure, if we assume that the threads of liquid move through the pipe in lines parallel to its axis.

Suppose we now attach to one of the tubes, **A″ B″**, a curved tube, shown in the sketch by dotted lines, having its open mouth at c normal to the axis of the pipe; conceive that this extension of the tube has no thickness, so that it will not disturb the normal flow through the pipe. The liquid in the vertical tube will then rise to the point **A″**, in the straight line **A D**, parallel to the line **a b**. Strictly speaking, if c is in the axis of the pipe, the liquid will rise higher than **A″**, as the velocity at c will be greater than the velocity due to the head **A a**, which represents the mean velocity head for the entire section of the pipe.

Now if we make $H =$ **A B** $=$ total head; $h' =$ **A a** $=$ velocity head; $h =$ a **B** $= H - h'$ $=$ "effective" or "frictional" head, which is absorbed in overcoming all sorts of resistance as the liquid flows through the pipe; $l =$ **B b** $=$ total length of pipe; $l' =$ length from **B** to any piezometric tube; $h_p =$ piezometric head in any tube. We have shown before that h' is pretty closely represented by $\frac{v^2}{2g}$, v being the mean velocity in the pipe.

---

* The point b in a circular pipe will be in the centre of pressure, which will be slightly different from the centre of the section, but this refinement can be neglected.

L L

Therefore, $h_p = h - \dfrac{h\,l'}{l}$; and approximately, $h_p = \left(H - \dfrac{v^2}{2\,g}\right) - \left(H - \dfrac{v^2}{2\,g}\right)\dfrac{l'}{l} = \left(H - \dfrac{v^2}{2\,g}\right)\left(1 - \dfrac{l'}{l}\right)$.

The Darcy experiments given in the last section of this chapter, have been reduced in accordance with the foregoing expression, allowance being made for effect of primary contraction.

Jacobson has made some interesting experiments with piezometers attached to a trumpet-shaped mouth piece, which are described by Dr. Lampe in Der Civilingenieur, Vol. XIX.

Mr. Mills has shown, that when proper care has been observed, piezometric attachments will accurately indicate the surface height in an open conduit, and one would naturally suppose that the same method would give equally satisfactory results, when applied to pipes. Entirely the reverse of this is true, for not a single series of pipe experiments thus far published, where h was determined by piezometers, gives thoroughly satisfactory results.

Of all such experiments, we regard the Lampe series as the most reliable; this accuracy resulted from the very great length of Dr. Lampe's experimental pipe, so that the piezometric errors were not relatively large enough to vitiate the final deductions; but, if in that pipe such considerable lengths as 2000 or even 3000 feet had been experimented upon, the deduced heads would often have been entirely untrustworthy.*

M. Darcy appears to have used great care in his pipe experiments, and it is quite unlikely that there were any errors of consequence in his given values of l, D, a, Q and c, as he used very proper methods for the determination of these elements. It will be shown in the next section of this chapter, that his work, when carefully analyzed, is most unsatisfactory and contradictory, if one considers the time and cost involved in the execution of his labors. It seems probable, therefore, that his errors chiefly resulted from erroneous values of h and c.

Messrs. Fteley and Stearns have shown the highest skill and accuracy as hydraulic experimenters; they were familiar with Mr. Mills' paper, and fully realized the necessity of observing the proper precautions in the connection of piezometric tubes to the experimental conduit. In their experiments with the Rosemary 4-foot pipe, Nos. 282-285, one of the four differs considerably from the other three, and this divergence is so considerable that we can only attribute it to misleading piezometric heights.†

It is not worth while discussing other defective data, where errors may be attributed to the use of a rough instrument like the Bourdon gauge, or where the bad results may have been due to errors in other elements than h.

Comparing the results obtained by Darcy, et cet., with those obtained by Couplet, Bossut and ourselves, we see for the latter vastly smoother curves, and curves which

---

* A description of Dr. Lampe's methods of experimentation, with a tabulated statement of his piezometric errors, has already been given. Pipe experiments, Nos. 275-278.

† It may be remarked that the experiments with the Rosemary pipe were made by the assistants of those gentlemen, neither of them being at the time able to personally superintend the execution of the experiments.

rarely contradict each other. Taking for instance, the author's New Almaden series, where Q was ascertained by absolute measurement of q, the curves are in all cases almost perfectly symmetrical, and the general harmony of the curves for the different series proves the substantial accuracy of the work.

With proper care, with h deduced from H, we are satisfied for velocities above 2 feet and diameters not less than 2 inches, that experiments can readily be made, which will show no comparative errors in n of over $\frac{1}{2}$ of one per cent. as a maximum. We are hence of the opinion that the discrepancies presented by the experiments of Darcy, Lampe and Stearns, must be attributed in a great measure to false piezometric heights, and not to unknown causes proceeding from slight changes in the water, such as we have conjectured affect the discharge through very small orifices, or for orifices and weirs with very low heads.

Piezometric errors can be attributed :

First.—To imperfect connections of the piezometric tubes with the conduit ; the effect of error from this cause should be constant, that is, always + or always — .

Second.—Accumulation of air in the tubes ; this seems difficult to avoid, although much care has been observed by Darcy and others to prevent the presence of air.

Third.—Obstruction of the small tubes, by sediment of various kinds ; a leaf clinging to the mouth of one of these tubes would considerably affect the height of the column.

Fourth.—When small glass tubes are used to directly show the heights of the columns, the uncertain or irregular amount of capillarity may be a source of error ; this, however, will not be the case when a mercurial column of proper size is employed.

In addition to these chances of error, there are probably others, which can only be ascertained by more careful investigation than has thus far been devoted to the subject.

In all ordinary cases, where h' is a small fraction of H, we feel assured that experiments with pipes, when h is deduced from H, are incomparably more reliable than when h is obtained by piezometers. We have shown that the expression $h = H - \frac{v^2}{2\,g\,o^2}$ cannot be very greatly in error, and any possible error from this source will in general have no notable effect upon the deduced value of n.[*]

CONCLUSIONS.

Variation in n, caused by Changes in Δ, D and v.

Taking the simple expression, $v = n\,(r\,s)^{\frac{1}{2}} = n\left(\dfrac{D}{4}\,s\right)^{\frac{1}{2}}$, we will now proceed to see what conclusions can be drawn from the foregoing experimental data, so far as the

---

[*] Error by the use of $h = H - \dfrac{v^2}{2\,g\,o^2}$ will be confined to a fraction—probably quite a small one—of $\dfrac{v^2}{2\,g}$, while with piezometers errors appear to be in some cases even greater than $\dfrac{v^2}{2\,g}$, when v is not large.

relative effects of D, v and Δ upon the co-efficient n are concerned.   In this discussion it will be more convenient for pipes to use D than r.   For open conduits r will be used, the ratio between these quantities being constant for circular pipes, i.e., 4 to 1.

Of the 363 selected experiments, we consider the New Almaden series—Nos. 287-339—much the most reliable.   The pipes employed were of small dimensions to be sure, but the mean areas were carefully obtained, and Q was ascertained by absolute measurement of q.   The only considerable danger of error in these series is in h; we have shown that h as deduced is probably slightly too great, and hence n probably too low, but errors from this cause will not largely affect the deduced values of n, except for Series IV., when h' was a large fraction of H.   Such possible errors for the other 9 series would hardly be appreciable for the smaller velocities, and as the lengths for 7 of the series are nearly identical, any slight errors in h will affect the several curves in a nearly similar manner.

From these New Almaden results, we will endeavour to deduce some general principles, which will enable us to more intelligently discuss the remaining data.

On Plate XIII. are plotted the New Almaden series, Nos. V. to X. inclusive, with v and n as co-ordinates; the values of n for the pipe used for Series III. are shown by a symmetrical curve which has been drawn from the data given on Plate XII., where the experiments constituting Series I., II. and III., have been plotted; the experimental points for these three series are very nearly in the same curve.   The seven curves on Plate XIII. represent the experiments made with seven distinct pipes; these pipes can be described as follows:

Δ *nearly Constant and very Low.*

Series III.   New and smooth wrought-iron, covered with an asphaltum varnish, and very smooth; $D = .087$; $l = 60.3$; $o = .98$.

Series VII.   New glass; almost perfectly clean; $D = .076$; $l = 63.9$; $o = .97$; composed of 12 joints.

Series VIII.   New glass; a few spots of dirt adhering on inner walls, but not enough to appreciably retard flow; $D = .062$; $l = 34.9$; $o = .82$; composed of 6 joints.

Series IX.   New glass; almost perfectly clean; $D = .042$; $l = 11.1$; $o = .82$; composed of 2 joints.   The area of this pipe less accurately obtained than for the others.

Series VI.   New wrought-iron, free from rust, with no varnish, hence Δ is very slightly higher than for the four preceding series; $D = .052$; $l = 60.1$; $o = .825$.

Higher Values of Δ.

Series V.   Composed of 4 joints of old pipe, of which the interior surface was covered with a thin hard scale, with occasional small nodules; and 1 joint of new varnished pipe. Total $l = 60.2$; of which 52.7 was rough, and 7.5 smooth. $D = .085$; $o = .98$.

Series X. Red-wood, bored by usual pipe-auger; surface as usual for such auger holes in moderately soft wood; $D = .105$; $l = 62.0$; $o = .80$.

By reference to Plate XIII. it will be observed that the experimental curves are nearly perfectly symmetrical; the only exceptions being Experiment No. 311, where n appears to be a very little low, and No. 338 where n appears to be fully 1 per cent. too high.* This symmetry proves that for each series H and Q were correctly determined.

Examining the 7 experimental curves, we see the following results due to changes in v, D and Δ:

Effect of Variation in v.—In all cases n increases with v; very rapidly for velocities below 2 feet, and more and more slowly as v further increases; the trend of all the curves as v diminishes is manifestly towards the axis of the co-ordinates, so we can roughly assume that the curves have a common origin at this axis. *Hence, with an infinitesimal velocity, n will be infinitesimal, no matter what values D and Δ may have.*

With Δ very nearly constant and D variable—Series IX., VI., VIII., and III. or VII.—as D increases the curves become somewhat more inclined for velocities above 2 feet. *Hence, it is probable that the effect of v upon n, increases with D.*

With D very nearly constant, and Δ variable—Series III. and V.—as Δ increases the effect of v upon n diminishes. This is shown still more forcibly by the curve for Series X., with D about one-fourth larger than for Series III.; for this curve, as v increases above 2.5 it has comparatively but little effect upon n. *Hence, with not very large values of Δ, and such small values of D, an increase in v above 2.5 will have but little effect upon n. Also the lower the value of Δ, the more will an increase in v increase n.*

Effect of Variation in D.—With v and Δ constant—Series IX., VI., VIII., and III. or VII.—n notably increases with D; with $v = 4$, n for $D = .042$ has a value of 89, and for $D = .076$ a value of 98; with the same velocity and $D = .062$, n has a value of 95.5. *Hence we can generally assume that n, v and Δ being constant, always increases with D.*

Effect of Variation in Δ.—With v and D constant—Series V. and III.—the smoother pipe shows much the larger value for n. For Series X., with a bored wooden pipe, $D = .105$; for a similar pipe with $D = .087$, or the same as for Series III., the values of n would be lower, and would be not over 62 for $v = 2$, and 64 for $v = 5$. With the same values of v, n for Series III. has the respective values of 90 and 100. *Hence*

---

* The value of n for this experiment should probably be 67.0 instead of 67.5 as plotted. *Vide* note in Table XCVII.

an increase in Δ, *with v and D constant, greatly diminishes n, and in these experiments for v above 1,* Δ *is undoubtedly much the most important of the three variables.*

The curves for Series III. and VII. almost absolutely coincide; the respective diameters were .087 and .076; this agreement of the two curves shows that the lower value of Δ for the glass pipe compensated for its smaller diameter.

We will now critically examine our other experimental data, taking them in the order in which they have been given, for the purpose of determining which of them shall be selected as authority.

Couplet. Versailles Pipe.—The experiments of Couplet were the first ever made with pipes of considerable length; they seem to us much the most reliable experiments extant, with very low velocities. *D* was constant for the 6 experiments selected—Nos. 1-6—and therefore any probable error in *D* will not notably affect our curve—certainly not its form. *H* seems to have been measured with sufficient care; *t* was quite short, being only $\frac{19}{10}$ seconds for No. 6; this experiment, however, was repeated with similar results; as the times were taken to $\frac{1}{2}$ seconds, it is probable that any error in *t* will not affect *n* more than 3 per cent.; now, as the values of *n* increase from 66 to 82, with a limited range of *r*, such an error will not vitiate the general accuracy of the curve.

Plotting the curve, with *v* and *n* as co-ordinates, we see it is symmetrical, *n* rapidly increasing with *v*; these 6 experiments will be accepted as authority.

Bossut.—The experiments of Bossut will be found plotted on Plate XII. By reference to that sheet, it will be seen that the curves for Series I. and II., with tin pipes having the respective diameters of .118 and .178, are fairly symmetrical, that for the larger diameter being somewhat the higher.* The curve for Series IV., with a lead pipe having a diameter of .089, is on the other hand a little higher than that for the tin pipe of .118 diameter. The point for Series III.—tin pipe, $D = .118$—is considerably lower than the curve for Series I. with the same diameter. Comparing the curves for Series I. and II., with that for the New Almaden Series III.—$D = .087$, it will be seen that the three curves fairly agree in form, but that the curves for the Bossut Series I. and II. are a little too high, making allowance for differences in diameters; the Bossut Series III. on the other hand is a little too low.

Dubuat.—On the same sheet will be found the curve of Dubuat's experiments with a tin pipe having a diameter of .089. This curve is almost identical with that of Bossut, Series I.

We can fairly deduce from these experiments of Bossut and Dubuat, that our first proposition in regard to the effect of *r* is correct. As to the effect of variation in *D*, the experiments are conflicting. As to the effect of Δ, that quantity was nearly constant.

---

* When we speak of an experimental curve or point being "higher" or "lower," it means that the ordinate representing the value of *n* is greater or less.

Scotch Authorities.—For Experiments Nos. 60-64, the great danger of error consists doubtless in the values of Q, and hence v. For two experiments with the Colinton pipe—Nos. 61 and 62—Mr. Leslie states that Q was the mean of quite a number of observations, the variation between maximum and minimum being in both cases about 6 per cent.. How these measurements were made does not appear.

Some weight will be attached to the Colinton series; the experiments, however, are of not much value, as nothing is known in regard to Δ; they indicate that for this pipe, increasing v from 5.3 to 14.5, increased n from 96 to 115.

The Loch Katrine experiment, No. 64, was with $D = 4$, and Δ evidently quite low. Probably H, a and l are given with reasonable accuracy; how Q was determined Mr. Gale does not state. This experiment is a very important one, as there have thus far been published the results of experiments with only one other iron pipe of the same size. These experiments—Nos. 282-285 of Mr. Stearns—with about the same velocity give n a value of 140, as against 112 given for the Loch Katrine pipe. Evidently, one or the other of these results is much in error; their respective claims or chances of accuracy will be weighed hereafter.

Darcy.—Examining the Darcy Series I., II. and III., with new wrought-iron; for Series I. we see a very ragged experimental curve, n for No. 67 apparently being as much as 20 per cent. in error, if Nos. 65, 66 and 70-77 are assumed as correct. The curves for Series II. and III. are much less contradictory. They indicate a rapid increase in n with D; n, for velocities of 3 feet, being about 66 for $D = .040$, 85 for $D = .087$, and 94 for $D = .130$. They generally indicate an increase of n with v, although contradictory in this respect.

Series IV., V. and VI. with new lead pipes, having the respective diameters of .046, .089 and .134, give curves practically identical, n being near 81 with $v = 1$, and 100 with $v = 5$. This would seem to show that with lead pipes n does not increase with D, but largely with v.

Series VII., VIII., IX. and X. were with sheet iron pipes lined with bitumen, having the respective diameters of .088, .271, .643 and .935. These 4 curves are fairly symmetrical; that for No. VIII. is considerably higher than the one for No. VII.; the curves for Nos. IX. and X. are nearly identical, and a little higher than the one for No. VIII.; for the curves IX. and X., n is 108 with $v = 2$, and 131 with $v = 11$. They indicate that n does not increase with values of D above .6.

Series XI. was with a new glass pipe, having $D = .163$. This curve is perfectly symmetrical, and from $v = 1.5$ to $v = 5$, closely agrees with the curve for the New Almaden glass pipe, with $D = .076$.

Series XII. and XIII. were with a cast-iron pipe, with D about .12. The pipe was an old one with deposits on its inner surface; in this condition n is nearly constant at 59; the pipe was then cleaned, its mean area again determined, and n is 80 for $v = .37$, and 99 for $v = 3.7$. The flow through this pipe was hence increased 1.6 times, with the same inclination, by the removal of the deposits.

Series XIV. and XV. were with an old cast-iron pipe, with D about .26, and was first experimented upon in its bad condition, and then again after being cleaned. The results were similar to those for the preceding series; with the same inclination the flow was increased about one-third.

Series XVI., XVII., XVIII. and XXII. were with new cast-iron pipes, having the respective diameters of .27, .45, .62 and 1.64. For velocities above 4, n for Series XVI. has a pretty constant value of 100; for Series XVII. a constant value of 113; and for Series XVIII. a constant value of 108. The curve for Series XXII. is exceedingly rough, and the experiments for this series are palpably worthless. The results for the first 3 series are contradictory, so far as D is concerned.

Series XIX. and XX. were with an old cast-iron pipe with $D = .80$, afterwards cleaned; n for Series XIX. has a pretty constant value of 75, and for Series XX. a pretty constant value of 97.

Series XXI. was with an old cast-iron pipe, with $D = .97$, very well cleaned, and n, with v from 1.8 to 10.4, has a constant value of 104.

Summing up these results:

Effect of Variation in v.—They generally indicate that with smooth pipes n increases with v, very rapidly with v less than 1, and more and more slowly as v increases from 1 to 16. There are, however, frequent contradictions to this conclusion for velocities less than 1.

With old cast-iron pipes covered with deposits, n is nearly constant, with values of v from 1 to 12.5.

In none of the series, excepting No. XXII., which is manifestly defective, is there any noteworthy lowering of n for high velocities.

Effect of Variation in D.—In the majority of cases n increases with D. The most notable exceptions are with the lead pipes.

Effect of Variation in Δ.—They show, without exception, that the lower the value of Δ, the greater is n. Contrasting Series XIV. with a cast-iron pipe in bad order, and Series VIII. with a pipe covered with bitumen, both pipes having about the same diameter, for velocities from 2 to 4 feet, the value of n for the smooth pipe ranges from 105 to 111, while for the rough pipe n is about 68.

We hence see that the Darcy series can only be considered entirely conclusive in regard to the great effect of changes in Δ. In other respects they are so frequently contradictory, that they can be of no service to us in the determination of our final curves for n. In the greater number of cases they confirm the general accuracy of the propositions deduced from the New Almaden series. As heretofore stated, we regard the very uneven results shown by the Darcy series to be largely due to the false indications of piezometric columns.

We have devoted a good deal of space to M. Darcy's pipe experiments, because they are the most elaborate investigations of the kind which have ever been executed, and because they are generally considered to be the highest authority.

M. Darcy deserves very great credit for his demonstration that the condition of the wetted surface has such an important, and often controlling effect upon the flow. This disproof of Dubuat's fallacious proposition is the most brilliant and valuable discovery in the science of Hydraulics during the present century.

Iben.—The Bonn experiments, Nos. 271-274, with $D = 1$, and Δ presumably low, show an increase in n from 89 to 106, with v from 1.6 to 3.1. The rough method employed for obtaining h, renders these experiments more or less unreliable.

Lampe. Danzig Pipe.—The dangers of error in these experiments, Nos. 275-278, have already been discussed. These experiments will be accepted as authority.

Kirkwood.—We regard these experiments as of very doubtful authenticity, and they will not be used.

Rochester Main.—This experiment appears to be entitled to considerable weight, as q was absolutely measured. It is a pity that Mr. Tubbs has not more fully described the methods followed in obtaining his experimental data.

Stearns. Rosemary Pipe.—These experiments, Nos. 282-285, are of much importance. We consider that in them there is no danger of considerable experimental error, except in the piezometric determinations of h. As the results of these experiments are flatly contradicted by the Loch Katrine experiment, No. 64, we will hereafter discuss their relative chances of error, using as a guide the open conduit experiments of Darcy and Bazin, and Fteley and Stearns.

Smith.—No. 286 of the author, with the Cherokee pipe, can only be regarded as an approximation, as there is too much uncertainty as to the proper values of the experimental elements. The given value of n may be in error either way 10 per cent., the chances perhaps being that n is a little too high.

The North Bloomfield series, Nos. 340-354, are entitled to a good deal of weight. The chief uncertainty in them arises from the indirect method of obtaining Q by weir measurement. The chances of error in the individual experiments will be fully stated in Chapter X. These experiments show that n increases with D, and with v, in the same manner as shown by the New Almaden series. For them, Δ was appreciably higher than for the New Almaden smooth pipes, and probably higher than for the Danzig and Rosemary pipes; the joints were roughly made, and the rivet heads also slightly added to the roughness. The curves in these pipes also slightly retarded the flow.

M M

No. 355 for the Humbug pipe, with $D = 2.15$, is more unreliable than the North Bloomfield series, on account of uncertainty about Q. The chances are that n is placed a little too high. Δ for this experiment was about the same as for Nos. 340-354.

No. 356, with the Texas Creek pipe, $D = 1.42$, is of great value, being by far the most authentic experiment on record, with a very high velocity, and a pipe of considerable size. The chances of error in this experiment are less than for the North Bloomfield series. Δ can be considered about the same, or perhaps a trifle higher, than for Nos. 340-355, and about the same as for No. 286. The curves for this Texas Creek pipe were sharp enough to appreciably retard the flow, possibly diminishing n one or two per cent. .

Darcy-Bazin.—The only experiments we have given with iron pipes having diameters larger than 2.4, are No. 64 with the Loch Katrine pipe by Mr. Gale, and Nos. 282-285 with the Rosemary pipe, by Mr. Stearns; each of these pipes was 4 feet in diameter, and the value of Δ appears to have been not very different. To determine which of these determinations is trustworthy, we will use the Darcy-Bazin Series XXIV., XXV. and XXVI., with semi-circular open conduits, each having a diameter of about 4 feet, and with Δ, a plaster of pure cement for No. XXIV., a plaster of cement mixed with one-third sand for No. XXV., and planks partly planed for No. XXVI.

In all these experiments h was directly measured, and Q obtained with a fair degree of accuracy. The general accuracy of these Darcy-Bazin experiments has been confirmed by the open conduit experiments of Messrs. Fteley and Stearns.

Some authorities, notably Humphreys and Abbot, have insisted that the air, even in calm weather, considerably retards the flow in open channels. This assumption has been proved by M. Bazin to be incorrect. In any event it seems clear, that comparing values of n for full circular pipes, with values of n for semi-circular open conduits, r, v and Δ being identical, n cannot be higher for the conduit than for the pipe. We are therefore on the safe side in using the Darcy-Bazin data. *

Fteley and Stearns.—Open Conduit Experiments, Nos. 444 and 445, were made with a section of the Sudbury conduit, where the brick-work was covered with a thin plaster of pure Portland cement; for these experiments the wetted surface was probably smoother than for the Darcy-Bazin Series XXV. (cement and sand), but on the other hand the semi-circular form of the latter conduit was more favourable for flow, than the oval form of the Sudbury conduit; hence the values of n for these two conduits should be about the same.

From the given data, we will now select those experiments with low values of Δ, which we regard as most trustworthy; they are as follows:

---

* The Darcy-Bazin Open Conduit (Series LI. and LII.) experiments showed that in a wooden rectangular pipe n, in $r^n (r s)^{\frac{1}{2}}$, had about the same value, as for a rectangular open conduit with the same values of Δ, s and r.

Name of Pipe.	Authority.	No. of Experiment.	Material.	Value of Δ.	D	v
New Almaden	Author	331-334	Glass	0	.042	2.1—4.4
„	„	316-321	Wrought-iron	0.2	.052	1.0—3.9
„	„	327-330	Glass	0	.062	1.4—4.4
		322-326	„	0	.076	2.0—5.0
		301-304	Wrought-iron	0.1	.087	2.2—5.4
„	„	293	„	0.2	.088	.90
		299-300	„	0.2	.088	1.05—1.4
Versailles	Couplet	1-6	Lead	0,	.44	.18—.47
North Bloomfield	Author	340-344	Riveted sheet-iron	1.5	.91	4.7—10.0
Bonn	Iben	271-274	Cast-iron coated	1.	1.00	1.6—3.1
North Bloomfield	Author	345-348	Riveted sheet-iron	1.5	1.06	4.0—10.7
„	„	349-354	„ „	1.5	1.23	4.4—12.1
Colinton	Leslie	61-63	Cast-iron	(?)	1.33	5.3—14.5
Danzig	Lampe	275-278	Cast-iron coated	1.	1.37	1.6—3.1
Texas Creek	Author	356	Riveted sheet-iron	1.6	1.42	20.1
Humbug	„	355	„ „	1.5	2.15	12.6
Cherokee	„	286	„ „	1.6	2.43 (?)	10.8
Rochester	Tubbs	281	Cast-iron, and riveted sheet-iron	1.2 (?)	{ 2.0 3.0	4.6 2.0
Loch Katrine	Gale	64	Cast-iron coated	1.1	4.0	3.5
Rosemary	Stearns	282-285	„ „	1.	4.0	2.6—6.2
		Conduit Nos.			r	
		223-234	Plaster of pure cement	...	.37—1.03	3.0—6.1
Conduit of semi-circular section	Darcy-Bazin	235-246	Plaster of cement and sand	1.	.38—1.04	2.9—5.7
		247-259	Partly planed plank		.39—1.15	2.6—5.5
Sudbury conduit	Fteley-Stearns	444	Plaster of pure cement	0.4	{ 2.05	2.7
		445	„ „		1.86	2.5

The foregoing experiments have been plotted on Plate XIV., with v and n as co-ordinates. It will be seen by an examination of this diagram, that the positions of the experimental curves and points, with a few exceptions, are in harmony with the conclusions drawn from the New Almaden results. That is to say; with v constant, n steadily increases with D; with D constant, n for small velocities rapidly increases with v, and this increase in n becomes less and less as the velocities become greater.

The exceptions to these general principles or laws are as follows :

No. 282 of the Rosemary pipe, which indicates that n increases with low velocities.

This experiment stands alone in this respect, so it should have but little weight as against the mass of other testimony.

No. 64, with the Loch Katrine pipe, which indicates that n for $D = 4.0$ is less than n for the Danzig pipe with $D = 1.37$, and only slightly greater than n for the North Bloomfield pipe with $D = 1.23$, the value of c in all three cases being the same. With the same velocity and diameter ($v = 3.5$ and $D = 4.0$), for the Rosemary pipe n is about 140., as against 112. for the Loch Katrine pipe. Looking at the three curves for the Darcy-Bazin semi-circular conduits, where r is a variable, we see that with $r = 1.0$, or $D = 4.0$;

For the surface of pure cement ...	...	...	...	$n = 154.$ with $v = 6.0$
,, ,, ,, ,, cement with one-third sand ...		...		$n = 141.$ with $v = 5.4$
,, ,, ,, ,, partly planed plank . .		...		$n = 128.$ with $v = 5.0$

A plaster of pure cement is doubtless a smoother surface than that of a cast-iron pipe covered with a coating of asphalt or tar ; probably a plaster of cement and sand has about the same character of surface as such a pipe. For the same velocity of 5.4 the Rosemary curve shows a value of n of 143, which closely agrees with its value for the conduit plastered with cement and sand. We hence see that the Loch Katrine experiment is directly contradicted by the Rosemary and Darcy-Bazin experiments, and indirectly by the Danzig and North Bloomfield curves ; we are therefore justified in rejecting it.*

The Colinton curve, Nos. 61-63 with $D = 1.33$, is considerably lower than the North Bloomfield curve with nearly the same diameter. Without looking for other reasons, we can fairly attribute this to roughness of surface in the Colinton pipe.

For Nos. 345 and 346 of the North Bloomfield series, n appears to be a little too low, compared with the other experiments of the three curves ; we have before stated that owing to the stoppage of a stone in this pipe, n for these two experiments has too small a value.

No. 286 for the Cherokee pipe and No. 355 for the Humbug pipe are somewhat contradictory ; neither of these experiments ranks high in point of accuracy.

Rejecting these exceptional experimental results, we can now proceed to draw curves on the diagram, representing the values of n for various diameters, the inner surface being in all cases quite smooth ; these curves are shown on Plate XIV. by symmetrical, solid, fine lines.

Δ for the four New Almaden experimental curves for small pipes, was exceedingly low ; in practice such smoothness will rarely be attainable. Hence for the diameters of .05 and .10 the curves for n will be drawn somewhat lower than is indicated as their proper position by the experimental curves ; they will also be made flatter, with velocities above 3., than the experimental curves, owing to the supposed effect of increasing Δ.

---

* Possibly Δ for this Loch Katrine pipe may have had a higher value than is indicated by the remarks made by Mr. Gale in regard to the condition of its inner surface.

The North Bloomfield curve for $D = 1.23$, is on the other hand for a riveted pipe, carelessly laid with poor joints, and having two rather sharp angles ; we will assume the curve for this pipe to nearly represent the curve of n for $D = 1.0$.

The datum point for the curve of n for $D = 4.0$ will be Open Conduit Experiment No. 244, with the Darcy-Bazin semi-circular conduit, covered with a plaster of cement and sand. The form of this curve is in harmony with the North Bloomfield and New Almaden experimental curves, following the law that, Δ being constant, the effect of v upon n increases with D ; that is to say, the larger the diameter, the more the curve for n should be inclined.

The curves of n with D between 1. and 1. have been interpolated from these two curves ; the curves with D from 5. to 8. have been drawn, so that they are harmonious with the other curves.

By reference to Plate XIV. it will be seen that the various curves of n agree very fairly with the experimental data considered reliable. The curves closely approximate to the given values of r for the Darcy-Bazin conduit (cement and sand), and those for the Sudbury conduit. The Rosemary curve (Nos. 283-285) is about 2 per cent. too high ; the Rochester compound pipe shows n about 4 per cent. too high ; the Danzig curve is slightly too high ; the experimental point for the Humbug pipe is slightly too low, and that for the Cherokee pipe about 9 per cent. too low.

Extending the curve for $D = 1.5$ from $v = 15$ to $r = 20$, it will be noticed that it agrees almost perfectly with the experimental point for the Texas Creek pipe. Δ for this pipe was somewhat higher than the standard assumed for the curves of n ; hence it is probable that the curve for $D = 1.5$ should have been more steeply inclined, or, in other words, that we have somewhat underestimated the effect of v in increasing n.

In the following table are given the values of n corresponding to the curves on Plate XIV., with D from .05 to 8., and c from 1. to 15. for diameters up to 4.. If it be required to find n and v, in our equation $r = n (r s)^{1/5}$, r and s being known, they can readily be obtained by two or more approximations. For instance, let $D = 4$. (hence $r = 1$.) and $s = .0001$; assume n to be 100. ; then $v = 1$.. Looking at the table or the diagram, we see that with $D = 4$. and $v = 1$., n should be 123. ; taking n as 125., we have $v = 1.25$. Again looking at the diagram we see that 125. is about the proper value for n with $D = 4$, and $v = 1.25$.

It must be kept in mind that the total head H must be corrected for losses of head in imparting velocity and contraction at the entrance ; this should be done by $h = H - \dfrac{v^2}{2 g o^2}$. For short pipes one or more additional approximations may be needed in order to obtain h, and hence $s = \dfrac{h}{l}$, before the proper value of n is determined.

The given values of n can, in our judgment, be used with entire safety for computing the flow of reasonably clean water, either through well-made cast-iron pipes, or through riveted sheet-iron or steel pipes, where the rivet heads do not form quite a notable portion of the area. The pipes must be properly coated with a varnish of asphaltum and coal tar, or some other preparation equally good; the joints must be smoothly united, and any curves must be well rounded. These remarks apply to diameters from 1. to 8.. For diameters less than 1. the given values of n are probably somewhat too high for either cast or riveted pipe; they are suitable for ordinary lap-welded pipe, which has also been coated.

These values of n will also apply to open semi-circular conduits, plastered with a mortar of cement and sand; for a hard smooth surface such as pure cement, n will be several per cent. larger. For open conduits of rectangular section n will be several per cent. smaller than our given values.

For values of D or v larger than those which are given, for the same degree of smoothness n will continually increase with D. For a riveted sheet-iron or steel pipe, with its inner surface properly coated, with D = 20. and v = say 5., n will probably have the very great value of 180, or perhaps even a higher value.

For velocities less than 1. the proper value of n is more or less uncertain; it can be approximately determined by reference to the curves on Plate XIV.

For values of D and v, intermediate to those given in the table, n can best be obtained by interpolation on Plate XIV; it must be remembered that Δ for the two least diameters, has a smaller value than for the other diameters.

TABLE XC.

Values of Co-efficient n, in v = n (r s)½ for Circular Pipes, or Semi-circular open Conduits, having quite Smooth Interior Surfaces, and no Sharp Bends.

{ Δ^{ss} for D= .05 and I.
{ Δ^{L} „ D=1. to 8.

Velocity v	DIAMETERS. $D = \frac{1}{4} r.$												
	.05	.10	1.	1.5	2.	2.5	3.	3.5	4.	5.	6.	7.	8.
1 (f)	...	80.0	96.1	102.8	108.8	112.7	116.7	120.2	123.0	127.8	131.8	134.8	137.5
2	77.8	88.9	104.0	110.9	116.2	120.3	123.8	127.0	129.9	134.3	138.0	141.0	143.3
3	82.4	93.7	108.7	115.6	120.8	124.8	128.3	131.4	134.2	138.6	142.3	145.4	147.6
4	85.6	97.0	112.0	118.9	124.0	128.1	131.5	134.6	137.4	141.9	143.5	148.6	151.0
5	87.6	99.3	114.4	121.3	126.5	130.6	134.1	137.1	140.0	144.7	148.1	151.2	153.6
6	89.1	101.0	116.3	123.2	128.6	132.6	136.3	139.4	142.3	146.9	150.5	153.5	
7	90.0	102.4	118.0	125.0	130.4	134.6	138.2	141.5	144.5	149.0	152.7		
8	90.6	103.3	119.3	126.4	132.0	136.3	140.0	143.3	146.3	151.0	154.9		
9	90.7	104.0	120.4	127.7	133.3	137.7	141.6	145.0	148.1	152.8	156.7		
10	90.8	104.5	121.4	128.8	134.5	139.0	142.9	146.4	149.7	154.6			
11	90.9	104.7	122.0	129.7	135.6	140.2	144.2	147.7	151.0				
12	91.0	104.8	122.5	130.4	136.4	141.1	145.2	148.8	152.2				
13	91.0	105.0	122.9	131.0	137.1	141.9	146.1	149.8	153.2				
14	91.0	105.0	123.2	131.5	137.6	142.5	146.7	150.5	154.0				
15	91.0	105.0	123.6	131.8	138.0	142.9	147.2	151.1	154.6				
20 (f)	...	...	123.9	132.9									

No attempt will be made to give the values of *n* for pipes having rough inner surfaces; the degree of roughness cannot be properly estimated, and any computation of the flow must necessarily to a great extent be guess-work. When it is required to compute the flow through such pipes, it must be remembered that for ordinary diameters, an increase in *v* above 2. or 3. has but little effect upon *n*. Also that Δ is a function of *D*; that is to say, a degree of roughness which would greatly lower *n* for small diameters, would have but little effect for great diameters.

There is no authentic evidence proving for pipes, that with large values of Δ, *n* decreases with an increase in *v*. Some of the Darcy-Bazin experiments with open conduits, where Δ was exceedingly high in relation to *r*, seem to prove that for velocities above 2. *n* does diminish as *v* increases; the Mississippi experiments of Humphreys and Abbot apparently show that this is true for rivers. If it be true for conduits and rivers, it is doubtless true for pipes. Some experiments made by Professor W. C. Unwin* show that the resistance encountered by a rotating disc, with a rough surface, is measured by a higher power than the square of the

* Proceedings, Institution of Civil Engineers, London, 1884-85. These experiments are more fully described at the close of this chapter.

velocity; this may indicate that in our Chezy formula, with high values of Δ, n decreases with an increase in v.

An iron pipe, unprotected by any surface coating, will soon become incrusted either by a hard scale, where water impregnated with lime flows through it, or by tubercular excrescences where soft water is used. These incrustations have the effect of not only diminishing the area, but also of increasing Δ; the flow is hence diminished in a double way. Some of the Darcy experiments, and those of Iben given in Table LXXVIII., illustrate the greatly diminished flow due to these two causes.

The author has had a large experience with riveted sheet-iron pipes in California, and has found no difficulty in protecting them both from rust and the formation of tubercles.[*] The Danzig and Rosemary cast-iron pipes had both been laid several years before the given experiments were made with them the high values of n for these experiments prove the efficiency of the interior coating.

FORMULÆ.

The Dubuat formula for pipes and streams is, in old French inches, $v =$
$$\frac{297\,(v^{\frac{3}{4}}-.1)}{\left(\frac{1}{s}\right)^{\frac{3}{4}}-L\left(\frac{1}{s}+1.6\right)^{\frac{3}{4}}} - .3\,(v^{\prime\prime}-.1),$$ in which L is the hyp. log.. Transforming this into English feet we have,

(A) $$v=\frac{88.51\,(v^{\frac{3}{4}}-.0298)}{\left(\frac{1}{s}\right)^{\frac{3}{4}}-L\left(\frac{1}{s}+1.6\right)^{\frac{3}{4}}} - .0894\,(v^{\frac{3}{4}}-.0298).$$

Prony proposed for pipes, in metrical measures, $\frac{1}{4}\,g\,D\,s = .000\,17\,v + .003\,416\,v^2$. Transforming this into English feet, we have,

(B) $$v=(.006\,65 + 9421\,v\,s)^{\frac{1}{2}} - .0816.$$

Prony suggests as a simpler form, but which must not be used for very low velocities, $v = 26.79\,(D\,s)^{\frac{1}{2}}$ metric, or,

(C) $$v=97.05\,(v\,s)^{\frac{1}{2}}.$$

Eytelwein proposed (English feet),

(D) $$v=47.9\left(\frac{H\,D}{\ell+54\,D}\right)^{\frac{1}{2}}.$$

Mr. Neville gives,

(E) $$v=140\,(v\,s)^{\frac{1}{2}} - 11\,(v\,s)^{\frac{1}{3}}.$$

Darcy's formula for new cast-iron pipes was in metrical measures, $R\,s = \left(.000\,507 + \frac{.000\,006\,47}{R}\right)v^2$, which in English feet becomes,

(F) $$v=\left(\frac{r\,s}{.000\,077\,26 + \frac{.000\,001\,62}{r}}\right)^{\frac{1}{2}}.$$

---

[*] The reader is referred to a paper by the author, published in the Transactions of the Am. Soc. of C E., February, 1884, for detailed information in regard to such pipes.

Weisbach gives,

(G)
$$v = \left(\frac{1}{\zeta} \frac{D}{l} 2gh \right)^{\frac{1}{2}},$$

in which the co-efficient ζ is greater for low than for high velocities. Weisbach recognised the fact that ζ diminishes as D increases, and also that ζ increases as the inner surface becomes rougher. For general use he makes $\zeta = .014\,39 + \frac{.009\,4711}{v^{\frac{1}{2}}}$ (metric); hence we have in English measures,

(H)
$$v = \left(\frac{1}{.003\,5975 + \frac{.004\,289}{v^{\frac{1}{2}}}} r s \, 2g \right)^{\frac{1}{2}}.$$

Hagen's formula* in metrical measures was,

(I)
$$v = \left[\frac{2g Ds}{\left(.023\,577 + \frac{.000\,115\,19 - .000\,004\,19 \, T_c + .000\,000\,092\,29 \, T_c^2}{v D} \right)} \right]^{\frac{1}{2}}.$$

In the foregoing expression T_c is the temperature in degrees of the Celsius scale.

Dr. Lampe proposes in metrical measure :

(a) $s = .000\,061\,341 \dfrac{v}{D^2} + .000\,793\,32 \dfrac{v^2}{D}$; and also suggests,

(b) $s = .000\,7555 \dfrac{v^{1.802}}{D^{1.25}}$. Transforming (b) into English measures we have,

(J)
$$v = 203.3 \; r^{.694} \; s^{.555}.$$

In the Kutter formula, given in Chapter VII., his co-efficient, n, of rugosity for a plaster of cement and sand is placed at .011 ; with this co-efficient his formula becomes in English measures,

(K)
$$v = \frac{206.3 + \frac{.002\,81}{s}}{1 + \frac{.011}{r^{\frac{1}{2}}} \left(41.66 + \frac{.002\,81}{s} \right)} (r s)^{\frac{1}{2}}.$$

D'Aubuisson, Saint-Venant, Lévy, Dupuit, Grashof, Gauckler, Boileau, Frank, Boussinesq, and Reynolds, have also proposed formulæ for the flow in pipes and streams. The formulæ of Bazin for open channels have been given in Chapter VII. Fanning, in his treatise on "Water Supply Engineering," gives tables, showing varying values of the co-efficient ζ, in $v = \left(\dfrac{1}{\zeta} 2g \, r s \right)^{\frac{1}{2}}$, in which he recognises the effect of changes in Δ, v and r, upon ζ.

---

* Druckhöhen-Verlust in geschlossenen eisernen Rohrleitungen.   Otto Iben, Hamburg, 1880.

In the following table are compared, with widely varying values of v and D, the resulting velocities given by some of the preceding formulæ. The values of v as deduced from our diagram on Plate XIV., are first given, and then the values computed by the several formulæ, in the order of their agreement with our results.

TABLE XCI.

Velocities in Circular Smooth Pipes, as deduced from Formulæ of Prony, et cet. —v and s being known.

D	s	Plate XIV.	Lampe (J)	Kutter (K)	Darcy (F)	Weisbach (H)	Prony (C)	Prony (B)	Dubuat (A)
.10	.005	.87	.83	.59	.94	1.01	1.09	1.01	.98
	.15	6.21	5.48	3.24	5.14	6.78	5.94	5.86	7.43
	.4	10.46	9.45	5.29	8.39	11.50	9.71	9.62	12.96
1.0	.000 4	.95	1.01	1.03	1.09	.89	.97	.89	.86
	.025	9.56	10.03	8.50	8.64	8.94	7.67	7.59	9.20
	.1	19.6	21.6	17.0	17.3	18.7	15.3	15.3	21.4
4.0	.000 07	1.03	1.01	1.09	.94	.72	.81	.73	.66
	.001 2	4.84	4.87	4.87	3.90	3.63	3.36	3.28	3.28
	.01	15.5	15.8	14.1	11.3	11.5	9.71	9.63	11.08
8.0	.000 025	.97	.92	1.03	.80	.59	.69	.61	.51
	.000 7	5.83	5.84	5.91	4.23	3.96	3.63	3.55	3.46
20.	.000 008	.90 (J)	.92	1.20	.72	.52	.61	.54	.36
	.000 155	5.0 (J)	4.77	4.83	3.16	2.85	2.70	2.62	2.37

An examination of the preceding table shows that the formulæ of Dubuat, Prony, Weisbach, and Darcy give results, especially for large diameters, widely varying from the experimental values indicated by the curves on Plate XIV. By giving n a constant value of 120, we would obtain considerably better results for these comparisons, than those deduced from any one of these five expressions. Darcy's formula gives very slightly better results than that of Weisbach.

The Kutter formula, for D less than 1., gives very much less than the true velocities; it agrees very well with our experiments for D from 3. to 8., except for very slight inclinations, when it gives results much too high. The Lampe formula agrees quite closely with our results, and can be used without very serious error; aside from its much closer approach to the truth, it is much preferable to the Kutter expression, on account of its greater simplicity.

From the experimental data, contained in this chapter and the preceding one, we could frame formulæ, giving the co-efficient N and the exponents $\frac{1}{m}$ and $\frac{1}{si}$,* in $v = N\,r^{1/m}\,s^{1/s}$, constant values for each assumed value of Δ; such expressions would give results approximating the truth. We prefer, however, to use the simplest form; the data we have given will enable the student to make a fair guess as to the value of n, where Δ is high in proportion to r; the assumption of such values of Δ must necessarily be a matter of judgment.

Professor Osborne Reynolds has published in the Transactions of the Royal Society, 1883, an account of experiments with glass and lead pipes, the diameters ranging from $\frac{1}{4}$ inch to 1 inch. With single joints of glass pipes up to 1 inch in diameter, by the aid of coloring matter in the water, he found that up to a certain limit of velocity for each diameter, the water moved in parallel threads or rings; the instant this limit was passed eddies formed, and parallelism disappeared.

He made a large number of experiments with two lead pipes, having the respective diameters of .0415 and .0202, with a very wide range of pressures; the "frictional" loss of head was determined by two piezometric tubes placed 5 feet apart. The experiments with the larger diameter showed that with the least pressure ($s = .000\,80$ and $v = .11$) n had a value of 39; with increasing pressures n increased to 91 with $v = .74$, and then dropped immediately to 86 with about the same velocity; this point Professor Reynolds considers to be that at which parallel movement ceases and eddying movement commences. The pressures were steadily increased beyond this limit, until s became 3.2; with this "inclination" v was 23, and n was 126; the increment of n with the increase in s was, however, quite irregular. With the smaller diameter the least pressure was equivalent to $s = .0086$, when $v = .26$, and $n = 39$; the limit at which parallel movement ceased was about $s = .0516$ and $v = 1.45$, when n was 90. Above this limit n was 92; then n decreased to 80 with $v = 2.1$, and with increasing pressures increased to 110, with $s = 3.9$ and $v = 15$; as with the larger pipe, the increment of n with s was quite irregular.

So long as parallel motion continued, these experiments agree roughly with the Poiseuille formula of $v = 52\,500\,(1 + .033\,68\,T, + .000\,221\,T,^2)\,s\,D^2$, the velocity increasing directly with the inclination and the square of the diameter. They indicate that as D increases, the limit of parallel motion diminishes; that is to say, with a considerable diameter, such as 1 foot, eddying movement will commence with a very small value of v. Professor Reynolds thinks this critical velocity is very nearly in the inverse ratio of the diameter; hence the critical velocity being .74 for $D = .04$, it should be about .03 for $D = 1.0$.

For low velocities the experiments of Jacobson and Reynolds confirm the general accuracy of the form of our curves on Plate XIV.; for very high velocities the experiments of Reynolds indicate that n continually increases with v, the pressure increasing with the 1.722 power of v, while we are of the opinion that n becomes nearly constant. Professor Reynolds appears to think that the effect of temperature is by far more notable when the motion is parallel, than when it is eddying; this agrees with our conjectures as to the effect of T.

One of the Reynolds series indicates that the curve for n, formed by using n and v as co-ordinates (as we have plotted our experimental data), has an abrupt break the instant parallel motion ceases.† Although this may be quite possible for a short smooth pipe, we do not regard it as definitely proved by Professor Reynolds; his methods for determining s and v can only be regarded as approximations; especially for the determination of s his piezometric measurements may have been quite misleading. The only trustworthy method of obtaining s and v for very low velocities is as follows: The pipe or pipes employed should be of lead, having a length of at the very least 100 feet for a diameter of one-half inch, and a still greater length for larger diameters. The discharge should be submerged, and the outlet and inlet tanks should be placed side by side; the respective heights of the surface in these tanks should be observed by hook-gauges attached to

---

* Eytelwein, we believe, first proposed the use of "fractional" exponents.

† It may be remarked that with the Darcy experiments, the irregularities in the curves for n are generally greatest when the velocity was less than .7.

the same vertical post, so that their zeros can be quickly and accurately compared; changes in inclination should be obtained by successively lowering the outlet tank. A trumpet-shaped mouth-piece should be attached to the inlet end, by which means the effect of contraction at the entrance can be nearly eliminated. The head absorbed in imparting velocity can be obtained with sufficient accuracy by $\frac{v^2}{2g}$. Q should be determined by the measurement of q in vessels of proper size; an iron pail in the form of a cone truncated near its apex, such as we used for our Greenpoint experiments, is the most convenient kind of vessel for such a purpose. D should be obtained by the weight of water contained in the pipe. Care must be observed to prevent accumulations of air in the pipe, and the curve or curves in the pipe should be smooth and gradual.

By such a method the value of n for a range of velocities from .01 to 2. can be obtained with the utmost precision. With such a pipe we hazard the guess that the curve for n will be almost perfectly symmetrical.

The question of parallel movement in pipes as contrasted with eddying movement, has but little practical interest, as with conduit-pipes of ordinary size parallel movement in all probability ceases at a very low limit of velocity. In fact with such a pipe composed of many joints, each of which forms more or less of an obstruction, it is difficult to conceive of parallel motion, unless at almost infinitesimal velocities.

Professor W. C. Unwin[*] has lately made some very interesting experiments in regard to the friction of discs having various conditions of surface, rotating in water having various temperatures; one experiment was also made with a polished disc rotating in a syrup formed by dissolving sugar in water. These experiments showed that the friction of a polished disc diminished 18 per cent. by raising T from 41° to 130°; the amount of friction increased rapidly as the surface of the disc was made rougher; for the smoother surfaces the resistance varied as the 1.85th power of the velocity; and for the rougher surfaces with a power of the velocity ranging from 1.9 to 2.1. How close the analogy is between the friction of such rotating discs, or the friction of vessels moving through the water, and the movement of water through conduits, we are hardly prepared to say.

---

[*] Proceedings Institution of Civil Engineers, 1884-85. The same experiments to which we have before referred.

CHAPTER IX.

EXPERIMENTS WITH ORIFICES AND WEIRS.

CALIFORNIA, 1874-1876.

SOON after the discovery of gold in California in 1848, associations or incorporated companies were formed for the purpose of building ditches and storage reservoirs for the supply of water to the placer mines. The amount of capital invested in these hydraulic works aggregated many millions of dollars, and a single company often sold water to hundreds of mining claims. The cost of water was by far the most important item in the miners' bill of costs, and hence it became necessary to have a standard of measure not only accurate, but also so simple that the amount of discharge could be readily computed by the common miner.

This was accomplished by the discharge of the stream of water sold to each customer through a rectangular, square edged, vertical orifice, with free discharge into the air, and having a constant head. In different parts of the State the standard opening varied ; the width varying from 2 to 4 inches, and the head above the top of the opening from 4 to 7 inches. Each square inch of the opening was called " a miners' inch ;" hence in a locality where the standard opening was 2 inches wide, if the miner wished a flow of 50 miners' inches the orifice was 25 inches long, and if only 10 inches was needed the length was reduced to 5 inches.

This method is analogous to the pouce d'eau used in Southern France, and was probably first introduced or suggested in California by some French or Mexican miner ;[*] the simplicity of this mode of measurement combined with a sufficient degree of accuracy, soon brought it into general use on the Pacific Coast, wherever water was sold for mining, irrigation, et cet. .

The standard, which had been in use since 1852 or 1853 in the mining districts supplied by the Eureka-Lake, Bloomfield and Milton water companies in Nevada County, California, was an opening 50 inches long, 2 inches wide, with constant head above opening of 6 inches ; the flow from this was called 100 miners' inches. If less water was needed, the opening was reduced in length by false pieces ; if more water was

---

[*] Mr. John Dunn adopted this system of measurement at Nevada City, California, in January, 1851 ; so far as we can learn, this was its first application in the State, for the purpose of gauging or selling water.

needed, as many openings were used as was necessary; for instance, if 350 miners' inches was needed, three standard openings and one-half, or 25 inches in length, of another opening were employed. Generally the miners bought water for 10 hours per diem at an agreed price per inch; for example, a miner using 350 miners' inches, for 10 hours each day, at the rate of 15 cents per inch, paid the water company 52.50 dollars per diem, and received the amount of water which would flow through orifices, having an aggregate length of 175 inches, a width of 2 inches, with a head of 6 inches above the top of the opening, during a consecutive period of 10 hours.

When water was used for the whole 24 hours of the day, the flow was termed "a miners' 24 hour inch," and of course meant 2.4 times the amount of discharge of "a miners' 10 hour inch."

The shallow placer mines became in time exhausted, and the water from these ditches was used upon a much larger scale in the deep placers, where the gravel, stones and earth were washed away by great streams of water discharging from pipes under high heads. The old system of water measurement became objectionable on account of the large size of the measuring tanks necessary to afford sufficient space for the openings, and a new standard or module was adopted, being the discharge from an orifice, 12 inches wide, $12\frac{3}{4}$ inches long, with a constant head of 6 inches above the top of the opening; the discharge from this orifice was roughly assumed to equal that from two of the old standards, or 200 miners' inches. It was desirable to know the exact discharge through each of these modules, and the author made at Columbia Hill in Nevada County, California, the experiments we are about to describe, the cost of the measuring apparatus being defrayed by the three companies before spoken of.

Apparatus at Columbia Hill.

A vertical section of the wooden canal, measuring tank, et cet., is shown by Fig. 7, Plate XV., and the horizontal plan, showing form of approach, et cet., by Fig. 8, Plate XV.

The feeding reservoir had an area of several acres, and was supplied by the flow from a higher reservoir with a nearly constant stream during each experiment. There was some little wind during the experiments, which caused more or less fluctuation in the surface of the approach to the orifices or weir. The hook-gauge, reading to .001 inch, was placed 7.6 feet from the weir, as shown by the plan, in a wooden box, having a number of small holes bored through its outer side from top to bottom, so that the interior surface level quite speedily responded to fluctuations in the channel of approach; the openings in the gauge box were, however, not large enough to permit the interior surface to be much disturbed by local wave movements.

The gauge was read at regular intervals of from 10 to 30 seconds, depending upon the total length of time of the experiment; means were deduced from these several readings, for orifices by the $\frac{1}{2}$ power of each observation, and for the weir by the $\frac{3}{2}$ power of each observation. This method was thought to be preferable to that sometimes followed of having one small avenue of communication between the feeding canal and the gauge box.

The inner depth below crest, G, for the weir was 3.8 feet; this was also very nearly the distance below the bottom of the several orifices experimented upon.

The feeding canal was 7.3 feet wide, at the dam where the various openings were placed.

The water, after its escape from the opening in use, passed into the lower wooden canal or flume, dropped down 2 feet at the step **A**, and discharged at the end of the flume **C**, until the commencement of an experiment. Gratings or racks were placed as shown by the section. There was an over-fall on each side of the flume, each 16.5 feet long, with the crests level and at equal elevations; these crests were 1.3 feet above the bottom of the flume. The stop-gate, **B**, was a sliding wooden gate, which could be put in place in about the space of two seconds.

The measuring tank was a rectangular box, made of narrow, well seasoned, planed pine plank, with an inner horizontal section of about 9×17.3, and a depth of 8.5. It was held by heavy outer timber frames well braced, and firmly tied each way by iron cross-rods. In spite of these precautions the tank became a trifle larger when it was filled than when it was empty; this was determined by several measurements on the outer sides from fixed points, and amounted to an average increased size of 2.14 cubic feet in 1874, and in 1876 to .86 cubic foot.

The leakage of the tank was ascertained by filling it with water, and noting by a hook-gauge the fall in its surface per minute. In 1874 this leakage amounted to an average of .078 cubic foot per minute while the tank was being filled; in 1876 this leakage amounted to an average of only .03 cubic foot per minute.

Low-water mark was fixed by two upright sharp steel points placed in opposite corners of the tank, and whose summits were at precisely equal elevations; before each experiment the water surface was adjusted by means of a draw-off cock with great exactness to these summits. Immediately above these points two wooden boxes were placed, open at the bottom, and having sharp steel hooks placed near the top of the tank; the points of these hooks were in a vertical line above the lower points, and the height of the water in the tank was the distance between the lower point and the upper hook. The horizontal section of the tank was very carefully measured in 1874 and again in 1876, by 8 sections one way and 5 sections the other way. The tank had been very carefully made, and the greatest difference between any two measurements for the length-wise sections was only .022.

The tank capacity, when filled, in 1874 was, $9.0319 \times 17.3494 \times 8.1075 = 1270.43$ cubic feet; less spaces occupied by tie-rods and boxes $= 2.31$ cubic feet; making net capacity 1268.12 cubic feet. In 1876 capacity, when filled, was $9.0442 \times 17.3620 \times 8.1100 = 1273.47$; less tie-rods, et cet., $= 2.31$; making net capacity 1271.14 cubic feet.

A second hook was placed between the lower point and upper hook, so that when desired, measurements of volume could be made with either the upper or lower section of the tank.

When the plate or frame forming the weir or opening at the dam, D, was in place, and the water in the reservoir brought to the desired level, the false-gates in the feeding canal were removed and the water passed through the opening, discharging at C. Low-water level was then accurately adjusted in the measuring tank, and the gate at B was closed; in a few seconds the incoming water filled the flume from A to B, and the time was taken when the water began to spill over the over-falls into the tank; the end of the experiment was the moment of time the water reached the fixed hooks near the top of the tank. An observer was stationed at each of these hooks; the times were determined to ⅕th second. The discharge was hence the net capacity of the tank, plus the amount of water above the over-fall level contained in the flume from A to B.* The height of the surface above the over-fall was measured at several points for each experiment; in Experiments (Weirs) Nos. 66-68 there was more or less uncertainty as to the exact mean height, because of the boiling of the water; any errors from this source even with No. 68 (Weirs), when Q was largest, could not have exceeded 2 or 3 cubic feet.

The boxes enclosing the hooks kept the surface of the interior water reasonably quiet, so that the time of filling to the hooks could be gotten very exactly. As the boxes were open at the bottom, the water level at the hooks fairly indicated the mean surface level in the tank.

An abstract of the notes for Experiment No. 68 (Weirs) will show the methods of observation followed:

Date, November 8th, 1876.

The total head, H, was 1.7327 taken from the mean of 7 readings, varying from 1.7296 minimum to 1.7362 maximum.

Time, 71.7 seconds.

Discharge.				
	Capacity of tank ...	...		1271.1
	Above over-fall in flume ...			+ 42.0
	Leak of tank ...	...	...	+ .0
	„ „ gate at B	...	...	+ 0.1
	„ „ dam at D	...	...	− 0.1
	Splashing over sides of tank	...		+ 0.3
				1313.4

$Q = \dfrac{1313.4}{71.7} = 18.318$ cubic feet per second.

Water approaching weir with surface velocity of .68 ft., from reservoir to beginning of surface curve at weir.

The small leakages, as just noted, were generally directly measured; they were

---

* The comparatively small quantity of water in the air, between the over-falls and the surface in the measuring tank, was not fully taken into account; this error was nearly balanced by the depression formed by the surface curves approaching the over-falls.

largest in Weir Experiment No. 66, when they amounted in the aggregate to 2.2 cubic feet.

In all the experiments with orifices the velocity of approach was too inconsiderable to be worthy of attention, and hence $H = h$. For Weir Experiments 64-68, effect of velocity of approach was doubtless slightly sensible for Nos. 64 and 65, and for Nos. 67 and 68 may have increased the discharge $\frac{1}{2}$ of one per cent.; as these last named experiments can only be considered as approximations, owing to the small capacity of the measuring vessel compared with Q, no attempt will be made to correct them for v_a.

The standard of measure was a "New York" level rod, which agreed closely with the standard in the U. S. Coast Survey Office in San Francisco.

EXPERIMENTS WITH ORIFICES.

Columbia Hill. Small Heads.

The discharge through the old module, 50 inches long and 2 inches wide, was determined by three experiments with an opening cut as true as possible in a pine plank, 3 inches thick, with the lower side or crest 1 inch broad, and then beveled at an angle of 45°. The inner corners were square with the inner face, and the escaping vein only came in contact with these corners. The length was 4.169, being .0023 in excess of the standard; the width for Experiments (*a*) and (*b*) was as nearly 2 inches as the means of measurement at hand—the broad end of a carpenter's square corresponding to the standard—permitted its determination; the swelling of the wood reduced the width of the opening a trifle for Experiment (*c*), as shown by the gauge fitting more tightly in the opening than it did after Experiments *a* and *b*. This diminution in size will be neglected.

The time, t, is the mean of the determinations by the two observers at the measuring tank.

TABLE XCII.

Determination of Value of Miners' Inch. Eureka Lake Standard, 50 Inches long, 2 Inches wide, 6-Inch Head above Top of Opening. Full Contraction.

$$(2 g)^{\frac{1}{2}} = 8.018$$

Temperature of water about 65°.

Orifice No.	Size of Opening.		$H = h$	t	Total Discharge.				Q	C	c
	l	w			Tank.	Flume.	Leaks.	Total.			
(a)	4.169	.166 67	.5827	488.7	1268.12	11.41	1.36	1280.89	2.621	.6163	.6169
(b)	,,	,,	.5855	487.7	1268.12	11.41	1.97	1281.50	2.626	.6165	.6170
(c)	,,	,,	.5888	488.0	1268.12	11.41	1.30	1280.83	2.625	.6140	.6146
Means										.6156	.6161
1	4.1667	.166 67	.5833						2.618	.6156	.6161

In the preceding table $Q = C (2 g h)^{\frac{1}{2}} l w$; and $Q = c \frac{2}{3} l (2 g)^{\frac{1}{2}} (H_s'^{\frac{1}{2}} - H_t'^{\frac{1}{2}})$; $H_s = h + \frac{w}{2}$; $H_t = h - \frac{w}{2}$. It is assumed that the value of c for module—No. 1—will be the mean of (a), (b) and (c), notwithstanding the slight variations in h.

The lower value of c for No. (c) than for Nos. (a) and (b), was doubtless chiefly caused by the slight diminishment of size before alluded to.

The chief danger of error in No. 1 is in the value of w; probably in this case the limit of error for the value of c is not far from $\frac{1}{300}$.

The discharge through the new and larger module, called the Bloomfield standard, was determined by two experiments. An opening with sides made as nearly square as was possible, was cut through a plank .12 thick, fastened to an outer plank of the same thickness having a still larger opening. The discharge was perfectly free into the air, the escaping vein only coming in contact with the inner corners. For No. (d) the length was 1.063, and the width 1.0005; No. (c) was made the day following, and the dimensions were then, $l = 1.0655$ and $w = 1.0018$.

The co-efficients and discharge for the exact module with $l = 1.0625$, $w = 1.$, $H_t = .5$, $H_s = 1.5$, and $h = 1.$ are obtained in the same manner as for the Eureka Lake standard (No. 1).

TABLE XCIII.

Determination of North Bloomfield Standard, called 200 Miners' Inches, 12¾ Inches long, 12 Inches wide, 6-Inch Head above Top of Opening. Full Contraction.

$(2 g)^{\frac{1}{2}} = 8.018$

Temperature of water about 50°.

Orifice No.	Size of Opening.		$H = h$	t	Total Discharge.				Q	C	c
	l	w			Tank.	Flume.	Leaks.	Total.			
(d)	1.0630	1.0005	1.0035	255.1	1271.1	17.8	0.7	1289.6	5.055	.5918	.5983
(c)	1.0655	1.0018	1.0029	253.7	1271.1	16.7	0.7	1288.5	5.079	.5926	.5992
Means										.5922	.5988
2	1.0625	1.	1.						5.045	.5922	.5988

In the foregoing experiments there was a surface current opposite the hook-gauge of .105; an allowance for this velocity of approach would not change the last decimal in the above value of c one point; *i.e.*, about $\frac{1}{10000}$.

Two indirect measurements by weirs were made by the author, of the discharge through a number of these Bloomfield modules placed side by side, the intervals separating them being about 6 inches long. In both cases there seemed to be a sensible increase in the discharge; that is to say 10 adjoining openings discharged more than 10 times the amount a single opening would discharge. These experiments were not made with sufficient care to absolutely determine this point, and hence will not be given here.

D'Aubuisson states that with three orifices placed side by side the co-efficient c was larger than when only one orifice was used. On the other hand, Francis's experiments with two 4-foot weirs, separated by an interval of 2 feet—Weir Experiments Nos. 50 and 56—show a slightly lower value of c, than would probably have been the case with one weir of the same length having full contraction, as will be seen by an examination of Plate VI.

Bazin's experiments with several adjoining orifices indicate a considerable increase in the co-efficient, when the number of openings was increased.[*]

The 4 following experiments were made with orifices nearly circular cut through thin sheet-iron plates; these plates were firmly bolted to an outer plank having a somewhat larger opening, so that there was no warping on account of the thinness of the iron. The edges were made as nearly square as was possible by dressing with a fine file.

Orifice No.	Thickness of Iron.	Diameters.		
		Maximum.	Minimum.	Mean.
3	.0090	.2542	.2525	.2533
4	.0085	.4195	.4177	.4185
5	.0085	.6637	.6615	.6627
6	.0060	1.0110	1.0070	1.0089

The mean diameters were deduced from 6 or 7 measurements for each orifice and were thought to be very near the truth.

The same care was taken in making these experiments, as was practised with the preceding experiments, but they cannot be considered as reliable, as they were single determinations.

TABLE XCIV.

Flow through Circular Orifices, with free Discharge into Air. Full Contraction.

$(2g)^{1/2} = 8.018$

Temperature of water 50° to 55°.

Orifice No.	Size.				Total Discharge.				Q	C	c
	Mean Diameter D	Mean Area a	$H = h$	t	Tank.	Flume.	Leaks.	Total.			
3	.2533	.05039	1.3315	1832.4	516.0	2.9	0.9	519.8	.2837	.6085	.6087
4	.4185	.1376	1.2045	1749.0	1271.1	5.4	1.1	1277.6	.7305	.6035	.6011
5	.6627	.3449	1.0887	750.4	1271.1	8.3	0.6	1280.0	1.706	.5912	.5929
6	1.0089	.7994	1.0457	333.9	1271.1	13.0	0.3	1284.4	3.847	.5869	.5913

The co-efficient C is obtained by $Q = C (2 gh)^{1/2} a$.

The co-efficient c is obtained from Table IV., giving the relative values of C and c for circular orifices.

---

[*] *Recherches Hydrauliques. Darcy and Bazin, p. 61.*

A small whirlpool or vortex kept forming and re-forming on the inner side in No. 6.   In all the other experiments with orifices, the surface of the water above the opening was quiet, or free from any apparent whirling movement.

The values of c for these 6 experiments will be found plotted on Plate III.

WEIR EXPERIMENTS.—*Columbia Hill.*

The weir experimented upon was a plate of boiler iron, .026 thick, with the crest .008 wide.   The opening was 1.754 deep; the length on top was 2.584 and on the bottom 2.587.   The edges were cut square with the inner face, and were made as true as could be done with a fine flat file.   These experiments were conducted with the same care used in making those with orifices.

The inner depth, G, below crest was 3.8, and the width of the feeding canal at the weir was 7.3.   Hence for Experiments Nos. 57 to 64 there was practically complete contraction, while for Nos. 65 to 68 there was a slight partial suppression, but hardly enough to warrant corrections being applied.

As before stated, velocity of approach will not be taken into account, although its effect was doubtless appreciable for Nos. 64-68.   For No. 68 the central surface current, from the reservoir to where the surface curve at the weir became sensible, was .68.

TABLE XCV.

Sharp-Crested Weir.   Free Discharge into Air.   Very nearly Complete Contraction.

$(2 g)^{1/2} = 8.0177$

Temperature of water 50° to 60°.

Weir No.	H	Mean Length. *l*	t	Total Discharge.				Q	C
				Tank.	Flume.	Leaks.	Total.		
57	.5659	2.586	358.4	1271.1	12.5	0.2	1283.8	3.582	.6087
58	.6163	2.586	318.7	1271.1	14.4	0.2	1285.7	4.034	.6032
59	.6470	2.586	296.8	1271.1	16.0	0.3	1287.4	4.338	.6030
60	.6703	2.586	281.9	1271.1	16.1	0.2	1287.4	4.567	.6020
61	.7072	2.586	260.3	1271.1	17.1	0.3	1288.5	4.950	.6021
62	1.0681	2.586	144.2	1271.1	24.3	0.6	1296.0	8.988	.5890
63	1.1063	2.586	137.3	1271.1	26.8	0.3	1298.2	9.455	.5878
64	1.2033	2.586	120.7	1271.1	30.1	0.3	1301.5	10.783	.5910
65	1.3257	2.585	106.6	1271.1	33.5	0.3	1304.9	12.241	.5804
66	1.5391	2.585	83.8	1271.1	35.2	2.2	1308.5	15.614	.5918
67	1.7195	2.585	72.1	1271.1	40.4	0.3	1311.8	18.191	.5840
68	1.7327	2.585	71.7	1271.1	42.0	0.3	1313.4	18.318	.5812

There was some little question as to whether termination of No. 62 was very exactly determined or not, and hence No. 63 was made with nearly the same head, which showed that no serious error had been made in No. 62.

Experiments Nos. 57-63 can be considered as reliable, and in these C is practically identical with c. Nos. 66-68 can only be considered as approximations for the reasons before given.

C is obtained by $Q = C \frac{2}{3} (2 g H)^{\frac{1}{2}} l H$.

All the foregoing experiments were made at Columbia Hill, California, in lat. 39° at an elevation of 2900 feet above sea level.

The apparatus employed was well suited for the exact measurement of Q up to a maximum of 9 or 10, and it was the intention of the author to have made a large number of additional experiments with weirs from .866 to 2. in length, and with square and round orifices of various sizes with heads up to 3 feet. Unfortunately, professional engagements prevented the completion of the contemplated investigations.

EXPERIMENTS WITH ORIFICES.

North Bloomfield and French Corral. Great Heads.

In July, 1874, the author made a series of experiments at North Bloomfield, California, and also one at French Corral in the same State, for the purpose of determining the effective duty of water-wheels, known as hurdy-gurdies, which are almost exclusively used for water-motors in California.[*] Incidentally measurements were made of the discharge through tapered nozzles of various forms, some of them having divergent adjutages, and through annular square-edged openings, where contraction was more or less perfect. The measurements of discharge and of the mean diameters were not made with the highest degree of accuracy, so that the following results may be in error as much as 3, or possibly 4 per cent. They have sufficient value, however, to warrant their insertion here, as the heads employed were exceptionally great.

The several nozzles and rings were placed at the end of a discharge-pipe, which was fed from a long main. The level of the water in the pen-stock at the head of the main was known, as was also the elevation of the nozzle. More or less water was being drawn from the main while the experiments were in progress, and hence a portion of the total head was absorbed in friction through the pipe. This consequent loss of head for Nos. (Orifices) 7-17 was determined by a new Bourdon gauge, placed at the upper end of the discharge-pipe at a known elevation.[†] The readings from this gauge were corrected by closing the openings in the main, and noting what difference there was between the pressure shown by the gauge, and that due to the head; these corrections varying from 1.6 to 3.2 lbs. were then applied to the gauge readings.

By this method the actual or effective heads at the nozzles were determined with comparative certainty to within 2 feet. As the least head employed was 312 feet, errors from this source would not amount to more than $\frac{1}{300}$.

The discharge was measured over a sharp-edged rectangular weir, with very true

---

[*] A detailed description of several forms of the hurdy-gurdy will be found in the Transactions of Am. Soc. of C. E., February, 1884.

[†] This gauge was placed at an elevation about 5 feet higher than the lower end of the discharge-pipe.

sides, made of $\frac{1}{4}$ inch boiler iron, edges beveled on the outer side, and .8660 long.   The water escaping from the nozzle was conducted by a square flume to the weir, the current being checked by a rack and cross boards; the weir was placed at the end of the canal, and there was nearly complete contraction. The height on the weir was read by a hook-gauge, reading to .001 inch, placed sufficiently far from the weir to be out of the influence of the surface curve. In Nos. 15, 17, 7 and 9, the discharge is thought to have been less accurately measured than in the other experiments, as the water in the feeding canal for the weir for these Nos.—15, et. cet.—was more turbulent than in the succeeding experiments, where the surface in the canal was quite quiet.

The forms of the nozzles used are shown by Figs. 1, 2, 3, 4, 5 and 6, Plate XV. The discharging nozzles A, B, C and D all screwed with a very nice fit into the larger nozzle H I, which screwed into the lower end of a long discharge-pipe of smooth sheet-iron, having a regular taper, with a maximum inner diameter of about .6.   The larger nozzle, H I, was of hard cast-iron, first accurately turned and then smoothly polished; its sides were very nearly straight—its inner form being nearly the frustrum of a cone—having a convergence of about 8° 25'.   The discharging nozzles A, B, C and D were also of polished cast-iron, very smooth, and with sides slightly curved.

The rings E, F and G were of thin saw-plate steel, held in place by annular screw-caps.   The pipes K L and M N, to which these rings were attached, were of smooth cast-iron, but not turned.

The sections of the large nozzle H I and of the discharging nozzles and rings, were as nearly circular as they could be made by a skilful mechanic.   They were kept perfectly free from rust by rubbing with oiled cloths, the oil being wiped off by a clean cloth at the beginning of each experiment.   The surfaces were hence probably smoother than would have been the case had the nozzles and rings been made of brass or any other comparatively soft material.

The diameters of the nozzles A, B, C and D were measured at their smallest sections by calipers, and then determined by a finely graduated scale; the holes through the rings were cylindrical in form, and were measured directly by the same scale.   This method of obtaining diameters of such small size was not sufficiently accurate, especially with A and E, where a slight error of measurement would sensibly affect the value of the co-efficient of discharge.

After C with its divergent mouth-piece or adjutage had been experimented upon, it was cut off close to its minimum section at *a*, and again used as C'.

The temperature of the water was not taken; it was not far from 60°·65°.   In the reductions a cubic foot of water is assumed to weigh 62.4 lbs. .

The discharge over the measuring weir has been re-calculated from the original notes, the co-efficient c' (in this case c_r) being taken from Plate VII.

The original notes are not at hand for No. 18; the total head for this experiment, however, has been corrected properly for head lost by friction in pipe.

TABLE XCVI.

Discharge through Nozzles and Rings. North Bloomfield, California. July, 1874.

$(2 g)^{\frac{1}{2}} = 8.018$

Orifice No.	No. of Nozzle	Measuring Weir				Size of least Section of Nozzle		Actual Velocity $\frac{Q}{a} = \frac{v}{c}$	Determination of Effective Head.						Theoretical Velocity $(2 g h)^{\frac{1}{2}}$	Co-efficient of Discharge $\frac{v}{V} = c$
		t	$H = h$	Plate VII c	Q	D	a		Total Head	Head in Feet	Pressure-Gauge in Lbs.	Indicated Head	Loss of Head	Effective Head h	v	
7	A	.9666	.2397	.6104	.3316	.0631	.002913	149.7	398.9	334.1	137.6	317.5	16.6	333.3	144.0	1.040
8	B	"	.4416	.5963	.8100	.0850	.005674	142.7	398.9	334.1	134.2	309.7	24.4	314.5	142.2	1.004
9	B	"	.4351	.5906	.7926	.0850	.005674	139.7	398.9	334.1	133.2	307.4	26.7	312.2	141.7	.986
10	C	"	.4538	.5936	.8484	.0868	.005917	143.4	398.9	334.1	134.9	311.3	22.8	316.1	142.6	1.006
11	C'	"	.4641	.5953	.8713	.0868	.005917	147.2	399.3	334.4	142.0	327.8	6.6	332.7	146.3	1.007
12	C'	"	.4634	.5952	.8740	.0868	.005917	147.7	399.4	334.5	143.5	331.1	3.4	336.0	147.0	1.005
13	D	"	.5682	.5932	1.1740	.1017	.008123	144.5	399.9	334.1	135.7	313.1	21.0	317.9	143.0	1.011
14	D	"	.5684	.5932	1.1727	.1017	.008123	144.4	399.9	334.1	131.7	310.8	23.3	315.6	142.4	1.014
	Rings															
15	E	"	.1950	.6157	.2454	.0597	.002799	87.68	358.9	334.1	135.0	311.5	22.6	316.3	142.6	.615
16	F	"	.3309	.6024	.5308	.0847	.005635	94.20	358.9	334.1	133.4	307.8	26.3	312.6	141.8	.665
17	F	"	.3300	.6025	.5287	.0847	.005635	93.84	358.9	334.1	133.2	307.4	26.7	312.2	141.7	.662
18	G	(Notes not at hand)			1.538	.1823	.02610	58.54	130.1				2.8	127.3	90.4	.647

The foregoing experiments were originally calculated by a modification of the formula of Mr. Francis for weirs,[*] which gives considerably lower values for Q with small heads, than does the interpolated curve for $l = .866$ taken from Plate VII.

We now make the co-efficient of discharge for nozzle A, $c = 1.040$, and for nozzle C', $c = 1.006$. It seems altogether improbable that c for either of these nozzles, or in fact for any of the nozzles, should be above .998; no probable error in the measurement of the diameter of A would account for this difference of 4 per cent. . It may be that the curve for c_e with $l = .66$, which has been drawn on Plate VII. from Lesbros' experiments, and from which the preceding values of c have been deduced, is too high, and that the lower values of c_e, as shown by the experiments of Poncelet and Lesbros (Weir Nos. 35-40) and of Castel (Weir Nos. 187-194), are nearer the truth than the determinations of Lesbros. Had these lower values of c_e been used in the determination of Q, the final results in the foregoing table would, apparently, have been much more satisfactory.

It is to be regretted that in these experiments Q was not obtained by absolute measurement. Any indirect calculation of Q is always more or less unsatisfactory, unless made over a weir, or through an orifice, whose co-efficients of discharge have been ascertained under precisely similar circumstances by measurement in a vessel of proper size.

Despite their imperfections the foregoing results are sufficiently accurate to show :

First.—With great heads and smooth converging mouth-pieces, the co-efficient of discharge, c, is nearly 1.

Second.—With great heads, short divergent adjutages have but little effect upon the discharge.

Third.—With great heads and small orifices, with complete contraction c will be about .60, and the size of the feeding channel probably need not be as great in proportion to the orifice, in order to avoid partial suppression, as is the case with small heads.

It may be incidentally observed that with equal quantities of water, the discharge from the rings gave a slightly higher useful effect upon the water-wheels than resulted when the nozzles were employed.

Experiment No. 18 was made at French Corral, California, at an elevation of about 1600 feet above sea-level ; all the others were at North Bloomfield, the elevation of the orifices being 2950 feet above sea-level.

The greatest danger of error in Experiments Nos. 1, 3 and 4 with orifices, undoubtedly consists in incorrect assumptions of the value of n. For Nos. 7-18 (Orifices) danger of error perhaps about equally rests in incorrect values of a and of Q.

The linear measurements for these areas were made with considerable care, by a finely divided ivory

---

[*] The original computation is given on page 18 Transactions Am. Soc. of C.E., 1884.

scale; this scale was compared with, and corrected by, a long box-wood rod, which in turn had been compared with a steel scale in the U.S. Coast Survey Office. This was a vicious method, as subsequent experience has fully shown us. Errors, with the use of several scales roughly compared, may very likely be all in one way: this is especially true of personal error. The only proper method in obtaining such small linear dimensions, where great accuracy is required to avoid notable comparative error, is by direct comparison of the orifice with the final standard.

It has been shown that the co-efficient c for No. 3 is apparently 1 per cent. too great, by comparing c for No. 3 with the Holyoke experiments. Supposing this error to be entirely due to an incorrect value of a for No. 3, it would involve an error of .0012 in the measurement of the mean diameter. This is a considerable quantity, but we now can conceive it to be a possible error, as we have of late seen more than one instance where errors of several comparisons have all been in one direction. Very likely the steel scale, which was not very finely divided, in the U.S. Coast Survey Office, may have been slightly different from the standard of Professor Rogers.[*]

With such large orifices, as those employed for Experiments Nos. 2, 5 and 6, the danger of comparative error in incorrect measurements of the dimensions of the orifice, is, of course, very much less than for the smaller orifices.

[*] A slight error in the reference standard would not affect the Columbia Hill experiments, nor in fact those at North Bloomfield. Because the measurement of g was made by the same standard.

CHAPTER X.

EXPERIMENTS OF FLOW THROUGH PIPES.

The following experiments were made by the author several years ago, the final results of which have been published in the Transactions of the American Society of Civil Engineers, 1883. A detailed description of the methods of observation employed will here be given, to enable other hydraulicians to determine what weight should be given to the several experiments.

NEW ALMADEN EXPERIMENTS.

In the month of April, 1877, 53 experiments were made at New Almaden, California, with small pipes of wood, old wrought-iron, new wrought-iron, and glass, with velocities from .9 to 6.9. The object in view was to determine the general laws governing the discharge through pipes, with widely varying conditions of the inner surface, and with varying velocities. It was thought that very exact measurements with small diameters would give more reliable data, than rougher experiments with large diameters, where it was impracticable to determine the discharge by absolute measurement, on account of the considerable cost of building measuring vessels of proper size.

The apparatus employed at New Almaden is shown by Figs. 1, 2 and 3, on Plate XVI., the discharge in all the experiments being perfectly free into the air. The measuring vessel was a rectangular wooden tank, of planed, tongued and grooved red-wood plank, held by outside wooden frames securely bolted together, so that there was no perceptible enlargement when the tank was filled. A cock was placed at the bottom, so that the water could be drawn off, and accurately adjusted to the low-water mark.

The measuring points for height were as follows:

At the bottom a vertical sharp steel point, A, to the summit of which the water was brought with great precision at the beginning of each experiment.

A sharply pointed steel hook, B, distant vertically 1.726 above the datum point.

A similar hook, C, near the top of the tank, and distant vertically about 2.15 above B.

The tank could therefore be divided into two sections.

The capacity of the lower section (L) was as follows, each dimension being the mean of 12 measurements;

(L) April 12, 1877   1.9712 × 1.9918 = 3.9263
  „ 16, „   1.9739 × 1.9955 = 3.9389

     Mean ... ...   3.932 × 1.726 = 6.787

Grooves in corners, .022 × .011 × 1.7 × 4 =   + .002
Space in water-cock, .17 dia. × .44 = .010, one half filled   + .005

    Capacity (L) =   6.794 cubic feet

The discrepancies between the measurements of April 12th and 16th, were partly due to slight changes in the tank, and partly because the sections were not taken at exactly similar points.

The capacity of the upper section (U) was determined in the same way, and was as follows;

(U) April 12, 1877   1.9814 × 1.9701 = 3.9036
  „ 16, „   1.9777 × 1.9722 = 3.9004

     Mean ... ...   3.902 square feet.

From April 10th to 14th the height of (U) was 2.157, and from April 15th to 17th 2.153; there was a + correction for grooves in corners of .028 × .011 × 2.2 × 4 = .003 cubic foot, hence;

  April 10-14 ... ...   (3.902 × 2.157) + .003 = 8.420 cubic feet
  „ 15-17 ... ...   (3.902 × 2.153) + .003 = 8.404 „ „

The total capacity (T) of the tank was therefore;

(T) April 10-14   6.794 + 8.420 = 15.214 cubic feet
  „ 15-17 ...   6.794 + 8.404 = 15.198 „ „

There was a slight leakage from the tank, which was carefully determined twice each day, and a table of corrections prepared, which it is not necessary to give here.

The maximum leakage was on April 10th, as follows;

  When (T) was used .0027 cubic foot per minute
  „ (L) „ „ .0011 „ „ „

The minimum leakage was on April 17th, as follows;

  When (T) was used .0010 cubic foot per minute
  „ (L) „ „ .0003 „ „ „

When everything was in readiness for an experiment, a tin trough, which carried the escaping water from the end of the pipe over and beyond the tank, was quickly removed, and the discharge allowed to enter the tank. In order to prevent commotion at the surface, when the water approached the measuring hook, an assistant caught the jet in a tin pipe of 4 inches diameter, the lower end of which was held some distance below the surface ; by this means the exact moment the surface level reached the hook could be very closely determined.

For the small pipes the discharge was sometimes measured by two buckets, one of which (B) held 94.27 lbs., the weight being ascertained by a Fairbank's platform bullion scale, sensitive to $\frac{1}{4}$th oz. This scale was also checked by a similar scale ; they were both known to be exact for a weight of $76\frac{1}{2}$ lbs. (flask of quicksilver). The temperature of the water was from 57° to 68° Fahr. ; hence weight of 1 cubic foot = 62.35 lbs., and capacity of (B) = 1.512 cubic feet. The other bucket (*b*) held 29.60 lbs. = .475 cubic foot. This bucket measurement was quite imperfect compared with the tank, but the limit of possible error with them was not over 1 per cent., with the chances of error (mean of three measurements) probably under one-half of 1 per cent.. This was determined by using first the tank, and then the bucket, for the same experiment ; for instance in No. 316, mean of two tank measurements gave $Q = .008\ 31$, while by bucket $Q = .008\ 35$.

The times were determined by a large stop-watch (T. S. Negus), which had a + rate of $\frac{1}{2500}$; hence a minus correction of .0004, *i.e.*, $t' \times .9996$, had to be applied ; the beats were 5 to a second, hence $\frac{1}{5}$th second was the limit of accuracy for time.

The water was fed to the pen-stock through a $2\frac{1}{2}$-inch hose, connected with a hydrant having a high and nearly constant head ; hence a very regular supply could be obtained. The disturbing effect of the water as it came from the hose was checked by the cross plank shown on the sketch. A waste-way or weir was placed on the side of the pen-stock, and a small quantity of water allowed to pass over it in all the experiments.

An engineer's level was set at a point equidistant from the pen-stock and the end of the pipe, and the difference in elevation determined before and after each experiment, the level rod being held first on the weir, and then on a sharp nick or bench-mark cut on the extreme end of each pipe. The elevations from the centre of the pipe to these points were measured, as also the height of the surface of the water in the pen-stock above the over-fall ; these last measurements being taken first at the commencement, and then at the ending of each experiment. The total head, H, hence was always the difference in elevation between the surface of the water in the pen-stock, and the centre of the discharge end of the pipe.

Two distinct measurements were made for each experiment when the tank was used, and three when the buckets were used. It would occupy too much space to give all these figures in detail ; a copy of the notes for Experiment No. 287 will show the methods of observation.

| | Head. | | | | | | Discharge |
No.	Over-Fall above Point at End of Pipe	Water Surface above Over-Fall.	Point above Centre of Pipe.	Total Head.	Time occupied in Filling Measuring Vessel.	Size in Cubic Feet of Measuring Vessel.	in Cubic Feet per Second.
287	8.444 8.446	.031			4' 39.8" 2' 59.3"	(T)	
	.196 .200	.034			12' 32.6" 10' 51.8"		
	8.248 8.246				7' 52.8" 7' 54.3"	15.214	
	8.247	.032	.051	8.330	472.6"	Leakage	
					Rate = 0.2"	8 × .0020 + .016	
					472.4"	15.230	.032 24

Whenever there was any appreciable difference in the times, as was the case in two or three experiments, the measurements were repeated.

The various inclinations were obtained by lifting the pipe, which rested on temporary brackets, the pen-stock not being moved. The glass pipe rested on planks, laid lengthwise.

The pipes were all straight, except the glass pipes which were slightly curved, as hereafter noted.

The standard of measure was a "New York" level rod, which agreed very closely with the standard scale in the office of the U.S. Coast Survey in San Francisco. A steel tape was used in measuring the lengths of the various pipes, which (with average temperature during the experiments and a constant strain) read .018 too long in 60 feet; thus making a constant minus correction of .0003, *i.e.*, $l' \times .9997$, for each foot in length.

The temperature of the water was from 57° to 68° Fahr. The water was almost perfectly limpid.

The diameters of the pipes—except for the $\frac{1}{2}$ inch glass pipe, and $1\frac{1}{2}$ inch wooden pipe—were computed from the weight of the water contained in them. This was determined by filling each joint, and weighing the total amount of water contained by the several joints constituting one pipe; the length being the sum of the lengths of the joints, and hence being somewhat shorter than the length of the pipe in place, on account of the spaces between the ends of the joints.

Description of Pipes.

Series I.—A lap-welded iron, gas or water pipe, about 1 inch in diameter, which had never been used, and was almost entirely free from rust. Its inner surface was quite smooth, except immediately at the ends of the joints, where there was a little roughness caused by the cutting of the threads for the outer screw coupling. The ends were a trifle smaller than the interior section of the pipe. Just at the weld the diameter was slightly smaller, than for the other axis. All the iron pipes used, were connected by the usual outer screw coupling, the joints not butting squarely together, but having intervals from .01 to .04 in length ; no account was taken of these trifling enlargements. The diameters of the two ends of each joint were taken at a point about .1 from the end.

The upper end of the pipe was just flush with the side of the pen-stock ; the co-efficient of contraction at the entrance was assumed at .800, on account of the slight roughness at the end.

The length, l, in place was (60.190 − .018) 60.172 ; the sum of the lengths of the five joints was (59.964 − .018) 59.946. The arithmetical mean of the end diameters was .086 85, and by weight of water contained by the several joints the mean diameter was .0878 (*vide* Series III., for method of determination), thus showing that the central parts of the joints were appreciably larger than the ends.[*]

Series II.—After the completion of the seven experiments with Series I., a funnel was screwed on the upper end of the pipe, projecting into the pen-stock as shown by Fig. 1, Plate XVI. This mouth-piece was 1. long; outer diameter .542 and inner diameter .086 ; it was of cast-iron not polished, but quite smooth; its form is shown by Fig. 3, Plate XVI.

The length of the pipe was measured from the beginning of the taper of the funnel (the funnel proper not being included in the length, l), and was (corrected) 60.247. The diameter was .0878, being unchanged from Series I. The co-efficient of contraction was assumed at .98.

Series III.—The pipe was then taken apart, and each joint immersed in a boiling bath of asphaltum and coal tar, thus giving the interior a very smooth and glassy surface, but which slightly diminished its diameter.

The measurements to obtain mean diameter were as follows:

Direct measurement.	Measurement by Weight.	
.086 25	Sum of lengths of five joints	59.964
.086 42	Error of tape	− .018
.086 25		———
.086 42		59.946
.086 50	Sum of the weight of water in five joints...	22.42 lbs.
.087 33	Let,	
.086 25	w = lbs. of water in 1 cubic foot	= 62.35
.086 08	l = length of pipe filled	= 59.946
.086 00	D = required diameter.	
Mean = .086 39	w = weight of water contained in pipe in lbs.	= 22.42
	Hence,	

$$D = \left(\frac{w}{\frac{\pi}{4} \, l w} \right)^{\frac{1}{2}} = .087\ 39$$

[*] The maximum end diameter was .0875, and the minimum .0865 ; the mean, computed from end sections, was .086 85.

The accuracy of the scales and of m (weight of cubic foot of water) had been verified by comparisons of the buckets (B) and (*b*) with the tank.

The mean diameter was assumed at .0873, thus giving a slight value to the direct measurement.

The length, *l*, of the pipe in place was,

Lower end of pipe to funnel	60.157	
Funnel, from lower end to beginning of flow	.125	
	———	60.282
Error of tape ...	...	–.018
		———
		l = 60.264

Co-efficient of contraction at entrance was placed at .98.

Series IV.—The first two joints of the pipe used in Series III., with the funnel mouth-piece.

The mean diameter, as deduced by weight of water, was .0876 ; *l*, including cylindrical portion of funnel, was (corrected) 16.685 ; co-efficient of contraction, .98.

Series V.—This pipe consisted of 4 joints of old lap-welded pipe, 52.7 feet long, and 1 joint at the upper end, for connection with the funnel, of new pipe, 7.4 feet long. The old pipe had been in use for 2 years, constantly filled with water, which had but little velocity. The water was from a clear spring, evidently containing lime. The interior surface was covered with a thin hard scale, with occasional small lumps, and was rather rough. It had never been coated with an asphaltum preparation; it was when new, doubtless of very nearly the same diameter as the pipe used in Series I. The short length of new pipe was smooth.

The diameter was substantially the same, .0853, both by end measurements, and by weight of water ; the diameter of the old pipe was from .0846 to .0850, and of the new joint, .0871. The length, including cylindrical portion of the funnel, was (corrected) 60.247. Co-efficient of contraction at entrance was assumed at .98, as in Series II.-IV.

Series VI.—A new ¾-inch lap-welded pipe, free from rust, but not coated with asphalt, consisting of 5 joints. The upper end of the pipe was flush with the side of the pen-stock, and was dressed smooth, with sharp clear-cut edges ; .825 was adopted as the co-efficient of contraction.

The mean diameter was, by end measurements .0515, and by weight of water .0524 ; .0523 was adopted. The measured diameters varied from .0508 to .0521. The length (corrected) was 60.127.

Series VII.—A glass pipe, formed of 12 joints as follows ;

Length.	Diameters.	
6.199	{ .0875 = entrance. .0747	The largest diameter was .0875 „ smallest „ „ .0719
6.342	{ .0750 .0844	The greatest difference between diameters at a
4.100	{ .0825 .0817	joint was (.0844 − .0825) .0019 ; generally they were pretty evenly matched.
6.062	{ .0822 .0800	The weight of water contained in the 12 joints was by two trials 18.22 and 18.22.
4.607	{ .0807 .0777	Hence $D = \left(\dfrac{w}{\frac{\pi}{4} \, l \, m} \right)^{\frac{1}{2}} = .07632$
4.014	{ .0771 .0779	
6.499	{ .0787 .0719	The measured length of the pipe in place (cor- rected) was 63.902 ; hence showing that the
4.644	{ .0735 .0765	intervals between the joints were very small.
5.747	{ .0760 .0746	
6.008	{ .0754 .0757	
4.918	{ .0746 .0746	
4.750	{ .0744 .0729 = discharge.	

	63.890 − .018	
Tape		.07751 = mean.
	63.872	

A diameter of .0764 was adopted, and $l = 63.902$. A funnel mouth-piece shown by Fig. 2, Plate XVI., having outer diameter of .422 and inner diameter of .0875, was fitted accurately to entrance end of pipe, the joint being very nearly absolutely exact; the co-efficient of contraction was assumed to be .97.

Series VIII.—A glass pipe, formed of 6 joints as follows ;

Length.	Diameters.	
6.070	{ .0717 = entrance. .0708	The largest diameter was .0717 „ smallest „ „ .0579
4.410	{ .0679 .0633	The greatest differences between diameters at a joint was (.0675 − .0633) .0042.
6.050	{ .0675 .0583	The weight of the water contained in the 6 joints
6.105	{ .0587 .0608	was by two trials, 6.61 lbs. and 6.59 lbs. ; mean = 6.60 lbs.
6.145	{ .0612 .0579	Hence $D = \left(\dfrac{w}{\frac{\pi}{4} \, l \, m} \right)^{\frac{1}{2}} = .06211$
6.163	{ .0587 .0596 = discharge.	The measured length of the pipe in place (cor- rected) was 34.941.

	34.943 − .009	
Tape		.06304 = mean.
	34.934	

The adopted diameter was .0522, with l = 34.941. The end of the pipe was flush with the side of the pen-stock ; hence co-efficient of contraction = .82.

Series IX.—A glass pipe, formed of 2 joints, as follows ;

	Length.	Diameters.	
	5.400	{ .0450 = entrance. .0125	The water was not weighed in this instance, the scales not being accurate enough for such
	5.730	{ .0404 .0404 = discharge.	small quantities. As with the other glass pipes the mean diameters were smaller than the mean
	11.130	.0421 = mean.	of the end diameters, it was assumed that the
Tape	− .003		same would be true of this series.
	11.127		

The diameter adopted was .0418. The measured length in place (corrected) was 11.127. The end of the pipe was flush with the side of the pen-stock, and hence .82 was again assumed as value of the co-efficient of contraction.

Series VII., VIII. and IX.—The glass pipes were all of " lead glass," of German or French manufacture. Their sections were not perfectly circular. Some of the pipes were curved ; the one having the sharpest curve, being a joint of Series VII., which in a length of 6½ feet, had an ordinate of .035. Hence it was not possible to lay the pipes in a perfectly straight line ; the general direction, however, was very nearly straight, and the curvature was so slight as not to deserve special consideration.

The ends of each joint were ground smooth ; the joints were butted together, and a water-tight connection made by placing over each connection a piece of tightly fitting rubber tubing. This tubing made a perfectly water-tight joint, except in one or two cases where a few drops of water leaked, but the leakage was too inconsiderable to deserve notice.

The inaccurate fits at the joints, amounting in one instance in Series VIII. to a difference in diameters of .0042, doubtless somewhat added to the friction. These irregularities of surface, however, made no eddies or cross currents perceptible to the eye, the water flowing through perfectly limpid, and showing no visible signs of commotion.

The pipes were used for the first time ; their interior surfaces were smooth, with no scratches. One or two of the joints in Series VIII. had a trifling amount of dirt adhering to the inner surface in spots ; not enough, however, to seriously increase the resistance. All the joints for Series VII. and IX. were almost perfectly clean.

Series X.—A pipe of 8 joints, made of heart red-wood, and bored by the usual pipe-auger. The interior surface was of the usual smoothness of such holes. The pipes were new, and had not been covered with any artificial coating. The connections were made by driving one joint into the other, an iron outer band preventing the wood from splitting.

Q Q

The diameters were quite uniform, varying from .1042 to .1062 ; the mean was .1052, which was adopted as the correct size.   The length (corrected) was 62.05.

Reductions.

The value of $(2\,g)^{1/2}$ was computed by the formula given by D'Aubuisson, in metrical measures,

$$g = 9.8051\,(1. - .002\,84\,\cos 2\,l)\left(1 - \tfrac{2}{r}\,e\right).$$

$$r = 6\,366\,407\,(1. + .001\,64\,\cos 2\,l).$$

$l =$ latitude ; $r =$ radius terrestrial spheroid ; $e =$ elevation above sea.

Reducing this to English measures,

$$g = 32.169\,54\,(1 - .002\,84\,\cos 2\,l)\left(1 - \tfrac{2}{v}\,e\right).$$

$$r = 20\,887\,510\,(1 + .001\,64\,\cos 2\,l).$$

$e = 400$, and $l = 37°\,10'$; hence $(2\,g)^{1/2} = 8.0179.$

In the following table, showing the results of these 53 experiments, l, D, q, t and H are experimental values.

$o = \dfrac{\pi}{4}\,D^2$. $Q = \dfrac{q}{t}$. $v = \dfrac{Q}{o}$. o is the assumed co-efficient of contraction at entrance ; the value of o of .980 for Series II., III., IV. and V., may possibly be a trifle too high, but in any event it would be .94, and this smaller value, if adopted, would not considerably change n. h is deduced by following formula ; $h' = \dfrac{1}{2}\,\dfrac{v^2}{g}\,\dfrac{1}{o^2}$, or, $v = o\,(2\,g\,h')^{1/2}$; hence $h = H - \dfrac{v^2}{2\,g\,o^2}$; h' being the assumed head required to give velocity to the water, taking into account contraction at entrance, and h being effective head absorbed in overcoming friction and adhesion, as the water passes through the pipe.

The final co-efficient n is obtained by the Chezy formula, $v = n\,(r\,s)^{1/2}$; r being hydraulic mean radius $= \tfrac{1}{4}\,D$, and s being sin of inclination $= \dfrac{h}{l}$; hence

$$Q = n\,\frac{\pi}{4}\,D^2\left(\frac{h\,\tfrac{1}{4}\,D}{l}\right)^{1/2} = .3927\,n\left(\frac{h\,D^5}{l}\right)^{1/2}.$$

TABLE XCVII.

Hamilton Smith, Jr.—Determination of a, in $v = n (r a)^n$. New Almaden, 1877.

No.	Length l	Mean Diameter D	Mean Area a	Capacity of Measuring Vessel g	Time Required to Fill Vessel t	Discharge per Second Q	Velocity v	Total Head H	Co-Efficient of Contraction c	h'	h	n	Remarks
287	60.172	.0878	.006 054	15.230	472.4	.032 24	3.335	6.330	.800	.689	7.641	100.9	Series I.;
288	60.172	.0878	.006 054	15.238	538.7	.028 29	4.673	6.590	"	.831	6.059	99.4	1 inch iron, new, uncoated, no funnel.
289	60.172	.0878	.006 054	15.243	637.9	.023 90	3.948	4.850	"	.379	4.471	97.8	
290	60.172	.0878	.006 054	15.230	793.4	.019 22	3.175	3.272	"	.245	3.027	93.5	
291	60.172	.0878	.006 054	15.268	1174.5	.013 04	2.154	1.663	"	.113	1.550	99.6	
292	60.172	.0878	.006 054	6.803	519.5	.008 60	1.421	.787	"	.049	.738	86.6	Water in No. 293 did not quite fill discharge end of pipe.
293	60.172	.0878	.006 051	.475	791.0	.003 80	.958	.473	"	.022	.451	74.7	
294	60.247	.0878	.006 054	15.230	467.1	.032 61	5.387	8.317	.980	.470	7.847	100.8	Series II.;
295	60.247	.0878	.006 054	15.235	534.5	.028 50	4.708	6.506	"	.359	6.147	99.5	Same pipe as Series I. with funnel.
296	60.247	.0878	.006 054	15.235	629.3	.024 21	3.999	4.817	"	.250	4.588	97.8	
297	60.247	.0878	.006 054	15.240	780.4	.019 53	3.226	3.276	"	.169	3.107	93.9	
298	60.247	.0878	.006 054	6.801	510.5	.013 31	2.199	1.669	"	.078	1.591	91.3	
299	60.247	.0878	.006 054	6.804	783.1	.008 69	1.135	.784	"	.033	.751	86.8	
300	60.247	.0878	.006 054	.475	74.6	.006 37	1.052	.491	"	.018	.473	80.1	Water in No. 300 did not quite fill discharge end of pipe.
301	60.264	.0873	.005 906	15.227	467.3	.032 58	5.443	8.353	.980	.480	7.873	101.9	Series III.;
302	60.264	.0873	.005 906	15.229	533.4	.028 50	4.761	6.597	"	.367	6.230	100.5	Same pipe as Series I. coated with asphalt, with funnel.
303	60.264	.0873	.005 906	15.236	589.3	.025 90	3.231	3.313	"	.168	3.145	93.3	
304	60.264	.0873	.005 906	6.797	511.3	.013 29	2.190	1.703	"	.080	1.623	91.6	
305	16.685	.0876	.006 027	15.204	366.5	.041 48	6.883	4.630	.980	.767	3.853	96.8	Series IV.;
306	16.685	.0876	.006 027	15.205	446.8	.033 88	5.621	3.179	"	.512	2.667	95.0	First two joints of above pipe, coated, with funnel.

TABLE XCVII.—*continued.*

No.	Length. l	Mean Diameter. D	Mean Area. a	Capacity of Measuring Vessel. q	Time Required to Fill Vessel. t	Discharge per Second. Q	Velocity. v	Total Head. H	Co-Efficient of Contraction.	Effective Head. K	Effective Head. h		Remarks.
307	16.685	.0876	.006 027	15.209 / 6.796	660.8 / 298.2	022.75	3.775	1.584	.960	.251	1.383	92.0	*Series IV.—continued.*
308	16.685	.0876	.006 027	15.216 / 6.796	1085.4 / 481.8	.014 06	2.333	.624	..	.088	.515	88.0	
309	60.247	.0853	.005 715	15.333	624.9	.024 38	4.266	8.115	.940	.295	8.020	80.1	Series V.;
310	60.247	.0853	.005 715	15.235	706.6	.021 56	3.775	6.563	..	.331	6.332	79.7	
311	60.247	.0853	.005 715	6.798	370.4	.018 35	3.211	4.846	..	.167	4.679	78.9	1 inch { 58.7 feet, old iron, uncoated.
312	60.247	.0853	.005 715	6.798	451.2	.014 97	2.619	3.278	..	.111	3.167	78.2	4 feet, old iron, coated, with funnel.
313	60.247	.0853	.005 715	6.802	651.2	.010 45	1.829	1.682	..	.051	1.639	76.1	
314	60.247	.0853	.005 715	.475	69.5	.006 83	1.190	.780	..	.023	.757	73.0	
315	60.247	.0853	.005 715	.475	91.3	.005 20	.910	.474	..	.013	.461	71.2	
316	60.127	.0533	.002 348	6.800 / 1.512	842.9 / 181.0	.008 33	3.878	8.354	.855	.314	8.040	92.9	Series VI.;
317	60.127	.0533	.002 348	1.512	207.9	.007 27	3.385	6.503	..	.262	6.341	91.2	⅜ inch new uncoated, no funnel.
318	60.127	.0533	.002 348	1.512	243.9	.006 15	2.862	4.867	..	.187	4.680	89.7	
319	60.127	.0533	.002 348	1.512	306.8	.004 93	2.295	3.306	..	.120	3.186	87.2	
320	60.127	.0533	.002 348	.475	140.2	.003 39	1.578	1.710	..	.057	1.653	83.2	
321	60.127	.0533	.002 348	.475	214.5	.002 21	1.029	.812	..	.024	.789	78.6	
322	63.902	.0761	.004 584	15.232	663.5	.022 96	5.009	8.670	.970	.115	8.555	100.8	Series VII.;
323	63.902	.0761	.004 584	15.231	756.2	.020 04	4.383	6.840	..	.318	6.522	99.3	⅜ inch glass, with funnel.

TABLE XCVII—*continued*

No.	Length. l	Mean Diameter. D	Mean Area. a	Capacity of Measuring Vessel. v	Time Required to Fill Vessel. t	Discharge per Second. Q	Velocity. v	Total Head. H	Co-Efficient of Contraction. c	N'	Effective Head. h	n	Remarks.
324	63.902	.0764	.004 584	15.238	902.2	.016 89	3.685	5.036	.970	334	4.812	97.3	*Series VII.—continued*
325	63.902	.0764	.004 584	6.797	503.6	.013 50	2.945	3.397	"	143	3.214	94.6	
326	63.902	.0764	.001 531	6.799	759.0	.008 96	1.956	1.661	"	.063	1.598	89.5	
327	34.941	.0622	.003 039	6.797	511.6	.013 29	4.373	5.072	.850	.442	4.630	96.3	Series VIII.: ¾ inch glass, no funnel.
328	34.941	.0622	.003 039	6.797	609.9	.011 14	3.666	3.660	"	.311	3.349	95.0	
329	34.941	.0622	.003 039	6.798	843.2	.008 06	2.652	2.039	"	.163	1.896	91.3	Water in No. 330 did not fill discharge end of pipe, for a distance of about .1 from end.
330	34.941	.0622	.003 039	.475	111.8	.004 55	1.398	.673	"	.045	.628	83.6	
331	11.127	.0418	.001 372	.475	78.0	.006 09	4.439	3.035	.820	.456	2.569	90.4	Series IX.: ⅜ inch glass, no funnel.
332	11.127	.0418	.001 372	.475	94.7	.005 02	3.659	2.155	"	.309	1.846	87.9	
333	11.127	.0418	.001 372	.475	127.5	.003 73	2.719	1.264	"	.171	1.095	84.8	
334	11.127	.0418	.001 372	.475	166.4	.002 85	2.077	.816	"	.100	.716	80.1	
335	62.05	.1052	.008 692	15.207	438.9	.034 60	3.946	8.324	.800	.386	8.138	67.9	Series X.: 1¼ inch wood, new, no funnel.
336	62.05	.1052	.008 692	15.208	497.3	.030 59	3.519	6.696	"	.201	6.395	67.6	
337	62.05	.1052	.008 692	15.210	581.7	.026 15	3.008	4.942	"	.290	4.722	67.2	In No. 338 there was a very slight loss by dripping between end of pipe and tank for one of the tests; hence t was taken as smallest value, or 709°; the other trial made t=714.1° which, if adopted, would give a value of 67.0 to n.
338	69.05	.1052	.008 692	15.212	709.	.021 46	2.469	3.399	"	.148	3.161	67.5	
339	62.05	.1052	.008 692	6.796	472.9	.014 37	1.653	1.567	"	.066	1.501	65.6	

Of the foregoing experiments, those of Series IX (short ½-inch glass pipe) are the least reliable. The others are believed to be very near the truth ; of course for the low velocities, where Q was measured by (b) having only a capacity of .475, the determinations of u are not quite as trustworthy, as when larger measuring vessels were used.

The method of ascertaining D, by the weight of water contained, gives far more accurately the mean areas for small pipes, than any practicable process of direct measurement.

The accuracy of the assumed values of o for Series I. and II. can easily be proved by a comparison of the values of n for the two series ; there being for Series I. full contraction, and for Series II. hardly any contraction.

The indicated results from these experiments have been fully analyzed in Chapter VIII.

The facilities for making these investigations were very kindly afforded by Mr. J. B. Randol, Manager of the Quicksilver Company, to whom the author here desires to express his obligations.

North Bloomfield Pipes.

The following experiments were made at North Bloomfield, California, in October, 1876, in behalf of the North Bloomfield Gravel Mining Company, of which the author was then the General Manager.

Three pipes were experimented upon, of about 15, 13 and 11 inches diameter respectively. They were made of sheet-iron, from .0054 to .0091 in thickness, single riveted. They had been several years (about 5 years) in use, and hence the rivet heads were somewhat smoother than when the pipes were first made ; no deductions for these heads will be made in computing areas. The pipes were made in riveted joints, each about 20 feet long, put together stove-pipe fashion, one end hence being slightly larger than the other. The pipes had been originally coated by immersion in a boiling bath of asphaltum and coal tar ; at the time of these experiments the interior surfaces were quite smooth, and almost absolutely free from rust.

The three pipes were laid side by side across a sharp ravine, the outlet tank being about 25 feet lower than the pen-stock (inlet tank) ; the discharge was conducted by a ditch some 600 feet long to a weir, over which the discharge for each experiment was measured. The co-efficient, c, for the same weir had been previously determined by absolute measurement, as we have already described. Figs. 4, 5 and 6 on Plate XVI. show the profile of the pipe-line, a plan of the pipe and ditch, and upon a larger scale a sketch showing position of the weir, hook-gauge, et cet. .

The pipes had each a funnel-shaped mouth-piece, shown by Figs. 10, 11 and 12, Plate XVI. These funnels were included in the measured lengths, l, but the diameters were obtained by measurements of the pipe proper ; the co-efficient of contraction at entrance was assumed at 1., and it was thought that this consequent neglect of the effect of contraction would be practically compensated for, by neglecting to consider the increased diameters of the mouth-pieces. Any errors produced by this hypothesis would be inconsiderable.

The flow through each pipe was first determined when they were in their normal positions—Experiments Nos. 340, 345 and 349 ; the total head, H, for Nos. 345 and 349 being the difference in elevation between the surface of the water in the pen-stock, and the surface in the outlet tank. In order to obtain lower velocities pieces of similar pipe were attached to the ends in the outlet tank, thus extending the several pipes on the general line of their inclination. The discharge was hence free into the air, and H was assumed as the difference in elevation between the surface in the pen-stock, and the centre of the discharge end of the pipe ; in Experiment No. 344 the escaping water did not quite fill the end of the pipe (top of jet being .04 below upper edge of pipe in the plane of the discharge end), and in this case H was assumed as the difference between surface in pen-stock and centre of the escaping jet (about .02 lower than centre of pipe end).

The difference in elevation between the pen-stock and the outlet tank was ascertained by two careful level-lines. The surface of the water in the pen-stock was read by an assistant from a gauge-rod ; his readings were to the nearest hundredth of a foot for Nos. 340-351, and for Nos. 352-354 to the nearest two-hundredth (.005). There was a maximum variation in the surface level at the pen-stock of .02 in Nos. 340, 345 and 347 ; in the other experiments there was no perceptible variation. In No. 349 there were a few small eddies—hardly vortices—slightly disturbing the surface just above the entrance of the pipe ; in all the other experiments the surface in the pen-stock was practically quiet.

In No. 349 the submerged jet of escaping water struck with much force the opposite side of the outlet tank, some water rising close to this side as much as .5 above the general surface level in the tank ; in No. 345 the submerged jet made less commotion. In both these cases the height of the surfaces in the outlet tank was measured on the north side of the jet in comparatively still water. In all the other experiments—except No. 344 as before stated—the top of the escaping jets rose higher than the top of the pipes, owing to their inclined position.

The limit of error in the determination of H could not have been over .02 for any one of these experiments, except Nos. 345 and 349, where, owing to the disturbed condition of the surface due to the submerged discharge, H might possibly be .05 or .06 in error.

The mean diameters and areas were obtained by measuring the circumference of each pipe with a steel tape at 14 points nearly equidistant ; the inner diameters were then computed by an allowance for the thickness of the iron, which for the mains varied from .0054 to .0069. No allowance, however, was made for the thickness of the asphaltum coating, which, had it been taken into account, would have reduced the stated mean diameters perhaps .001.

The maximum variations in the inner diameters shown by these measurements were for the mains (*i.e.* pipes in place, not including funnel mouth-pieces) as follows ;

	Maximum.	Minimum.	Mean.
11 inch pipe	.923	.889	.9105
13 ,, ...	1.061	1.049	1.056
15 ,, ...	1.240	1.221	1.230

The above mean diameters were slightly changed by variations in the sizes of the joints used in extending the pipes at the discharge end. The diameters can fairly be considered as being correct within .0035.

The lengths were carefully measured ; the values given of *l*, should be within .2 of the truth.

The standard of measure was the same as that used for the pipe experiments at New Almaden, and for the measurements with the 2.6 weir; agreeing closely with the U.S. Coast Survey standard.

The course of the ditch feeding the pipes was at right angles to their line, and at the junction of the pen-stock the width of the ditch was 10.5 feet ; there was hence no approaching velocity of any moment at the entrance, to be taken into account.

During the experiments the mouths of the two pipes not in use were tightly closed, so that the flow was entirely confined to the pipe being experimented upon.

Measurement of Q.

By reference to Figs. 4 and 5, Plate XVI., it will be seen that the arrangements for the measurement of Q at the weir, were somewhat defective. The surface of the water, from the outlet tank to the hook-gauge at the weir, had a considerable inclination in the ditch ; for instance in Experiment No. 349, the surface at the tank was 77.90, while at the gauge it was (75.36 + 1.46) 76.82, thus having a fall of 1.08 in about 600 feet horizontal, and giving a considerable initial velocity to the stream, which is perhaps not fully taken into account by the mean velocity, r_a, where it was measured at the section **a b**, Fig. 4. As there were no screens across the ditch, this initial velocity was not at all checked. Unfortunately the surface velocity from **E** to **H** was not measured ; this would have much more satisfactorily shown the effect of the approaching velocity, than its computation from v_a at the section **a b** as hereafter given.

The position of the hook-gauge was also objectionable, being placed on the down-stream side of the weir in comparatively dead water, where the surface was in all probability slightly elevated by the force of the current, as the stream passed by towards the weir.

The hook-gauge, reading to .001 inch, was placed about .7 north-west from the side of the flume in a vertical box, extending some distance below the crest of the weir ; small holes were bored through the sides of this box from top to bottom, so that the level of the interior water very speedily became the same as that in the exterior canal.

For each experiment the water was allowed to run for some 20 minutes, until a nearly perfect regimen was established in the ditch, and then the hook-gauge was read every minute for a period of from 10 to 30 minutes, until the readings became practically constant. The mean of these last readings for a period of from 6 to 13 minutes was taken as the correct value of H at the weir. The largest fluctuation—difference between maximum and minimum—in the final gauge readings was in No. 347, being in that experiment .0014 (.8118 − .8132); the average fluctuation in these readings for the 15 experiments was .0007.

The measured value of H, aside from the defects of position, can therefore be considered as having been very accurately ascertained. There was but little wind at the time, which accounts for the steadiness of the water.

There was a slight leakage at the stop-gate **A**, and also under the weir; also a very slight leakage from the pipes. This was in part measured, and in part estimated; by reference to the following table it will be observed that the total leakage was an inconsiderable fraction of Q.

The mean velocity of approach, v_a, as measured at the section **a b**, Fig. 4, Plate XVI., has been reduced by the formula $h = H + b\,\dfrac{v_a^2}{2\,g}$, b having a constant value of 1.4.

The inner depth below crest was 3.08, so contraction can be considered as complete, except with No. 349, where there was probably a slightly increased discharge, due to partial bottom suppression. Hence for the following reduction, c will be considered as c_s, and its value will be taken from curve on Plate VII., for $l = 2.6$.

TABLE XCVIII.

Hamilton Smith, Jr.—Determination of Volumes of Water passing through Pipes.   15 North Bloomfield Experiments Measured over Sharp-Edged Iron Weir ;   Full Contraction.

$$G = 3.08 \qquad (2\,g)^{\frac{1}{2}} = 8.018 \qquad Q' = c\,\tfrac{2}{3}\,(2\,g\,h)^{\frac{1}{2}}\,l\,h$$

No.	H	v_a	h_a	h	l	Plate VII. c_i	Q'	Leakage.	Q
340	.8486	.32	.0022	.8508	2.586	.5969	6.475	.050	6.525
341	.7702	.28	.0017	.7719	2.586	.5989	5.614	.030	5.644
342	.6604	.24	.0013	.6617	2.586	.6018	4.478	.037	4.515
343	.6062	.22	.0011	.6073	2.586	.6034	3.947	.025	3.972
344	.5092	.18	.0007	.5099	2.586	.6065	3.053	.018	3.071
345	1.0913	.42	.0038	1.0951	2.586	.5908	9.359	.017	9.376
346	.9414	.38	.0031	.9445	2.586	.5944	7.542	.030	7.572
347	.8124	.30	.0020	.8144	2.586	.5977	6.072	.025	6.097
348	.6116	.22	.0011	.6127	2.586	.6032	3.999	.025	4.024
349	1.4611	.58	.0073	1.4684	2.585	.5828	14.329	.036	14.365
350	1.3344	.55	.0066	1.3410	2.585	.5855	12.563	.024	12.587
351	1.1442	.45	.0044	1.1486	2.586	.5897	10.034	.020	10.054
352	1.0349	.41	.0037	1.0386	2.586	.5923	8.666	.024	8.690
353	.9889	.38	.0031	.9920	2.586	.5934	8.104	.024	8.128
354	.7292	.27	.0016	.7308	2.586	.6000	5.181	.018	5.199

Stones weighing from 2 to 18 lbs. were sent through the pipes, during Experiments Nos. 340, 342, 349, 350, 346 and 345, and the time of passage determined by a stop-watch beating ½ second.* After No. 347 had been made with the 13-inch pipe, two stones were put in the entrance, but were caught by a projecting edge of a joint, and remained lodged in the pipe during Experiments Nos. 346 and 345. Some time after these investigations the pipe was taken apart, and the stones were found lodged as stated. These obstructions doubtless retarded the flow in Nos. 346 and 345.

* The moment of the escape of such a stone from the lower end of a pipe, can be very well determined by sound.

The times of passage of these stones were as follows;

Pipe.	No.	Times.		Weight of Stone in Pounds.	Specific Gravity.	Description of Stone.
		Maximum.	Minimum.			
11-inch	340	77.	...	3	...	Oval quartz pebble.
		...	72.7	17	2.3	Rough edged. Volcanic tufa.
	342	126.5	...	8	2.5	Round quartz boulder.
		...	123.3	8	2.1	Rough edged. Volcanic tufa.
15-inch	349	63.3	...	6	2.3	Smooth edged. Volcanic.
		...	87.6	18	2.3	Nearly round. ,,
	350	76.1	...	5	2.3	Rough edged. ,,
		...	73.6	10	2.7	Black quartz boulder, nearly round.
13-inch	346		94.	8		This experiment unreliable, as after Experiment No. 347 for flow had been made (No. 347 preceding No. 346 in point of time), two stones had lodged in pipe.

The water was nearly clear for Nos. 340 to 343; it was muddy for Nos. 344 and 354, and more or less discolored for the other experiments.

The temperature of the water was not observed. It did not, however, vary much, and must have been not far from 55° Fahr.

The final reductions of these 15 experiments will be found in the following table. As before stated, the co-efficient, o, of contraction at the entrance will be assumed as 1.; hence the effective head, h, is obtained by, $h' = \frac{v^2}{2g}$ and $h = H - h'$. As heretofore,

$$r = n \,(r\,s)^{\frac{1}{2}} = n \left(\frac{h \frac{1}{4} D}{l}\right)^{\frac{1}{2}}.$$

$$Q = n\, D^2 \frac{\pi}{4} \left(\frac{h \frac{1}{4} D}{l}\right)^{\frac{1}{2}} = .3927\, n \left(\frac{h\, D^5}{l}\right)^{\frac{1}{2}}.$$

$$(2\,g)^{\frac{1}{2}} = 8.018.$$

No corrections will be made either for the rivet-heads, or for the two main angles in the pipe line, which were respectively 9° and 11°. Probably the projections or irregularities at the "stove-pipe" joints retarded the flow more than the rivet-heads and curves combined; this, however, can only be a matter of conjecture.

TABLE XCIX.

Hamilton Smith, Jr.—*Determination of n, in c = n (r s)½. North Bloomfield, October, 1876.*

No.	Length			Mean Diameter D	Mean Area a	Table XCVIII Q	Velocity $\frac{Q}{a}=v$	Elevations		H	Head Absorbed in imparting Velocity k'	Effective Head		Maximum Velocity of Stones sent through Pipes.	Trans. Am. Soc.C.E.,1883 Value of n there given.
	Funnel	Main	l					Pen-Stock.	Discharge End.			h	n		
340	7.8	677.0	684.8	.9305	.6511	6.535	10.091	102.40	78.18	24.220	1.562	22.658	113.5	9.42	115.8
341	7.8	689.2	697.0	.911	.6518	5.644	8.659	102.22	83.215	19.005	1.166	17.839	113.4		113.5
342	7.8	706.1	713.9	.911	.6518	4.515	6.927	102.29	89.44	12.850	.746	12.104	111.5	5.79	111.9
343	7.8	713.5	721.3	.911	.6518	3.972	6.094	102.34	92.14	10.200	.578	9.622	110.6		111.0
344	7.8	722.8	730.6	.911	.6518	3.071	4.712	102.30	95.745	6.555	.345	6.210	107.1		108.1
345	12.0	672.9	684.9	1.056	.8758	9.376	10.706	102.37	77.76	24.610	1.783	22.757	114.4		114.9
346	12.0	687.6	699.6	1.056	.8758	7.573	8.646	102.42	85.73	16.690	1.163	15.527	113.0		113.4
347	12.0	697.2	709.2	1.056	.8758	6.097	6.962	102.74	91.855	10.885	.754	10.131	113.4	7.4 (?)	113.7
348	12.0	706.4	718.4	1.056	.8758	4.024	4.595	102.65	97.52	5.136	.328	4.803	109.4		109.8
349	14.8	669.6	684.4	1.230	1.1882	14.365	12.090	102.29	77.90	24.390	2.274	22.116	121.3	11.88	123.6
350	14.8	680.8	695.6	1.230	1.1882	12.587	10.593	102.31	83.385	18.925	1.746	17.179	121.6	9.98	123.5
351	14.8	690.2	705.0	1.230	1.1882	10.054	8.462	102.08	89.365	12.715	1.114	11.601	119.0		119.8
352	14.8	695.9	710.7	1.230	1.1882	8.690	7.414	102.550	93.000	9.550	.832	8.718	119.1		119.4
353	14.8	697.6	712.4	1.230	1.1882	8.128	6.841	102.610	94.065	8.545	.728	7.817	117.8		118.1
354	14.8	705.1	719.9	1.229	1.1863	5.199	4.383	102.450	98.535	3.915	.299	3.616	111.6		112.0

These 15 experiments were reduced in 1876, and the results published in the Trans. Am. Soc. of C.E., 1883; the values of n thus deduced are given in the last column of the preceding table. They have been here again reduced from the original notes, in accordance with the views expressed in our chapters on Velocity of Approach and Weirs.

By comparison it will be observed that our present values of n appreciably vary from those first given. These discrepancies nearly altogether result from a much smaller value being now attached to the effect of velocity of approach at the measuring weir, than was thought by the author in 1876 to be a proper correction for that cause. He was then a believer in the Bernoulli theorem "that no force is lost," but from a careful study of the experimental data which has before been fully discussed, has largely changed his views in this respect.

Owing to the initial velocity in the ditch feeding the weir, it is probable that in these later computations we have underestimated h_a', but this is perhaps sufficiently compensated for by the faulty position of the hook-gauge. The diameters, as we have before explained, are probably very slightly underestimated. Owing to slight partial bottom suppression at the measuring weir, which has been disregarded, the chances are that n for No. 349, and perhaps for No. 350, is placed a little too low.

The results here given, with the exceptions of Nos. 345 and 346, can be regarded, however, as being nearly as accurately determined, as is practicable with large pipes, where Q is obtained by indirect measurement. In point of accuracy this series cannot be considered as comparable with the preceding experiments with small pipes made at New Almaden.

The foregoing remarks well illustrate the uncertainties attending the weir measurement of water, unless the investigations are conducted by one thoroughly familiar with the minor laws governing the flow over weirs. It is hardly to be expected that an engineer, busied with important works of construction, can devote sufficient time to master a subject, which has perplexed the minds of such great physicists as Bernoulli and Dubuat.

Humbug Pipe.—*26 inches diameter.*

The main water supply for the North-Bloomfield mine is conducted across Humbug Cañon by two sheet-iron riveted pipes, each 26 inches in diameter, and having a united delivery capacity of about 90 cubic feet per second.

In 1873 the discharge through one of these pipes was determined by the flow through apertures under a constant head, but under conditions not favorable for very exact measurement. It was not practicable to measure Q directly.

The length, mean diameter, and head (H) were obtained with accuracy, the methods employed being the same as with the experiments just described.

The pipe had been coated with a preparation of asphaltum and coal tar when first laid in 1868, and its interior surface was very smooth at the time of the experiment. The thickness of the iron was .0054 ; as it was single riveted with light rivets, the rivet heads formed no obstruction worthy of consideration. Neither were the bends in the pipe sharp enough to perceptibly retard the flow.   The stove-pipe connections were similar to those employed for the three North Bloomfield pipes.

The discharge was submerged in the outlet tank.   The course of the ditch into which the water escaped, was at right angles to the line of the pipe, as was also the case with the feeding ditch at the pen-stock ; hence there was no disturbing element at either end from velocities in the ditches.

The pipe had a short funnel-shaped mouth-piece, which is included in the given length; co-efficient of contraction is hence assumed at 1., and $h = H - \dfrac{v^2}{2\,g}$.

TABLE C.

Hamilton Smith, Jr., 1873.—North Bloomfield 26-Inch Sheet-Iron Pipe.   Determination of n,

$$in\ v = n\,(r\,s)^{\frac{1}{2}} = n\,\left(\frac{h\,\frac{1}{2}\,D}{l}\right)^{\frac{1}{2}}.$$

No.	Length.	Mean Diameter.	Mean Area.	Q	$\frac{Q}{a}=v$	Total Head.	Head Absorbed in Imparting Velocity.	Effective Head.	n	Maximum Velocity of Stones passing through Pipe.
	l	D	a			H	h'	h		
355	1193.8	2.154	3.643	45.92	12.605	22.067	2.471	19.596	134.1	11.24

Cubical wooden blocks were also sent through the pipe, and occupied a very slightly larger space of time in passing through, than did stones weighing 10 to 15 lbs.

The element of uncertainty in this experiment is the value of Q.   It is thought that the quantity given should be within 4 per cent. of the truth, with the chances that it is stated somewhat in excess of its real value.   The reason for this supposition is, that while the velocity of stones passing through the pipe was only 11.24, the calculated mean velocity of the water was 12.60, which appears to be too great a difference.   In other experiments these differences were as follows :

Experiment.	Diameter of Pipe.	Velocity of Stones or Wooden Blocks.	Mean Velocity of Water.
No. 340	.91	9.43	10.02
„ 342		5.79	6.93
„ 349	1.23	11.88	12.09
„ 350		9.58	10.59
„ 355	2.15	11.24	12.60
„ 356	1.42	20.94	20.14

The inclination of the Bloomfield pipe, No. 355, was not quite as steep as that of Nos. 340 and 349, and not nearly as steep as No. 356 ; nor were the rivet heads in it any larger than those in Nos. 340 and 349, while they were much smaller than those in the Texas Creek pipe, No. 356.

It may be also observed that the velocity of the stones in No. 355, was slightly greater going down the incline to the synclinal part of the pipe, than in ascending the hill on the opposite side.

TEXAS CREEK PIPE.—17 *inches diameter*.

The experiment with this pipe is most interesting, owing to the great head of 300 feet, and consequent velocity of over 20 feet per second ; we will therefore describe it in considerable detail.

The pipe experimented upon was laid across the Big Cañon branch of the South Yuba river, Nevada County, California, in 1878 ; the following described measurements were made in 1878 and 1879. The elevation of the pen-stock is about 5500 feet above sea-level.

The reader is referred to Trans. of Am. Soc. of C.E., February, 1884, for a profile of the pipe line, and for mechanical details which are not here pertinent.

The pipe is made of wrought-iron sheets, double riveted on the long seam, in joints of about 20 feet in length ; these joints for a linear distance of 1350 feet are put together stove-pipe fashion, and for the remaining 3090 feet by an inner sleeve, with an outer band, the spaces between the pipe and the band being filled with lead.

The templates used in the shop where it was built, show the following dimensions:*

---

* The thicknesses of the iron as given here, are slightly different from those stated in the paper in the Trans. Am. Soc. of C.E.

1350 linear feet, iron	.0069 thick, internal diameter	= 1.4156
220 ,, ,, ,,	.0079 ,, ,, ,,	= 1.4146
240 ,, ,, ,,	.0091 ,, ,, ,,	= 1.4175
250 ,, ,, ,,	.0104 ,, ,, ,,	= 1.4137
320 ,, ,, ,,	.0114 ,, ,, ,,	= 1.4243
610 ,, ,, ,,	.0133 ,, ,, ,,	= 1.4233
1450 ,, ,, ,,	.0138 ,, ,, ,,	= 1.4227

4440 feet total length { Diameter by arithmetical mean = 1.4188
                        {   ,,   ,, geometrical ,,   = 1.4196

The mean of a large number of end measurements, made after the pipe was delivered, was 1.4166, and is probably more accurate than the geometrical mean of 1.4196 obtained from the templates, as this latter size was probably a trifle diminished by the draw of the rivets.

The pipe had an unusually heavy coating of asphaltum and coal tar, which diminished its internal diameter fully .001. Hence 1.416 can be assumed with safety as representing with much exactness the mean diameter, D.

The rivet heads for the larger thicknesses of the iron formed noteworthy obstructions for, say, one-half the length. The curves were made comparatively easy, but with the very high velocity of 20, doubtless somewhat retarded the flow. Neither of these two retarding influences will, however, be taken into account in our computations, as there are practically no data of value to enable one to even approximately estimate their effects.

Comparing the condition of the interior surface of the Texas Creek pipe with those used in Experiments Nos. 340-354, it can be considered as being appreciably rougher.

There was a short funnel-shaped mouth-piece at the entrance, which is included in the stated length of the pipe; the co-efficient of contraction for such an entrance would be in the neighborhood of .90; as the length of the funnel is insignificant compared with the total length, this co-efficient, o, will be placed at .92, and the increased diameter of the funnel will be neglected.

The discharge was submerged, the pipe terminating at the bottom of the outlet tank. Neither the velocity of the water entering the pen-stock, nor the velocity of the water escaping from the outlet tank, practically had any effect upon the flow through the pipe.

The total head, H, was the difference in level between the surface in the two tanks. It was determined by the mean of two lines of levels, and, as stated (303.62), is within .35 of the truth.

The length of the pipe as laid was twice measured, and $l = 4438.7$ as given, cannot be in error more than 1.5.

The standard of measure was the same as that used in all the preceding experiments—U.S. Coast Survey.

This pipe line is across a very precipitous mountain cañon, as is shown by the profile before referred to, which accounts for the rather large limits of error as above given.

The temperature of the water during the several trials varied from about 60° to about 50° Fahr.. The water was from a clear mountain stream, and was perfectly limpid.

The volume of water passing through the pipe was measured in May, 1879, first by the flow through orifices 1×1.0625, and then over a weir 5.48 long. These measurements were made at the lower end of a wooden box or canal, situate just above the upper mouth of the pipe. The canal was horizontal, 16 feet long, $13\frac{1}{2}$ feet wide, with its bottom 1.83 feet below crest of weir (G). The water entered the upper end of this canal with considerable force, which in these first two experiments was only partially checked by a grating or rack placed across the upper end of the canal.

These two experiments, after allowances for wastage at the pen-stock and a small loss by leakage, gave the following results;

$$Q \text{ by weir measurement} = 31.69$$
$$Q \text{ by orifices} \qquad = 30.90$$

Neither of these experiments, however, was regarded as reliable owing to the large initial velocity in the canal before spoken of, and for which in both cases a considerable correction was applied to H.

There was much more suppression of contraction at the bottom with the orifices than with the weir, which doubtless accounts for the smaller value of Q deduced from the flow through the orifices.

After greater precautions had been taken to check the primary current in the canal, the experiment was repeated on June 1st, 1879, the total head, H, being almost exactly the same as it was in the former trials; we will use this last determination for our data.

The weir was 5.500 long, vertical, with square corners; the crest was a plank 1 inch wide, planed true, and with perfectly square corners. The discharge was hence free into the air, the escaping vein only coming in contact with the inner corners. The inner depth below crest, G, was 1.83. The weir was placed in the centre of the canal, with each end distant 4 feet from the side of the canal, so that there was almost perfect end contraction. As H was 1.50, with $G = 1.83$, there was an appreciable bottom suppression, which we will assume added $1\frac{1}{4}$ per cent. to the discharge, this correction being in accordance with the experimental data for partial suppression we have given in the first portion of this volume.

The velocity of the surface in the central line of the canal opposite the hook-gauge was very nearly 1.00, while the mean velocity, v_a, at the gauge was .736; we will adapt the former velocity with the factor of correction, $b = 1.$; hence $h_a' = \dfrac{1^2}{2\,g} = .0155.$* The

---

* With $v_a = .736$, and $b = 1.4$, $h_a' = b \dfrac{v_a^2}{2g} = .0118.$

primary or initial velocity as the water entered the canal was not entirely checked, as is indicated by the considerable difference between the surface and the mean velocities. This value of h'_a, .0155, is probably a little too low.

The measured head, H, was 1.5050; the water in the little gauge-box, in which the hook-gauge was placed, was quite steady, the fluctuations in level being minute.

The co-efficient c_c (full contraction) on Plate VII. for $l = 5.5$ and $h = 1.50$, is .5932; adding 1½ per cent. to this for partial bottom suppression, for this weir $c = .6011$. Then, with $Q' = c \frac{2}{3} (2 g h)^{\frac{1}{2}} l h$, and $(2 g)^{\frac{1}{2}} = 8.017$, we have;

H^*	h_a'	h	l	c	Q'
1.5050	.0155	1.5205	5.500	.6011	33.129

Of this volume there was wasted at the pen-stock the flow over a weir with full contraction, and with very little velocity of approach, as follows;

$H = h$	l	c , Plate VII.	Q''
.417	1.50	.605	1.306

There were also losses by leakage at the pen-stock and by small leakages from the pipe, aggregating about $.102 = Q'''$. Hence $Q' - (Q'' + Q''') = Q = 31.721$, being the volume passing through the pipe.

Hence we have by the formulas,

$$ v = n \, (r \, s)^{\frac{1}{2}} = n \left(\frac{h \frac{1}{4} D}{l} \right)^{\frac{1}{2}} ; $$

TABLE CI.

Hamilton Smith, Jr., and H. C. Perkins, 1879.—Texas Creek 17-Inch Shot-Iron Pipe.

No.	Length. l	Mean Diameter. D	Mean Area. a	Q	$\frac{Q}{a} = c$	Total Head. H	Losses of Head. h'	Effective Head. h	n	Maximum Velocity of Wooden Blocks.
336	4438.7	1.416	1.5748	31.721	20.143	303.62	7.46	296.16	131.1	20.94

Six blocks of wood, each 4 inches × 6 inches × 6½ inches, with corners rounded, and loaded so they would just sink—specific gravity 1.05—were sent through the pipe; the time of passage varied from 211 seconds to 219½ seconds; the mean of the two least times was 212 seconds, showing a velocity of 20.94. Two similar blocks, unloaded—specific gravity of, say, .8—passed through the pipe in 213 seconds and 216 seconds respectively. A small round stone weighing ½ lb. came through in 231 seconds, and another weighing 3 lbs. in 232 seconds, thus having a velocity of 19.2.

---

* With a trifle less water passing over the weir, and with its length 5.480 instead of 5.500, the measured head, H, was in the first experiment in May, 1879, 1.4830. This shows the effect of the initial velocity spoken of.

The above experimental value of n[*] compares favourably in point of accuracy with Experiments Nos. 340-354.

---

[*] The same value for n is given in the Transactions of the American Society of Civil Engineers, although the values of h, and e for the weirs are somewhat different, one discrepancy balancing the other. The computations as here given have been made from the original notes, in conformity with the views expressed in the first portion of this volume.

CHAPTER XI.

EXPERIMENTS WITH ORIFICES, 1884-5.

HOLYOKE.

BEING especially desirous of accurately determining what differences there were between the co-efficients of efflux for submerged vertical orifices, and those for the same orifices with a free discharge, the author during the summer of 1884 made a large number of experiments at Holyoke, Massachusetts. The experiments with a free discharge, were repeated in 1885. The general arrangement of the apparatus employed is shown by Fig. 1, Plate XVII. The supply of water was admitted to the pressure tank, **A**, by an iron pipe, with a stop valve just above the tank ; this pipe received its supply from an upper reservoir, not shown by the sketch, where the surface of the water was maintained at a nearly constant level.

The orifices to be experimented upon, were pierced in brass plates of $\frac{1}{4}$ inch thickness ; those for the two round and two square orifices, were .71 square, the orifice proper being slightly divergent, with a thickness of $\frac{1}{12}$ inch, as shown by Fig. 2, Plate XVII. The plates were made of this considerable thickness, in order to secure entire rigidity, so that the form of the orifice could not be changed by any possible unequal strains, due either to the wood screws by which the plate was fastened to the side of tank **A**, or to changes in the plank to which it was screwed. A rubber gasket, between the plate and the wood, formed a water-tight joint.

There were 5 brass plates used ; two circular, having the respective diameters of about .05 and .1 ; two square, having the respective sides of about .05 and .1 ; and one oblong, being a rectangle having a length of about .3 and a width of .05. A plate of boiler iron, with a round orifice having very nearly the same diameter as the .1 round orifice in the brass plate, was used for a few experiments in 1885, to determine whether or not there was any difference of discharge for orifices in different metals. The holes for the round orifices when examined by a microscope of high power proved to be almost absolutely perfect ; the edges of the rectangular openings were not so perfect.

The sizes of the orifices were determined with great care by Prof. W. A. Rogers, of Cambridge, Mass. The means of 3 sets of observations gave the following results, with $T = 62°$ for the 6 plates ;

Round	.05	Brass	Diameter	.049 802	Area	.001 9480
"	.1	"	"	.099 994	"	.007 8531
"	.1	Iron	"	.099 999	"	.007 8538
Square	.05	Brass		.049 756 × .049 800	"	.002 4778
"	.1	"		.100 100 × .100 089	"	.010 019
Oblong	.3 × .05	"		.299 736 × .049 892	"	.014 954

The measurements of the brass plates were made after the completion of the series of experiments made in 1884; after the experiments in 1885, the orifice in the brass plate, $D = .10$, was carefully remeasured and found to be .099 9964, or a trifle larger than the original determination. This enlargement may have been due to wear, or possibly to a little rough usage in June, 1885, by the mechanic who used this plate as a standard in making the orifice of nearly the same diameter in the iron plate.

The apparatus for exact measurement, employed by Prof. Rogers, is exceedingly perfect, and the foregoing sizes are doubtless within a very small fraction of the truth; the bronze yard of Prof. Rogers, which had twice been compared with the Imperial yard in London, was his standard of reference.

// The standard of measurement for capacity and heads was a graduated box-wood rod, whose slight errors had also been determined by Prof. Rogers.

The tank B had two discharge orifices, at a and b; when the discharge was submerged, the orifice at b was closed, and the water flowing through the experimental orifice, o, escaped at a: when the discharge was free into the air, the water escaped at b from the lower tank. These tanks A and B were very solidly built, with the bottom 5 inches thick, and sides 3 inches thick. With the foregoing arrangement it was essential for accuracy that there should be no leak from A into B, and no leak from B. It was thought that these conditions had been almost perfectly complied with. This arrangement was used in the year 1884.

The measuring tank C was made of pine, 3 inches thick, with outer clamping frames of wood, held together by iron rods. Its inner section was nearly 2.5 × 2.5, with a height of 4.5. The inner surface was smoothly planed, and the joints of the plank very neatly united. In this tank were placed a steel point at B, and three fixed steel hooks at M, U and T. Just before the commencement of an experiment, the surface of the water was accurately adjusted to either B, M or U, depending upon which vertical section of the tank was to be used for q. The water escaping from the tank B, was then introduced by a bent tin pipe into the measuring tank C, and the time taken, by either one or two stop-watches, the instant the assistant caught the jet at a or b in the tin pipe; when the water in the measuring tank had risen to nearly the summit of the desired hook,—either M, U or T—the tin pipe was quickly removed, and the time again taken. In 1884 the difference between the surface of the water and the standard hook, was measured by a small hook-gauge, and the capacity computed from this measurement; in 1885 the much more accurate method was employed of determining the capacity between the surface of the water and the summit of the fixed hook, by using a wooden rod, .25 × .25 and 2.5 long, graduated so that each division represented

.001 cubic foot ; this rod was depressed in a vertical position until the surface of the water rose just to the summit of the fixed hook ; the depth of the submerged portion of the rod instantly gave the correction to be applied.*   By waiting until the water in the tank was quiescent, these adjustments could be made to within .001 of a cubic foot.

The tank C was measured 5 times, by means of 7 equidistant horizontal sections, with 8 measurements to each section.‖ These measurements gave the following results, for the entire tank, B—T ;

Aug. 12, 1884	Tank nearly new	27.286
„ 27, „	„ 2½ weeks in use	27.306
Sept. 3, „	„ 3½ „ „	27.303
May 4, 1885	„ during the past winter had been somewhat enlarged by action of ice	27.333
„ 15, „ ...	At termination of May, 1885, series...	27.331

The enlargement from August 12-27, 1884, was due to swelling of the wood.   The tank was so firmly made, that there was no difference appreciable between its size when empty, and when full.

The tank was also measured on May 4th, 1885, by an iron vessel, which will be described hereafter, whose capacity had been determined to be by weight of water .9310, and by capacity of cylindrical tubes .9312 ; the mean of these measurements being .9311 with T = 46°.   By this vessel the capacity of the tank—B to T—was 27.345†.   Assuming the rod measurement of 27.333 to be correct, the capacity of the iron vessel was .9307, which was assumed to be its size when used for q at Holyoke in 1885.

The contents of the several vertical sections or divisions of the tank were determined, by assuming that the total capacity, or B—T, was 27.333 ; the results of the flow through the orifices, with nearly uniform heads, filling first one section and then another, were compared, and the sub-division of the total capacity thus made ;

$$\begin{array}{l} \text{B—M} = 16.968 \\ \left.\begin{array}{l} \text{M—U} = 4.881 \\ \text{U—T} = 5.484 \end{array}\right\} 10.365 = \text{M—T} \\ \overline{\text{B—T} = 27.333} \end{array}$$

Computed from the rod measurements these capacities were as follows ;

$$\begin{array}{l} \text{B—M} = 16.975 \\ \text{M—U} = 4.875 \\ \text{U—T} = 5.483 \\ \overline{\text{B—T} = 27.333} \end{array}$$

---

* In 3 or 4 experiments the water was above the hook ; in these cases the correction was ascertained by a hook-gauge.

† Tank C adjusted at B, and 29 pails of water poured in, the pail being held by the handle when filled ; the surface of the water was then .0555 below hook T.   The horizontal area of the tank at this point is 6.1837 ; hence,

29 × .9311=27.002

.0555 × 6.1837 = .343

(T of water 48°)   27.345=capacity tank C, from B to T.

Errors in direct measurement of the small sections were necessarily much larger, proportionately, than for the whole tank. Hence the indirect method of segregation was preferred; very likely 4.879 would have more accurately represented the capacity of M—U, than 4.881, the quantity adopted; this difference would represent a comparative error of $\frac{1}{2400}$ in the value of M—U, $\frac{1}{5000}$ in M—T, and $\frac{1}{6000}$ in B—M.

The vertical elevations B—M et cet. remained constant during the experiments of 1885, showing that the tank remained unchanged in its vertical plane.

// The leakage of the measuring tank in 1884 was quite appreciable, ranging from .0025 cubic foot, to .000 05 cubic foot per minute. In 1885 the tank when filled to T, lost a maximum of only .0026 cubic foot per hour; as the longest experiment in this year was only about 30 minutes, this loss was not appreciable, being probably due entirely to evaporation. We therefore had in 1885, a measuring vessel *without leakage* and of unchanging form, a desideratum rarely obtained.

The stop-watches used beat to $\frac{1}{4}$ seconds. In 1885 only one watch was used, whose rate when compared with a good second pendulum clock, varied from +.0019 to +.0009; part of this maximum variation was doubtless due to personal error; a uniform + rate of .001 25 was assumed as a correction for May, 1885, being the mean of a large number of determinations, with times varying from 60 to 1800 seconds. For June, 1885, a constant + rate of .001 was adopted.

When the experiments made in 1884 had been computed, the results presented were slightly inharmonious. The curves for each orifice were symmetrical, but they did not agree comparatively close enough, one with the other. Determined to know the reason of these discrepancies, in 1885 we made a new series of experiments, all with free discharge, using every possible precaution to prevent error. These new series of experiments revealed the fact, that in 1884 the slight known leak from the tank B was large enough to affect the results; this leakage was so small as not to notably —hardly appreciably—affect the co-efficients from the three larger orifices with the larger heads, but it became quite notable with the two smaller orifices with small heads. The tanks A and B rested upon a plank floor, which was entirely open below; during 1884 the leakage from both A and B seemed to be only a few occasional drops falling from this floor; doubtless much the larger portion of the leakage from B was absorbed by evaporation. //

The possibilities of error for the series of 1885 will be critically discussed hereafter.

Details of the 1885 series will only be given; we will hereafter state the general results of the 1884 series.

FREE DISCHARGE, 1885.

For heads of less than 1., the experimental plates were placed at the lower end of the tank B at o'; the centre of each orifice being .75 above the floor and distant 1.5

from each side of the tank **B**. A hook-gauge was placed at f, and the elevation of its zero-point determined at least three times each day by a delicate spirit level, extending from the bottom of the orifice to the summit of the hook. The maximum variations of these determinations were very slight, as the wood of the tank **B** was thoroughly damp. This gauge was read to about .0002.

The water came to the orifices through the racks **g** and **g'** very evenly ; the perfection of the jets escaping at o' proved that the supply was evenly distributed, this being a wonderfully delicate test. The surface of the water at f was so steady, that no enclosing box was used for the hook-gauge.

The iron pail was sometimes used for *q* for the .05 round orifice with small heads. When this vessel was employed, it was quickly shoved upon a horizontal plank, into position to catch the jet ; this instant was the beginning of *t*. When nearly full the pail was held by the handle about half an inch above the plank, and the time ended when the pail was brimming full.

We may remark that in all these experiments, both of 1884 and 1885, the quantity *q* received in a measuring vessel was as nearly as possible the true quantity discharged from the orifice during the observed time, *t*.*

For heads above 1.7, the experimental plates were placed at **o**, on the side of the tank **A**, and, as at **o'**, with their centres .75 above the floor, and 1.5 from the sides of the tank. The heads for this tank were measured by the hook-gauge at **c**, which was read to .0005; the elevation of the zero point of this gauge was determined, by first measuring the height from the bottom of the orifice to the fixed hook **d**, which was exactly vertical above the centre of the orifices; the water in the tank was then brought to the level of the hook **d**, and a reading taken of the gauge at **c** ; these measurements were generally repeated three times each day. The variations of this zero-point were appreciable, being as much as .003 during the first period of twelve days, when nearly all the experiments were executed, and being probably due to unequal swelling of the wood mounting of the gauge, and the tank. The possible error from this source could not be over .0015, if indeed as much, for any one experiment; such a possible error for *H* over 1.7, would be barely appreciable.

When an orifice was placed at **o**, the lower end or side of the tank **B** was removed, and the water was conducted from **o** to the end of **B** through a small wooden square pipe or trough, which was perfectly tight, not a single drop escaping from it. From the end of this wooden pipe, the water was at will directed into the measuring tank by the bent tin pipe before described.

The use of the screen in the tank **A**, enclosing the iron feeding pipe, did not result in a *perfectly* uniform supply of water to the larger orifices, when the highest heads were employed. This was indicated by the slight imperfections of the .1 round jet.

----

* Excepting the leakage from the tank B in 1884, which will be discussed hereafter, and slight losses with the .3 × .05 orifice, when dividing inner plates were attached.

Experiments were made, without any screen around the feeding pipe, to see whether or not this uneven supply of water had any appreciable effect upon the discharge.

The experiments were made at an elevation of 90 feet above sea level, in latitude $43° 12'$; hence $(2 g)^{1/2} = 8.0210$, *vide* Table II.

In the following table, the exact time of the beginning of each experiment is given, in order to chronologically trace variations in c. That is to say, month, day, hour and minute. Hours from 8 to 12 are a.m., and those from 1 to 6 are p.m.

The given temperature, T, is that of the water. During the experiments made in May, the temperature of the air did not differ more than $8°$ from that of the water. During the experiments made in June the temperature of the air varied only $3°$ from that of the water.

The corrected time, or t, is only given; the noted time, t', was always $\frac{1}{500}$ larger than t in May, and $\frac{1}{1000}$ larger than t in June.

H is the mean head during each experiment from the centre of the orifice. O is the maximum variation or oscillation of the head during each experiment. This variation was largely due to a slight leakage of air into the feeding pipe; this leakage caused a variation in the amount of vacuum in the iron pipe between the valve and its lower end, sometimes causing a slightly irregular flow; after the cause of this trouble was discovered, and the pipe made perfectly tight, there was but little difficulty in maintaining a nearly perfectly uniform head for each experiment. The hook-gauge was read either every 30, or every 60 seconds, depending upon the length of the experiment. H is the arithmetical mean of the gauge readings; where O was largest the means of the $\frac{1}{2}$ power of the various readings were also calculated, but they in no case differed appreciably from the arithmetical means.

The co-efficient c, representing the arithmetical mean of several experiments with nearly identical heads, has been calculated from C by Tables III. and IV.

The experiments were commenced on May 6th, and concluded on June 6th, 1885.

TABLE CII.

Holyoke, Massachusetts, May and June, 1885.—Flow through Vertical Orifices, with
$(2 g)^{\frac{1}{2}} = 8.0210$ $Q = C a (2 g H)^{\frac{1}{2}}$

Brass. .05 round.

No.	Date, 1885.	T	t	q Vessel.	q Correction.	q	Q	H H	H O
19	May 9,	2-35	50"	212.2	.9307		.9307	.004 386	.1849	.0010
	" "	2-39		212.3	"		.9307	.004 384	.1856	.0005
	" "	2-43		212.0	"		.9307	.004 390	.1855	0
	" "	2-47		212.0	"		.9307	.004 390	.1855	0
20	" 12,	4-11	52"	1072.3	4.881	— .028	4.853	.004 496	.1901	.0066
	" "	4-33		1078.1	"	— .036	4.845	.004 494	.1889	.0003
21	" 8,	1-53	49"	205.9	.9307		.9307	.004 520	.2009	.0030
	" "	1-57		205.8	"		.9307	.004 522	.1996	.0008
	" "	2-02		205.6	"		.9307	.004 527	.1999	.0005
	" "	2-06		205.9	"		.9307	.004 520	.1996	.0005
22	" 13,	9-24	51"	204.2	.9307		.9307	.004 558	.2028	.0003
	" "	9-28		204.3	"		.9307	.004 556	.2026	0
	" "	9-32		204.1	"		.9307	.004 560	.2029	.0001
	" "	9-36		204.0	"		.9307	.004 562	.2029	0
23	" 13,	9-02	51"	188.6	.9307		.9307	.004 935	.2404	.0008
	" "	9-06		188.8	"		.9307	.004 930	.2403	.0001
	" "	9-09		188.9	"		.9307	.004 927	.2399	.0001
	" "	9-13		189.1	"		.9307	.004 922	.2398	.0001
	" "	9-17		189.2	"		.9307	.004 919	.2396	.0003
24	" 9,	11-11	50"	918.7	4.881	— .048	4.833	.005 261	.2830	.0005
	" "	11-31		1033.8	5.484	— .045	5.439	.005 261	.2821	.0010
25	" 9,	2-15	50"	173.7	.9307		.9307	.005 358	.2832	.0005
	" "	2-19		174.3	"		.9307	.005 340	.2814	.0015
	" "	2-22		174.1	"		.9307	.005 346	.2812	.0005
	" "	2-26		173.2	"		.9307	.005 374	.2827	.0010

TABLE CII.

Free Discharge into Air. Contraction perfect. Jets only touching Inner Edges of Orifices.
c from Tables III. and IV.

$D = .049\,802 \qquad a = .001\,9480$

No.	C		Means.		REMARKS.
	Experi ment.	Mean.	H	c	
19	.6528 .6513 .6523 .6523	.6522	.185	.6525	Jets for all experiments with this orifice, beautifully perfect from the orifice to the floor. When orifice was placed at O′, this vertical fall of the escaping jet was over 7 ft.
20	.6600 .6617	.6608	.190	.6611	
21	.6454 .6478 .6480 .6475	.6472	.200	.6475	
22	.6477 .6474 .6479 .6482	.6478	.203	.6481	
23	.6441 .6436 .6438 .6433 .6432	.6436	.240	.6438	
24	.6329 .6340	.6334	.283	.6336	
25	.6444 .6442 .6452 .6468	.6451	.282	.6453	

TABLE CII.—*continued.*

No.	Date			T	t	Vessel.	Correction.	q	Q	H	O
	May 8,		2-18		173.5	.9307		.9307	.005 364	.2823	.0005
	,,	,,	2-21		173.2	,,		.9307	.005 374	.2823	.0005
26	,,	,,	2-25	49°	173.1	,,		.9307	.005 377	.2825	0
	,,	,,	2-29		173.2	,,		.9307	.005 374	.2825	0
	,,	9,	8-51		173.7	.9307		.9307	.005 358	.2830	0
27	,,	,,	8-55	49°	173.4	,,		.9307	.005 367	.2830	0
	,,	,,	8-59		173.3	,,		.9307	.005 370	.2830	0
	,,	8,	11-47		162.6	.9307		.9307	.005 724	.3355	0
	,,	,,	11-51		162.5	,,		.9307	.005 727	.3351	.0005
28	,,	,,	11-54	49°	162.6	,,		.9307	.005 724	.3350	0
	,,	,,	11-57		162.7	,,		.9307	.005 720	.3345	0
	,,	13,	8-44		161.1	.9307		.9307	.005 777	.3378	.0017
29	,,	,,	8-48	51°	160.9	,,		.9307	.005 784	.3359	.0006
	,,	,,	8-51		161.5	,,		.9307	.005 763	.3353	.0001
	,,	,,	8-55		161.1	,,		.9307	.005 777	.3356	.0001
30	,,	12,	5-01	52°	862.0	5.484	−.043	5.441	.006 312	.4034	.0091
	,,	,,	5-21		861.3	,,	−.054	5.430	.006 304	.3995	.0011
	,,	8,	11-31		143.3	.9307		.9307	.006 495	.4349	.0010
31	,,	,,	11-34		143.3	,,		.9307	.006 499	.4365	.0010
	,,	,,	11-37	49°	142.7	,,		.9307	.006 522	.4373	.0005
	,,	,,	11-40		143.2	,,		.9307	.006 499	.4375	0
	,,	8,	11-10		130.2	.9307		.9307	.007 148	.5359	.0010
	,,	,,	11-15		129.7	,,		.9307	.007 176	.5359	.0005
32	,,	,,	11-18	49°	129.8	,,		.9307	.007 170	.5370	0
	,,	,,	11-21		129.8	,,		.9307	.007 170	.5365	0
	,,	,,	11-24		129.7	,,		.9307	.007 176	.5355	0
	,,	9,	9-08		2050.0	16.968	−.065	16.903	.008 245	.7201	.0055
33	,,	,,	9-49	49°	585.3	4.881	−.061	4.820	.008 235	.7186	.0010
	,,	,,	10-04		655.8	5.484	−.094	5.390	.008 182	.7220	0

TABLE CII.—*continued.*

No.	C	Mean.	H	c	REMARKS.
					Brass. .05 round.—*continued.*
26	.6461 .6473 .6474 .6470	.6470	.282	.6472	
27	.6446 .6457 .6461	.6455	.283	.6457	
28	.6324 .6332 .6329 .6330	.6329	.335	.6330	
29	.6362 .6387 .6369 .6382	.6375	.336	.6376	
30	.6368 .6384	.6376	.401	.6377	
31	.6303 .6296 .6312 .6289	.6300	.437	.6301	
32	.6249 .6274 .6262 .6265 .6276	.6265	.536	.6265	
33	.6219 .6217 .6162	.6199	.720	.6199	For this determination expiration of *t* was not taken very exactly

TABLE CII.—*continued.*

No.	Date.		T	t	Vessel.	Correction.	q	Q	H	O
	May 9,	10-23		524.3	4.881	−.048	4.833	.009 218	.9132	.0035
34	,, ,,	10-36	50°	588.3	5.484	−.080	5.404	.009 186	.9122	.0020
	,, ,,	10-51		592.5	,,	−.063	5.421	.009 149	.9059	.0015
	,, 13,	9-47		1814.6	16.968	−.038	16.930	.009 330	.9289	.0010
	,, ,,	10-26		1816.7	,,	−.022	16.946	.009 328	.9288	.0012
35	,, ,,	11-01	52°	521.3	4.881	−.018	4.863	.009 328	.9303	.0015
	,, ,,	11-15		519.3	,,	−.034	4.847	.009 334	.9284	.0022
	,, ,,	11-28		585.1	5.484	−.032	5.452	.009 318	.9292	.0008
	,, ,,	11-43		584.8	,,	−.030	5.454	.009 326	.9278	0
	,, 13,	4-13		432.7	5.484	−.037	5.447	.012 588	1.7383	.0030
36	,, ,,	4-26	54°	431.4	,,	−.049	5.435	.012 599	1.7376	.0030
	,, ,,	4-39		431.2	,,	−.049	5.435	.012 604	1.7423	0
37	,, 13,	3-11	54°	659.2	10.365	−.047	10.318	.015 652	2.7231	.0100
	,, ,,	3-27		659.2	,,	−.026	10.339	.015 684	2.7349	.0040
	,, 13,	2-25		576.2	10.365	−.049	10.316	.017 904	3.5790	.0080
38	,, ,,	2-40	54°	576.2	,,	−.050	10.315	.017 902	3.5668	.0050
	,, ,,	2-54		578.5	,,	−.031	10.334	.017 864	3.5641	.0040
	,, 13,	1-35		506.1	10.365	−.037	10.328	.020 407	4.6581	.0100
39	,, ,,	1-49	54°	508.7	,,	−.046	10.319	.020 385	4.6152	.0140
	,, ,,	2-03		506.8	,,	−.044	10.321	.020 365	4.6276	.0010

Brass. .1 round.

No.	Date.		T	t	Vessel.	Correction.	q	Q	H	O
	May 12,	8-31		722.2	10.365	−.031	10.334	.014 309	.1296	.0025
40	,, ,,	8-47	50°	724.5	,,	−.027	10.338	.014 269	.1291	.0007
	,, ,,	9-04		726.9	,,	−.003	10.362	.014 255	.1268	.0011
41(?)	,, 12,	9-42	51°	507.3	10.365	−.044	10.321	.020 345	.2644	.0090
42	,, 11,	2-48	51°	394.6	10.365	−.033	10.332	.026 183	.4563	.0030
	,, ,,	2-58		393.5	,,	−.042	10.323	.026 214	.4583	.0010

TABLE CII.—*continued.*

No.	C	Mean.	H	c	Remarks.
	.6174				Brass. .05 round.—*continued.*
34	.6155	.6160	.910	.6160	
	.6152				
	.6196				
	.6195				
35	.6190	.6194	.929	.6194	For this series, adjustments and corrections for *q,* were made with the utmost precision.
	.6200				
	.6187				
	.6197				
	.6111				
36	.6117	.6113	1.74	.6113	
	.6111				
	.6071				
37	.6070	.6070	2.73	.6070	
	.6057				
38	.6067	.6060	3.57	.6060	
	.6056				
	.6051				
39	.6043	.6051	4.63	.6051	
	.6059				

$D = .099\,994$ $a = .007\,8531$

No.	C	Mean.	H	c	Remarks.
	.6310				Jet enlarges and diminishes three times in falling 7 feet.
40	.6305	.6307	.129	.6337	The section of the jet is uniform; the swell looking in front, being just opposite the stricture looking sideways.
	.6306				
41 (f)	.6281	.6281	.264	.6286	Head for this experiment jumping about badly.
	.6154				A little manifestation of alternate swelling and contraction of the jet, as above noted. Jet perfect at orifice.
42	.6152	.6153	.457	.6155	

TABLE CII.—*continued.*

No.	Date.		T	t	Vessel.	Correction.	q	Q	H	O
43	May 11,	2-16		329.1	10.365	−.044	10.321	.031 361	.6613	.0005
	" "	2-26	51°	329.6	"	−.042	10.323	.031 320	.6596	.0005
	" "	2-34		329.6	"	−.042	10.323	.031 320	.6618	.0020
44	" 11,	1-35	51°	748.0	27.333	−.065	27.268	.036 454	.8993	.0025
	" "	2-02		750.2	"	−.036	27.297	.036 366	.8998	.0023
45	" 14,	4-13		542.6	27.333	−.065	27.268	.050 254	1.7460	.0070
	" "	4-35	49°	545.6		−.027	27.306	.050 048	1.7283	.0010
	" "	4-54		545.3	"	−.032	27.301	.050 066	1.7300	0
46	" 6,	2-57	49°	402.9	27.333	−.064	27.269	.067 68	3.1806	.0030
	" "	3-14		403.4	"	−.047	27.286	.067 64	3.1776	.0005
47	" 6,	2-09		336.6	27.333	+.002	27.335	.081 21	4.5969	.0015
	" "	2-21	49°	335.0	"	−.113	27.220	.081 25	4.6019	.0025
	" "	2-42		336.3	"	−.035	27.298	.081 17	4.5948	.0020
48	" 6,	8-38	49°	335.9	27.333	−.030	27.303	.081 28	4.6023	.0090
	" "	8-55		336.6	"	0	27.333	.081 20	4.6019	.0005
49	" 6,	9-13	49°	402.2	27.333	−.035	27.298	.067 87	3.1985	.0010
	" "	9-29		402.2	"	−.031	27.302	.067 88	3.2007	.0005
50	" 5,	2-19	48°	401.5	27.333	−.037	27.296	.067 99	3.1946	.0010
	" "	2-37		402.1	"	−.006	27.327	.067 96	3.1941	.0010
51	" 5	3-10	48°	400.7	27.333	−.033	27.300	.068 13	3.2213	.0050
52	" 5,	10-08		336.0	27.333	−.031	27.302	.081 26	4.5968	.0025
	" "	10-24		336.5	"	0	27.333	.081 23	4.5965	.0015
	" "	10-40	48°	335.6	"	−.062	27.271	.081 26	4.5971	.0015
	" "	11-00		337.8	"	+.064	27.397	.081 10	4.5933	.0040
	" "	3-48		337.0	"	−.012	27.321	.081 07	4.5906	.0210

TABLE CII.—*continued.*

No.	C	Mean.	H	c	Remarks.
					Brass. .1 round.—*continued.*
	.6122				
43	.6122	.6119	.661	.6120	Perfect jet for the entire fall of 7 feet.
	.6112				
44	.6103	.6096	.900	.6096	Perfect jet as above.
	.6090				
	.6038				
45	.6044	.6042	1.73	.6042	Jet *very* nearly perfect for 2.5 feet; perfect at orifice.
	.6043				
46	.6025	.6025	3.18	.6025	Jet very nearly perfect, and sometimes apparently perfect, from orifice to where it strikes trough.
	.6024				
	.6013				
47	.6013	.6013	4.60	.6013	Jet seems to be perfectly true for .2 from orifice; then it sometimes begins to twist a little, and at other times is perfect for a distance of 1.5 from orifice.
	.6012				
48	.6015	.6012	4.60	.6012	For Nos. 48 and 49, no rack or protection of any kind around discharge end of iron feeding pipe. Jets for both heads are exceedingly ragged and twisting, being very far from smooth even immediately at orifice.
	.6010				
49	.6025	.6024	3.20	.6024	
	.6024				
50	.6039	.6038	3.19	.6038	For No. 50, one vertical plank, .93 high, placed across tank, A, just in front of feeding pipe; open on top. Jet ragged and twisting even at orifice; not apparently as bad a jet, however, as for Nos. 48 and 49.
	.6037				
51	.6026		3.22	.6026	For No. 51, two vertical planks, 1.86 high, placed across tank, A, just in front of feeding pipe; open on top. Jet ragged and twisting.
	.6017				
	.6015				For Nos. 52 and 53, four vertical planks, 3.92 high, placed in front of feeding pipe, open on top.
52	.6017	.6013	4.59	.6013	Jet twists a good deal, and is perceptibly not absolutely perfect near the orifice; very much better, however, than in Nos. 48 to 51, but not as good as No. 47.
	.6008				
	.6007				

TABLE CII.—*continued.*

No.	Date			T	t	Vessel.	Correction.	q	Q	H	O
	May	5,	11-21		403.7	27.333	0	27.333	.067 71	3.1883	.0340
53	„	„	11-38	48°	403.4	„	−.013	27.320	.067 72	3.1900	.0020
	„	„	11-53		403.5	„	−.003	27.330	.067 73	3.1875	0
54	„	5,	4-24	48°	337.4	27.333	−.012	27.321	.080 98	4.5973	.0020
55	„	5,	4-41	48°	336.1	27.333	−.017	27.316	.081 27	4.6190	.0140
	June	6,	11-43		524.3	27.333	−.040	27.293	.052 056	1.8767	0
56	„	„	12-01	62°	524.9	„	−.055	27.278	.051 968	1.8767	0
	„	„	12-19		526.3	„	−.041	27.292	.051 856	1.8637	0
57	„	6,	2-28	62°	501.4	27.333	−.040	27.293	.054 434	2.0487	0
	„	„	2-45		500.7	„	−.074	27.259	.054 442	2.0487	0
	„	6,	3-07		337.6	27.333	−.038	27.295	.080 85	4.5583	.0010
58	„	„	3-22	62.5°	337.5	„	−.054	27.279	.080 83	4.5517	0
	„	„	3-36		337.6	„	−.054	27.279	.080 80	4.5477	0

Iron. .1 round.

No.	Date			T	t	Vessel.	Correction.	q	Q	H	O
	June	6,	10-02		532.4	27.333	−.058	27.275	.051 230	1.8156	0
59	„	„	10-30	61.5°	531.7	„	−.106	27.227	.051 207	1.8232	.0030
	„	„	10-51		532.5	„	−.060	27.273	.051 217	1.7966	0
	„	6,	4-33		531.8	27.333	−.043	27.290	.051 316	1.8070	.0040
60	„	„	4-51	63°	531.6	„	−.077	27.256	.051 272	1.8028	.0010
	„	„	5-09		533.4	„	−.060	27.273	.051 130	1.7936	0
	„	6,	9-08		428.9	27.333	−.037	27.296	.063 64	2.8046	0
61	„	„	9-24	61°	430.0	„	+.057	27.390	.063 70	2.8076	0
	„	„	9-39		428.7	„	−.045	27.288	.063 65	2.8076	0
	„	6,	8-21		332.4	27.333	−.051	27.282	.082 08	4.6766	0
62	„	„	8-36	60.5°	332.2	„	−.067	27.266	.082 08	4.6766	0
	„	„	8-51		332.2	„	−.053	27.280	.082 12	4.6766	0

TABLE CII.—*continued.*

No.	C	Mean.	H	c	REMARKS
					Brass. .1 round.—*continued.*
53	.6020 .6020 .6023	.6021	3.19	.6021	Jet apparently twisting a little more than in No. 52.
54	.5996		4.60	.5996	Orifice well oiled with sperm oil, before No. 54.
55	.6004		4.62	.6004	After No. 55, orifice found to be pretty nearly free from oil.
56	.6033 .6022 .6030	.6028	1.87	.6028	}
57	.6038 .6038	.6038	2.05	.6038	} Nos. 56 to 58 with normal conditions every way.
58	.6012 .6014 .6015	.6014	4.55	.6014	}

$$D = .099\,999 \qquad a = .0078538$$

No.	C	Mean.	H	c	REMARKS
59	.6035 .6022 .6066	.6041	1.81	.6041	Jet nearly perfect; perhaps twisting a trifle more than with brass plate, $D = .10$, with same head.
60	.6060 .6062 .6060	.6061	1.80	.6061	
61	.6033 .6035 .6030	.6033	2.81	.6033	Jet twisting more than for Nos. 59 and 60.
62	.6025 .6025 .6028	.6026	4.68	.6026	Jet twisting a good deal, after distance of .2 from orifice; perhaps a little more than with brass plate, $D = .10$, and same head.

TABLE CII.—*continued.*

Brass. .05 × .05

No.	Date.			T	t	Vessel.	Correction.	q	Q	H	O
63	May	9,	3-30	51°	681.4	4.881	-.027	4.854	.007 124	.3117	.0100
	"	"	3-46		760.7	5.484	-.057	5.427	.007 134	.3150	.0030
64	"	9,	4-14	51°	660.4	4.881	-.062	4.819	.008 599	.4619	.0070
	"	"	4-28		640.1	5.484	-.060	5.424	.008 474	.4523	.0035
65	"	9,	4-49	51°	479.8	4.881	-.046	4.835	.010 077	.6504	.0040
	"	"	5-02		536.7	5.484	-.071	5.413	.010 066	.6518	.0010
66	"	9,	5-19	51°	413.2	4.881	-.060	4.821	.011 668	.8863	.0080
	"	"	5-32		468.7	5.484	-.054	5.430	.011 585	.8734	.0025
	"	"	5-44		469.3	..	-.053	5.431	.011 573	.8706	.0005
67(?)	"	14,	8-53	52°	648.2	10.365	-.032	10.333	.015 941	1.7017	.0190
68	"	14,	9-59		630.7	10.365	-.052	10.313	.016 352	1.7887	.0040
	"	"	10-15	52°	631.0	"	-.042	10.323	.016 360	1.7919	.0130
	"	"	10-34		628.6	"	-.046	10.319	.016 416	1.7921	0
69	"	14,	10-55		504.7	10.365	-.044	10.321	.020 450	2.8171	0
	"	"	11-11		505.3	"	-.052	10.313	.020 410	2.8121	0
	"	"	11-30	52°	530.2	16.968	-.006	16.962	.020 431	2.8121	0
	"	"	11-53		831.7	"	-.021	16.947	.020 376	2.8021	0
70	"	14,	1-48		440.8	10.365	-.038	10.327	.023 428	3.7045	.0020
	"	"	2-01	51°	443.9	"	-.009	10.356	.023 330	3.6941	0
	"	"	2-14		442.7	"	-.028	10.337	.023 350	3.7012	.0010
71	"	14,	2-33		650.3	16.968	-.015	16.953	.026 070	4.6298	.0020
	"	"	2-48	51°	396.4	10.365	-.031	10.334	.026 070	4.6266	0
	"	"	3-00		396.0	"	-.046	10.319	.026 058	4.6266	0

TABLE CII.—*continued.*

.049 756 × .049 800 = .002 4778 = area.

No.	C	Mean.	H	c	Remarks.
63	.6420 .6396	.6408	.313	.6410	Jet very regular, with diamond-shaped facets, soon after leaving orifice ; not cruciform.
64	.6366 .6340	.6353	.457	.6354	Jet as above.
65	.6287 .6286	.6286	.651	.6286	Jet as above.
66	.6236 .6237 .6240	.6238	.877	.6238	Jet as above.
67 (?)	.6149		1.70	.6149	Head jumping about for this experiment.
68	.6152 .6149 .6170	.6157	1.79	.6157	Jet very nearly perfect, and cruciform for .6 from orifice.
69	.6130 .6124 .6130 .6125	.6127	2.81	.6127	For this series, adjustments and corrections for q made with great precision. Jet as in No. 68.
70	.6124 .6107 .6107	.6113	3.70	.6113	Jet very nearly perfect ; cross continues for 1.2 from orifice.
71	.6096 .6098 .6096	.6097	4.63	.6097	Jet very nearly perfect ; cross continues for 1.5 from orifice.

TABLE CII.—*continued.*

Brass. .1 × .1

No.	Date		T	t	Vessel.	Correction.	q	Q	H	O
	May 11,	11-09		484.6	10.365	−.042	10.323	.021 302	.1790	.0050
72	" "	11-22	50°	480.9	"	−.032	10.333	.021 487	.1814	.0005
	" "	11-34		479.6	"	−.067	10.298	.021 472	.1814	0
	" 11,	10-23		792.5	27.333	−.030	27.303	.034 452	.4804	.0005
73	" "	10-45	50°	795.1	"	−.020	27.313	.034 352	.4789	0
	" 11,	9-41		671.3	27.333	−.032	27.301	.040 669	.6762	.0035
74	" "	10-02	50°	670.5	"	−.026	27.307	.040 726	.6774	.0005
	" 11,	8-49		572.3	27.333	0	27.333	.047 760	.9371	.0035
75	" "	9-08	50°	570.8	"	−.048	27.285	.047 801	.9412	.0035
	" "	9-28		215.7	10.365	−.041	10.324	.047 863	.9399	0
	" 14,	5-44	51°	426.7	27.333	−.030	27.303	.063 99	1.7155	0
76	" 15,	8-31	49°	427.4	"	−.004	27.329	.063 94	1.7095	0
	" "	8-48		426.9	"	−.031	27.302	.063 95	1.7095	0
	" 15,	9-05		337.5	27.333	−.036	27.297	.080 88	2.7455	0
77	" "	9-20	49°	337.8	"	−.005	27.328	.080 90	2.7455	0
	" "	9-34		337.7	"	−.007	27.326	.080 92	2.7455	0
	" 15,	9-49		289.8	27.333	−.034	27.299	.094 20	3.7375	0
78	" "	10-03	49°	289.6	"	−.075	27.258	.094 12	3.7375	0
	" "	10-17		289.8	"	−.050	27.283	.094 14	3.7375	0
	" 15,	10-30		261.7	27.333	−.002	27.331	.104 44	4.5865	0
79	" "	10-43	50°	261.5	"	−.035	27.298	.104 39	4.5865	0
	" "	10-55		261.7	"	−.030	27.303	.104 33	4.5865	0

Brass. .3 long × .05 wide.

No.	Date		T	t	Vessel.	Correction.	q	Q	H	O
	May 12,	2-44		687.6	27.333	−.042	27.291	.039 690	.2609	.0001
80	" "	3-04	52°	688.5	"	−.051	27.282	.039 625	.2608	.0003
	" "	3-24		687.2	"	−.060	27.273	.039 687	.2611	.0015

TABLE CII.—*continued.*

.100 100 × .100 089 = .010 019 = area.

No.	C	Mean.	H	c	Remarks.
72	.6265 .6278 .6273	.6272	.181	.6292	Jet very short. No facets as with .05 × .05 with small heads. No signs of any interior vortex.
73	.6185 .6177	.6181	.480	.6184	Jet considerably rippled.
74	.6154 .6157	.6155	.077	.6157	Jet perfect, except a little rippling.
75	.6139 .6131 .6143	.6138	.939	.6139	Jet very nearly perfect; sometimes perfect. Shape cruciform.
76	.6079 .6086 .6087	.6084	1.71	.6084	Jet very nearly perfect; a little uneven 1. from orifice.
77	.6074 .6076 .6077	.6076	2.75	.6076	Jet as above.
78	.6063 .6058 .6060	.6060	3.74	.6060	Jet more perfect than with preceding head.
79	.6068 .6066 .6062	.6065	4.59	.6065	Jet almost absolutely perfect; more perfect than with any other head for this orifice.

.299 736 × .049 892 = .014 954 = area.

No.	C	Mean.	H	c	Remarks.
80	.6478 .6469 .6475	.6474	.261	.6476	Jet nearly perfect; close to orifice, entirely free from ripple. No signs of any interior vortex above orifice.

TABLE CII.—*continued.*

No.	Date.		T	t	Vessel.	Correction.	q	Q	H	O
81	May 12,	2-07	52°	540.3	27.333	+ .065	27.398	.050 718	.4421	.0007
	,, ,,	2-24		536.7	,,	− .087	27.246	.050 766	.4425	.0010
82	,, 12,	12-20	52°	444.0	27.333	− .056	27.277	.061 435	.6568	.0095
	,, ,,	1-35		440.6	,,	− .038	27.295	.061 950	.6713	.0015
	,, ,,	1-51		440.8	,,	− .064	27.269	.061 862	.6674	.0015
83	,, 12,	11-30	51°	377.6	27.333	− .046	27.287	.072 26	.9172	.0065
	,, ,,	11-46		379.4	,,	− .027	27.306	.071 97	.9156	.0025
	,, ,,	12-01		378.5	,,	− .032	27.301	.072 13	.9178	.0010
84	,, 15,	4-25	52°	272.1	27.333	− .015	27.318	.100 40	1.8240	0
	,, ,,	4-38		272.8	,,	+ .008	27.341	.100 22	1.8120	0
	,, ,,	4-51		272.7	,,	− .007	27.326	.100 21	1.8120	0
85	,, 15,	3-59	52°	323.7	27.333	+ .006	27.339	.122 21	2.7180	0
86	,, 6,	4-55	49°	219.7	27.333	+ .057	27.390	.124 67	2.8270	.0020
	,, ,,	5-08		218.8	,,	− .017	27.316	.124 84	2.8300	0
87	,, 6,	4-18	49°	190.2	27.333	− .056	27.277	.143 41	3.7475	.0030
		4-30		189.8	,,	− .125	27.208	.143 35	3.7500	0
	,, ,,	4-42		190.6	,,	0	27.333	.143 40	3.7425	.0020
88	,, 15,	3-21	52°	169.8	27.333	− .095	27.238	.160 41	4.6980	0
	,, ,,	3-34		170.1	,,	− .068	27.265	.160 29	4.6970	0
	,, ,,	3-45		170.5	,,	0	27.333	.160 31	4.6940	0

Brass. .3 long × .05 wide.

No.	Date.		T	t	Vessel.	Correction.	q	Q	H	O
89	May 7,	2-42	49°	218.6	27.333	− .056	27.277	.124 78	2.8335	.0020
	,, ,,	2-54		219.2	,,	− .007	27.326	.124 66	2.8246	.0015
90	,, 7,	3-12	49°	191.3	27.333	− .052	27.281	.143 61	3.7193	.0120
	,, ,,	3-25		190.6	,,	− .077	27.256	.143 00	3.7390	0
	,, ,,	3-36		190.9	,,	− .045	27.288	.142 94	3.7340	0

TABLE CII.—*continued.*

No.	C	Mean.	H	c	Remarks.
					Brass. .3 long × .05 wide.—*continued.*
81	.6359 .6362	.6360	.443	.6361	Jet as for No. 80.
82	.6320 .6304 .6313	.6312	.665	.6312	Jet a little rippled.
83	.6291 .6271 .6277	.6280	.917	.6280	Jet regular, but not quite perfect
84	.6197 .6207 .6206	.6203	1.82	.6203	Jet fairly true.
85	.6180	.6180	2.72	.6180	
86	.6182 .6187	.6184	2.83	.6184	Jet twisting a little.
87	.6176 .6171 .6180	.6176	3.75	.6176	Jet twists a little.
88	.6170 .6166 .6169	.6168	4.70	.6168	Jet fairly true ; twists a little.

With vertical brass sheets of various thicknesses, placed across centre of orifice ; the sheet being normal to the plane of the orifice, and projecting into feeding reservoir.

No.	C	Mean.	H	c	Remarks.
89	.6197 .6200	.6199	2.83	.6199	Brass sheet .74 × .75, and .0008 thick ; hence area of orifice = (.299 736 − .0008) × .019 892 = .014 914 for Nos. 89 and 90. Jets somewhat truer than for experiments with this orifice without the sheet ; the form of the jet apparently unchanged
90	.6181 .6182 .6183	.6182	3.73	.6182	

TABLE OIL.—*continued.*

No.	Date.		T	t	Vessel.	Correction.	q	Q	H	O
91	May 8,	8-35	49°	189.4	27.333	-.089	27.244	.143 84	3.8107	.0110
	,, ,,	8-46		189.6	,,	-.007	27.326	.144 12	3.8230	0
92	,, 8,	8-59	49°	220.1	27.333	-.016	27.317	.124 11	2.8226	.0005
	,, ,,	9-11		220.6	,,	0	27.333	.123 90	2.8229	.0005
93	,, 15,	2-21	52°	233.3	27.333	-.047	27.286	.116 96	2.7620	0
	,, ,,	2 33		233.4	,,	(?)		(?)	2.7668	.0010
94	June 5,	9-43	64°	312.4	27.333	-.068	27.265	.087 28	1.8120	0
	,, ,,	9-56		312.7	,,	-.083	27.250	.087 14	1.8010	0
	,, ,,	10-10		312.5		-.086	27.247	.087 19	1.8010	0
95	,, 5,	10-32	63°	254.7	27.333	-.080	27.253	.107 00	2.7120	0
	,, ,,	10-45		255.4	,,	-.023	27.310	.106 93	2.7120	0
	,, ,,	10-58		255.5	,,	-.024	27.309	.106 88	2.7120	0
96	,, 5,	11-13	63°	195.4	27.333	-.078	27.255	.139 48	4.6600	0
	,, ,,	11-25		194.8	,,	-.096	27.237	.139 82	4.6720	0
	,, ,,	11-37		195.4	,,	-.052	27.281	.139 62	4.6700	0

TABLE CII.—*continued.*

No.	C	Mean	H	c	REMARKS.
					Brass. .3 long × .05 wide.
91	.6160 .6162	.6161	3.82	.6161	Same sheet, and same area as Nos. 89 and 90. Sheet, however, projects through orifice (plate ¼" thick) and extends .085 in length on outer side of plate, thus dividing contracted vein. A very small loss of water in these experiments, by dropping from this projection, which is not caught by wooden pipe or trough.
92	.6175 .6164	.6170	2.82	.6170	The jets follow along the dividing sheet their forms unchanged.
93	.6246 ...	.6246	2.76	.6246	Brass sheet .67 × .64, and .0182 thick; hence area of orifice = (.299 736 − .0182) × .049 892 = .014 046. This sheet does not project through orifice. A perfect fit was not made for this sheet against the plate, and hence a little water comes out in spray underneath the edges of the sheet, which was chiefly lost, i. e., not going into measuring tank. The jet unites after its escape from the orifice ; perhaps not perfectly, but nearly so. The correction for *q* for the second determination was not noted ; it was not very different from the first.
94	.6242 .6251 .6255	.6249	1.80	.6249	
	.6255				Brass sheet, .67 × .67, and .040 17 thick ; hence area of orifice = (.299 736 − .040 17) × .049 892 = .012 9503. This sheet does not project through orifice, and fits perfectly to the face of the brass plate. For Experiments Nos. 94 to 96 there was a very slight loss of water by spray, which did not go into measuring tank ; this loss was probably greater for No. 96 than for the other two experiments.
95	.6251 .6248	.6251	2.71	.6251	The escaping jets for these experiments united at a distance of about .05 from the plane of the orifice ; from this point the jet had the same appearance as when no sheet was employed—Nos. 84 to 88.
96	.6220 .6227 .6220	.6222	4.67	.6222	

The results obtained in May, 1885, under normal conditions for the five brass plates, have been plotted on Plate III., with values of c as ordinates, and values of H as abscissæ. From these experimental points, a curve of the most probable value of c, with T from 48° to 54°, has been drawn for each one of the five orifices. Except for the values of H below 1. for the .05 round orifice, and Nos. 41 and 67 which were known to be unreliable, these curves do not as a maximum differ more than .000 75, or $\frac{1}{800}$th part, from the given values of c, and rarely more than .0005, or $\frac{1}{1200}$th part. The symmetry and harmony of these curves almost conclusively demonstrate the general accuracy of the experimental data.

In Chapter III. we have fully discussed these results.

Chances of Error and Variation.

The methods employed in 1885 can be considered perfect, except in the measurement of t; this could have been more accurately obtained by an automatic chronograph, with the beginning and ending of each experiment graphically registered by means of electric currents. The chances of error and variation in detail were as follows:

Orifices.—The dimensions of the orifices were measured by Professor Rogers with the greatest care. The given areas of the .1 round and .1 × .1 orifices are doubtless free from appreciable error; the areas of the .05 round and .05 × .05 orifices may perhaps be slightly appreciably in error. Such errors will of course be constant.

The edges of the orifices both in 1884 and in 1885 were always free from the slightest tarnish; before each series of experiments they were wiped perfectly free from any trace of oil or grease, except in two experiments with the .1 round orifice, where the edges of the orifice were purposely wet with oil.

Professor Rogers found by experimentation with the .1 round orifice, that its diameter increased with increased temperature pretty closely with the general co-efficient of expansion for brass. The temperature of the water for the May, 1885, series only varied 6°, so that enlargement by expansion was not appreciable.

Measuring Vessels.—The standard of capacity was the value of 27.333, for the measuring tank from B to T. We have already described the several measurements of this tank. We consider this value as having been determined certainly within $\frac{1}{500}$th part of the truth. Any error in this value will be constant for the complete series of experiments in 1885.

The segregation of the capacities of the several vertical sections of the measuring tank may be slightly in error, especially for the section M-U, which may possibly be in error .004, or $\frac{1}{300}$th part. Such errors will only appreciably affect values of c for the two smallest orifices. For the three other orifices the entire tank was nearly always employed.

The errors of adjustment to the lower datum point or hook were very slight.

These adjustments could be made with wonderful precision, and for the point B with rapidity, as the water at this point soon became quiescent.

The errors of correction for capacity between the surface of the water and the upper datum hook, could also be obtained within a limit of about .001 cubic foot, by waiting until the water became almost absolutely quiescent. This required considerable time, and when the entire tank was used, this correction was made generally with a limit of accuracy of about .003; such an error would not be appreciable. When the smaller sections of the tank were used, greater care was observed in making this correction.

As before stated, the losses from the tank when filled, only amounted to about .003 cubic foot per hour. Its form also remained unchanged.

Taking all the chances of comparative error in obtaining Q by the measuring tank, we may assume that for the smaller sections, an aggregation of errors all in one way might amount to $\frac{1}{600}$; for the entire tank such errors could not exceed $\frac{1}{1600}$.

When Q for the .05 orifice was obtained by the iron pail, with the capacity of .9307, there might be a possible error of say $\frac{1}{500}$, or .0019; the capacity of this pail was known certainly to .0005; any error would chiefly arise from incorrect observation of the instant when the pail was filled.

Time.—Errors in t arose from an uneven rate for the watch and from personal error. Probably the rate did not vary beyond the limits of $+.001$ and $+.0015$; $+.001\,25$ was the rate adopted; error in rate hence probably did not exceed $\frac{1}{700}$th. Personal error had a probable limit of about .25° counting errors both at beginning and ending; for short times, such an error would be quite appreciable, being $\frac{.25}{12} = \frac{1}{48}$ for No. 32.

Head.—When the plates were placed at O′ for heads below 1 foot, the united errors of zero determination for the hook-gauge, error in the gauge-rod, and errors of observation probably did not exceed .0012, and could not possibly have exceeded .002; the readings were taken by an assistant who had had long experience in such work, and his determinations were frequently verified by the author. The surface of the water was nearly absolutely quiescent, and the position of the hook could be readily fixed within a limit of .0003 or less; when the height of the water varied during an experiment the error from this source was roughly in the proportion of $\frac{O}{H}$; in the few cases when these oscillations of the surface were irregular, a special note to that effect is given in Table CII.

When the pressure tank **A** was used, united errors of zero-point, rod and observation probably did not exceed .002, and could not possibly have exceeded .004.

Errors in H would of course be greatest with the smallest heads; for heads above 2.5 feet they could have had hardly any appreciable effect upon c. The given value of H represents the mean of a number of observations, which probably nearly eliminated errors of observation; errors of zero-point and rod would be nearly constant for the experiments from which the mean value of c was deduced.   Hence for small heads the mean errors in H for such a series were probably less than .001, and we can hardly conceive that they could have exceeded .0015.   A mean error in H of .0015 for $H = .3$ would affect c about $\frac{1}{200}$th part, and for $H = .9$ about $\frac{1}{600}$th part.

Variation in Temperature.—The range of T in May, 1885, was only 6°; such a variation we surmise did not appreciably affect c for the .1 round and .1 × .1 orifices, unless possibly for Nos. 10 and 72, when H was quite small.   For the .05 round and .05 × .05 orifices with heads less than 1., we conjecture that even this slight variation in T appreciably affected c.   An increase in T having the effect of diminishing c.

Impurities in the Water.—The water used in all these experiments was drawn from the Connecticut River, above the City of Holyoke.   There are many manufacturing establishments on the river above this point, and the water doubtless contains small quantities of grease or fatty matter.   The water of the Connecticut can hardly be called muddy, even in times of flood; its color is dark, in common with most of the streams of New England, probably due to vegetable acids.   In May, 1885, the river was at a pretty high stage, and the water was less limpid than in June, 1885, when the river was at a lower stage.   The action of the water had only a very slight tarnishing effect upon the faces of the brass plates, and none whatever upon the edges of the orifices; this latter result was doubtless due to the fact that these edges were frequently rubbed with oil, when not in use.

Barometric Variation.—The following table gives the height of the barometer, reduced to freezing point and sea level, as observed at the U. S. Signal Station, at the City of Springfield, about 12 miles distant from Holyoke.

Barometric Elevations.   National Armory, Springfield, Massachusetts.   Reduced to Sea Level.
(Station 215 feet above Sea.)

Date, 1885	7 A.M.	2 P.M.	9 P.M.	Date, 1885	7 A.M.	2 P.M.	9 P.M.
May 6	30.191	30.267	30.188	May 11	30.104	30.133	30.202
,, 7	30.199	30.117	30.076	,, 12	30.296	30.299	30.336
,, 8	30.099	30.038	30.099	,, 13	30.341	30.293	30.330
,, 9	30.010	29.858	29.887	,, 14	30.221	30.138	30.080
,, 10	29.887	29.866	29.894	,, 15	30.057	30.108	30.337

SUBMERGED AND FREE DISCHARGE.   1884.

From August 13th to September 3rd, 1884, 185 experiments were made with the 5 orifices in brass plates, of which 114 were with free discharge, and 71 with submerged discharge.

As before stated there was during these experiments a slight leakage from the tank **B**. For free discharge the amount of this leakage can be determined closely, by a comparison of the results of 1884 with those of 1885 which are free from error of this kind. For submerged discharge the leakage must have been somewhat greater than for the free discharge, as the water in the tank **B** then stood at a considerably higher level. We will assume a constant leakage for each of the 10 series of experiments ; that for the 5 series with free discharge being assumed a quantity which will cause the results of 1884 to agree pretty closely with those of 1885 ; for the submerged discharge we will assume that the leakage was increased nearly one-third.

The 1884 experiments were made in the following order ; x is the assumed rate of leakage per second for each series ; T is the observed temperature of the water.

Date, 1884	Series.		x	T
August 13-14	.10 round, free		.000 10	72.5°
„ 14-15	„ „ submerged		.000 15	
„ 16-18	.05 × .05 „		.000 13	78°
„ 18-21	„ free		.000 08	80°
„ 21-22	.05 round, „		.000 18	82°
„ 22-25	„ „ submerged		.000 20	82° and 72°
„ 27-Sept. 2	.10 × .10 „		.000 17	68.5° and 71°
September 2-3	„ free		.000 12	70.5°
„ 3	.3 × .05 „		.000 15	
„ 3	„ submerged		.000 20	

In the following table are given the results of the 1884 experiments, corrected by the foregoing values of x. H, Q and c' are the mean values of the head, flow and co-efficient c, as determined in 1884 ; these means generally represent the results of 3 experiments ; c'' is obtained by $\frac{x\,c'}{Q} + c' = c''$, being equivalent to $Q + x = c''\,a\,(2\,g\,H)^{\frac{1}{2}}$; hence c'' represents the true co-efficient for the 1884 series. c is the co-efficient, as determined in 1885, and has been obtained by the curves on Plate III.—most probable curve for c for each of the 5 orifices. For the submerged orifices, the value of c for free discharge is also given, to illustrate effect of submergence.

TABLE CIII.

Holyoke Experiments with Orifices in Brass Plates, 1884. Discharge Free and Submerged. Corrected for losses by leakage, by experiments of 1885; this correction, or z, constant for each of the 10 series.

.05 Round.

	Free Discharge. z=.000 18						Submerged Discharge. z=.000 20			
H	Q	c'	c''	1885 c	No.	A	Q	c'	c''	1885 c=free.
4.64	.0292	.5990	.6044	.6050	97	4.08	.0188	.5553	.6016	.6054
3.76	.0181	.5990	.6050	.6036	98	2.16	.0137	.5954	.6041	.6090
2.80	.0157	.5998	.6067	.6069	99	.437	.0062	.5980	.6183	.6204
1.97	.0132	.6008	.6090	.6099						
1.02	.0096	.6094	.6208	.6166						
.412	.0062	.6159	.6338	.6305						

.1 Round.

	Free Discharge. z=.000 10						Submerged Discharge. z=.000 15			
H	Q	c'	c''	1885 c	No.	A	Q	c'	c''	1885 c=free.
4.64	.0814	.5998	.6005	.6014	100	3.97	.0750	.5980	.5992	.6018
4.14	.0768	.6001	.6009	.6017	101	3.57	.0711	.5974	.5987	.6020
3.94	.0750	.6003	.6010	.6018	102	2.99	.0650	.5975	.5989	.6025
3.64	.0721	.6008	.6016	.6020	103	2.58	.0605	.5982	.5997	.6028
3.14	.0671	.6012	.6021	.6024	104	2.00	.0533	.5989	.6006	.6035
2.64	.0616	.6022	.6032	.6027	105	1.51	.0462	.5987	.6006	.6051
1.94	.0529	.6037	.6048	.6036	106	.985	.0375	.6001	.6025	.6084
1.40	.0451	.6054	.6067	.6035	107	.648	.0304	.5997	.6027	.6122
1.02	.0384	.6055	.6071	.6030	108	.250	.0189	.6000	.6048	.6229
.502	.0273	.6132	.6154	.6148						
.315	.0216	.6131	.6149	.6200						

.05 × .05

Free Discharge. z = .000 08					Submerged Discharge. y = .000 13					
H	Q	c'	c''	1885 c	No.	h	Q	c'	c''	1885 c = free.
4.65	.0261	.6083	.6102	.6097	109	4.06	.0242	.6036	.6068	.6103
3.94	.0240	.6082	.6102	.6105	110	2.21	.0179	.6048	.6092	.6140
3.36	.0222	.6090	.6112	.6113	111	.350	.0072	.6091	.6201	.6390
2.55	.0194	.6105	.6130	.6129						
1.39	.0144	.6154	.6188	.6178						
.397	.0079	.6333	.6399	.6372						

.1 × .1

Free Discharge. x = .000 12					Submerged Discharge. x = .000 17					
H	Q	c'	c''	1885 c	No.	h	Q	c'	c''	1885 c = free.
4.63	.1050	.6059	.6066	.6061	112	3.95	.0963	.6037	.6048	.6062
3.89	.0960	.6061	.6069	.6063	113	3.11	.0856	.6040	.6052	.6070
3.21	.0872	.6061	.6069	.6069	114	2.32	.0737	.6026	.6040	.6079
2.70	.0802	.6074	.6083	.6074	115	1.52	.0598	.6038	.6055	.6095
2.21	.0725	.6077	.6087	.6080	116	.771	.0425	.6029	.6053	.6149
1.68	.0633	.6076	.6088	.6089	117	.410	.0312	.6058	.6091	.6205
1.15	.0524	.6079	.6093	.6117	118	.207	.0222	.6070	.6117	.6277
.774	.0430	.6091	.6108	.6148						
.538	.0360	.6114	.6134	.6179						
.374	.0301	.6131	.6155	.6214						

.3 × .05

Free Discharge. x = .000 15					Submerged Discharge. w = .000 20					
H	Q	c'	c''	1885 c	No.	h	Q	c'	c''	1885 c = free.
4.60	.1583	.6156	.6162	.6170	119	2.77	.1232	.6178	.6188	.6180
3.80	.1441	.6163	.6169	.6175	120	1.63	.0949	.6194	.6207	.6213
3.00	.1280	.6169	.6176	.6179	121	.614	.0582	.6198	.6219	.6321
2.16	.1090	.6184	.6192	.6192						
1.34	.0862	.6223	.6234	.6232						
.830	.0681	.6235	.6249	.6283						
.557	.0562	.6275	.6292	.6333						

The preceding comparison of the results of 1884 and 1885, with free discharge, shows a pretty fair accordance, except for heads less than 1 foot. With the .1 × .1 and .3 × .05 orifices, with heads below 1, the corrected co-efficients for 1884 appear to be notably lower than those for 1885. Possibly this may be in part due to the much higher temperature of the water in 1884. These lower heads were, however, less accurately determined in 1884 than in 1885, and the lower the head the greater the danger of error in incorrect assumptions of the value of the leakage x.

Submerged Discharge.

The height of water in the tank **B**, above the centre of the submerged orifices, varied from .57 to .73.

The head in the tank **B** was determined by a hook-gauge at **e**. The zeros of this gauge and the one in the tank **A** at **c**, were compared by filling the tanks **A** and **B** to about the height of the hook **e**, as shown in the sketch; the supply of water was then shut off, and simultaneous readings made of the two gauges. The gauge at **e** was enclosed in a wooden box, with small orifices pierced in its bottom and sides. There was some commotion in the surface of the water in the tank **B**, with the three largest orifices, with heads above 2 feet, but not enough to notably affect the general surface elevation in **B**.

For submerged discharge, with heads above 1., with the three largest orifices, the chances of error in c'' (including leakage) are probably not over $\frac{1}{500}$; for the two smallest orifices, with the same heads, the probable errors in c'' are not over $\frac{1}{250}$. The smaller the head, and the smaller the orifice, the greater is the danger of error in c''.

The 25 determinations for submerged discharge are plotted on Plate V. The heavy solid curved line represents the most probable curve for c, for the orifice with free discharge, as shown on Plate III. The dotted curve below the curve for free discharge, is the most probable curve for c, for submerged discharge.

The experimental curves or lines for the 5 series appear to be fairly symmetrical, except No. 99 for the .05 round orifice; this experimental point, judging from the other experimental curves, seems to be about .003 too high; the dangers of error in this experiment are greater than in any of the others.

By reference to Plate V., it will be noticed that the experimental submerged curve for the .3 × .05 orifice, is slightly higher, for h above 2, than the curve for free discharge. This was doubtless caused by the divergent sides of this orifice, which were .021 in length, being the full thickness of the brass plate;[*] when the orifice was submerged these divergent sides doubtless notably increased the flow. These comparatively long divergent sides for one orifice, were purposely used, to determine their effect. The short divergent sides for the .05 round and .05 × .05 orifices, probably

---

[*] In the other four plates, the length of the divergent sides of the orifices was only .0009.

appreciably increased the submerged flow; this increase was probably hardly appreciable for the .1 round and .1 × .1 orifices, when the length of the divergent sides was only $\frac{.000}{.100}$ths of the diameter or side of the orifice.

The results for submerged discharge, as indicated by Plate V., have been already discussed in Chapter III.

In the foregoing table the discrepancies between the results of 1884 and 1885 have been attributed almost entirely to the leakage from the tank B in 1884. It was known at the time that there was a slight loss from this source, but some of the corrections given seem to us to be improbably large.

The temperature of the water in 1884 ranged from 68.5° to 82°, while in 1885 it was from 48° to 54°. We conjecture that an increase in T of 25° or 30° will very slightly, but still appreciably, diminish the co-efficient c, even for the .1 round and .1 × .1 orifices with the largest heads. By "appreciably" we mean a change in c of .0005. This surmise, however, is not substantiated by a comparison of Nos. 47 and 58 with the .1 round orifice, where a variation in T of 13.5° does not appear to have affected c; this can hardly be considered as conclusive, as such a variation in T, according to our conjecture, would only change c .0002 or .0003, a quantity barely appreciable. For the three other orifices we feel confident that for heads less than 1 foot, the higher temperature in 1884 notably diminished c.

We hence regard it probable that the given corrections for *leakage* are excessive.

On Plate III. have also been plotted, three experimental values of c, for the Greenpoint circular orifice of about .02 diameter, and the six experimental values of c obtained by the author in his California experiments with orifices; all of these being with free discharge, and full contraction.

In conclusion it may be observed, that for all the Holyoke experiments H is always equal to h. That is to say, the area of the feeding canal or tank, a_s, was always so very much larger than the area of the orifice, a, that the velocity of approach, v_s, was almost inappreciable, and the head, h_s, imparting this velocity was absolutely inappreciable.

The Holyoke Water Power Company constructed at its own expense the tanks employed at Holyoke, and also furnished without charge the services of all the needed assistants. This was done at the instance of its hydraulic engineer, Mr. Clemens Herschel. We here desire to express our appreciation of the company's liberal aid in behalf of the science of Hydraulics, and also our personal obligations to Mr. Herschel, for his valuable advice and co-operation during these investigations.

GREENPOINT.

The experiments at Greenpoint were all made with a circular orifice, having a diameter of about .02, pierced through a hardened steel plug. The inner face was true; the orifice proper had divergent sides .016 in length. The diameter of the orifice was measured by Professor W. A. Rogers, of Cambridge, Mass.; the mean of several determinations was .020 15, with $T = 50°$. The orifice was almost absolutely circular, with nearly true edges; this orifice was, however, not quite as perfect as the circular orifices in brass plates, used at Holyoke; under high magnifying powers, in some places on its edge, there could be seen a very slight rounding of the corner.*

The steel plug, through which the orifice was pierced, was .042 in diameter, and was very firmly "bushed" in the centre of a rectangular cast-iron plate, .04 thick, and .64 × .69. On the outer side of the orifice an iron plug, fitted with a handle, could be screwed to the iron plate; a soft leather washer secured a perfectly tight joint, when the plug was thus attached.

The feeding or pressure tank employed was a solid cast-iron vessel, 3.65 high. The upper 3 feet of this vessel was cylindrical; this tube was carefully bored, and had a nearly circular and uniform section with a mean diameter of .504 51, with T about 50°, The lower portion of the vessel was a parallelopipedon, having an inner base of .52 × .58, and an inner altitude of .58. The iron plate before described, formed one of the vertical sides of the parallelopipedon; it was attached to the vessel by many screws, a rubber gasket forming a tight joint. Near the bottom of the vessel, and on the opposite side to the plate, a hole was pierced; a short pipe with a right bend was screwed in this hole, and to the pipe was attached a glass tube, having an inner diameter of about .0315, with a nearly uniform circular section; the axis of the glass tube was parallel with the axis of the iron vessel (tank); a tight connection for the glass tube was made by putting rubber rings around it at the bottom, which were tightly pressed by an iron "follower."

During all the experiments made with the iron tank, there was no leakage whatever, all the joints being perfectly tight.

The iron tank was firmly fixed in an exactly vertical position, and a vertical wooden rod, with a graduated scale drawn on white paper glued to the face of the rod, firmly placed immediately behind the glass tube. Horizontal "sight-bars" were placed in front of the tube, so that the height of the liquid column in the tube could be read at several points, with a very fair degree of accuracy.

All linear measurements were carefully reduced to the standard of Professor Rogers, being the same standard employed at Holyoke.

As in some of the experiments T had a wide range, it became important to determine the effect of changes in temperature upon the diameter of the orifice. This was determined by Professor Rogers, with a maximum variation in T of 86°. He found that the diameter increased with T, but with a somewhat irregular movement; this was probably in part due to the composite character of the plate. In one experiment the increase in T of 76°, resulted in an increase in D of $\frac{1}{130000}$ for each degree, being a larger rate than the usually accepted co-efficient of expansion for either

---

* The orifices in the five brass plates, and the one used at Greenpoint, were made by Messrs. Buff and Berger, of Boston, who are skilful workmen. The circular orifices were entirely satisfactory; the rectangular orifices were not as perfect as was desired. It may be remarked that it is exceedingly difficult to make perfect rectangular openings; they can only be finished up properly, by the aid of a microscope with pretty high power.

cast-iron or steel. These determinations proved that no notable error would ensue, by the assumption that with an increase in T, the diameter of the orifice would increase in the same proportion as the diameter of the cylinder of the iron tank ; therefore the increase in the areas would also be directly proportional.

The measuring vessel employed was a bucket or pail made of best quality " galvanized " sheet-iron, having the form of a truncated cone, with a bottom diameter of 1.25, a top diameter of .5,* and an altitude of about 1.45. A curved handle was attached to the pail ; when supported by this handle the capacity of the pail was increased from .0006 to .0007, although the bottom was stiffened by light cross-bars ; the amount of this enlargement was determined by filling the pail exactly to the rim when it rested on its bottom ; the pail was then lifted by the handle, and the dropping of the surface level noted ; this was about .0033, showing the gain in capacity to be .000 66.

The capacity of the pail was determined in two ways :

First.—By filling the iron tank and its attached glass tube with water ; the screw plug closing the orifice was then removed,† and in a few seconds the pail was quickly placed into position to catch the jet ; at this instant the height of the column of water in the glass tube was noted ; when the water had fallen in the iron cylinder to near the bottom, the pail was quickly removed, and the height of the column in the glass tube again noted. The pail not being filled, the experiment was repeated, the concluding instant being the moment when the pail, supported by its handle, was brimming full. The mean of three determinations, with T from 46° to 49°, showed that the capacity of the pail, when supported by its handle, was equivalent to 4.640 in length of the sum of the mean areas of the iron cylinder and the glass tube. The sum of these areas was .200 69 ; the capacity of the pail, held by the handle, was hence .9312. This method was not entirely satisfactory, owing to the varying amount of capillarity in the glass tube, which will be commented upon hereafter. Mr. Ross E. Browne has found with very small orifices, that when a plug is removed quickly, the jet is at first abnormally long, quite an interval of time being required for it to diminish to a permanent length. In our experiments we could hardly notice such a phenomenon ; the jet with us for all liquids became constant, certainly inside of a period of 2 seconds.

Second.—By ascertaining on very accurate scales the weight of water contained in the pail, the pail resting on its bottom. The gold bullion scales in the U.S. Assay Office, in New York, determined the weights, with various temperatures, showing the following results. The given weight is the net weight of the water, with the pail brimming full ;

---

* The top diameter should not have been more than .35 or .4 ; this smaller top size would have permitted more accurate determinations of the instant when the pail was full.

† It is almost unnecessary to observe that the female thread into which this plug screwed had a much larger diameter than the orifice.

the weight of a cubic foot of water at the given temperature, has been taken from Table I. ; the capacity of the pail is deduced from these values ; T was determined by a standard thermometer.

Experiment.	T of Water.	Net Weight.		Weight of a Cubic Foot of Water.	Capacity of Pail.
		Troy Ounces.	Avoirdupois. Pounds.		
1st	46°	846.82	58.068	62.418	.9303
2nd	100°	841.67	57.715	61.998	.9309
3rd	167°	827.02	56.710	60.862	.9318

The temperature of the room was about 75°. The water used was from the Croton supply of the city, which is very pure, being nearly free from mineral salts and sediment.

If we assume that the iron of the pail expanded $\frac{1}{180.000}$th part in length for each degree of temperature, and that the temperature of the iron was identical with that of the water, as the pail expanded in three dimensions, the relative capacities of the pail should have been;

1st	By supposition, 1.		Result, as above, 1.		
2nd	"	"	1.001 08	"	"	1.000 64
3rd	"	"	1.002 42	"	"	1.001 61

The last two determinations were quickly made, and the mean temperature of the iron was in all probability less than that of the water.

These results can be considered as very satisfactorily agreeing with the values assumed by Rossetti, and embodied in Table I.

With $T = 46°$, the pail resting on its bottom had a capacity as above of .9303 ; hence when supported by its handle a capacity of .9303 + .0007 = .9310.

For the following experiments, .9311 will be assumed its capacity, with $T = 40° - 46°$.

The capacity of this same pail, with $T = 48°$, was afterwards found to be .9307, by comparison with the large wooden vessel used at Holyoke. The most probable capacity of the pail, with this temperature, is .9308 or .9309.

The times were taken by a stop-watch, whose rate varied from + .0021 to + .0013, including personal error, when compared with a standard second pendulum clock. A uniform rate of + .0017 was adopted.

In all the experiments the inner edges (or corner) of the orifice were perfectly free from rust.

The jets, both with water and with quicksilver, were, when not otherwise expressly noted, always beautifully perfect, and very steady. Not a flaw nor the sign of the slightest twist could be observed, from the orifice to where the jet finally struck.

The water used was drawn from the supply of the City of Brooklyn, which is very limpid, and remarkably free from impurities. This water, however, passes through pumps, and doubtless contains a very slight quantity of oily matter.

NORMAL DISCHARGE. *Constant Heads.*

Nos. 122 and 123 were obtained by the use of the iron tank. The supply of water was obtained by a rubber hose, with its discharge end slightly immersed in the water in the tank; this supply was rather irregular. The heads were read by the scale back of the glass tube; a reading was taken every 15 seconds, and H is the arithmetical mean of these readings; O gives the maximum variation in H for an experiment. No. 124 was obtained by attaching the plate to an open wooden tank of considerable size; the heads were read every minute by a hook-gauge; the supply of water was obtained from an upper reservoir. The edges of the orifice were in all cases wiped free from any trace of oil or grease.

We had a queer experience in making No. 124, which is worth relating. The wooden tank employed had been used for the workmen of the Continental Works, as a wash-stand, and was pretty thoroughly saturated with grease of various kinds. Before it was put in place, the inside was scoured with sand, and most of the surface grease removed. The plate was then put in place, and the tank filled with water; the jet was slightly imperfect, clinging at one point on the divergent side of the orifice. Attributing this phenomenon to greasy matter in the tank, its surface was carefully scoured with turpentine; the jet still remained imperfect, always clinging to some one minute point whose position could be changed by rubbing the inner edge of the orifice. After bothering from early in the morning until late in the afternoon, rubbing the orifice with naphtha and alcohol, and the jet still remaining imperfect, we made a final trial, and, as if my magic, the jet was perfect, and we had no further trouble. We cannot suggest any plausible reason for this phenomenon; it seems to us that it certainly was not due to any imperfection in the orifice.

In the following table; q in all cases was .9311, the pail being suspended by the handle just before the termination of the experiment; C in all cases is identical with c; T is the temperature of the water; the temperature of the air was 49° for Nos. 122 and 123, and 36° for No. 124.

Nos. 122 and 123 were made on March 6th, 1885, and No. 124 on March 12th, 1885.

For Nos. 122 and 123 correction has been made for excess of indicated height in glass tube, due to capillarity. H is of course the corrected height from the centre of the orifice.

TABLE CIV.

Flow of Water through a Vertical Orifice, in Steel, with Free Discharge into Air, and Full Contraction.

D = .020 15   a = .000 3189   q = .9311   (2 g)$^{1/2}$ = 8.020

No.	T	t	q	H	O	C	Means	
							H	c
122	46"	324.2	.002 8720	3.211	.032	.6267		
	46"	324.8	.002 8667	3.205	.045	.6261	3.19	.6264
	46"	325.5	.002 8606	3.191	.025	.6261		
	46"	326.5	.002 8518	3.164	.064	.6269		
123	46"	366.8	.002 5385	2.486	.052	.6295		
	46"	369.4	.002 5206	2.453	.049	.6293	2.43	.6298
	46"	373.9	.002 4902	2.386	.034	.6303		
	46"	374.7	.002 4850	2.376	.024	.6303		
124	40"	650.3	.001 4318	.7428	.0045	.6496		
	40"	652.4	.001 4272	.7375	.0045	.6498	.739	.6495
	40"	655.6	.001 4246	.7365	.0075	.6491		

NORMAL DISCHARGE.   *Dropping Heads.*

The following experiments were made by first screwing in the stop-plug, and filling the iron tank higher than the desired summit level; the plug was then removed, and the time began when the column of water in the glass tube dropped to the upper "sight" line; the time ended when this column reached the lower "sight" line. Corrections for *H* have been made for the average amount of capillarity, at the time of the several experiments. These experiments were made March 5th and 6th, 1885. As a matter of course, during an experiment the tank was receiving no water.

TABLE CV.

Flow of Water through a Vertical Orifice, with Full Contraction and Free Discharge. The Supply of Water coming from Vertical Tubes, having very nearly Circular and Uniform Sections.

No.	Height from Centre of Orifice.		*T*	*t*		Remarks.
	At Beginning.	At Ending.		Expt.	Mean.	
125	3.157	.559	51.5° 48°	254.6 254.3 253.8 255.2 254.6	254.5	Jet perfect.
126	3.157	1.758	47°	113.3 112.9 112.8 112.8 112.9	112.9	Jet perfect.
127	1.758	.559	47°	140.9 140.4 140.2 139.8 140.9	140.4	In some of these experiments when the head was dropping from .66 to .559 (at the close of the experiment) the jet was a little ragged, showing a slight adherence to divergent sides.
128	3.159	.661	52° 50° 48° 46°	242.0 241.1 242.0 240.1	241.3	Jet perfect.
129	3.159	1.760	47° 46°	113.7 113.5	113.6	Jet perfect.
130	1.760	.661	46° 46°	126.6 126.3	126.5	Jet perfect.
131	3.163	.665	46°	239.2 238.3 238.3	238.6	Jet perfect.

Experiments, forming No. 131, were made immediately after experiments with hot water. The plate was then taken off (after completion No. 131), and a little rusty matter found on inside of face of plate ; the edges of the orifice, however, were perfectly clean and sharp.

The tube of the iron tank had a mean diameter of .504 51 ; hence $.I' = .199\,91$. The mean diameter of the glass tube was .0315; hence $A'' = .000\,78$; $A' + A'' = A = .200\,69$, being the united areas of the two tubes.

The sum of the means of t for Nos. 126 and 127 is 253.3 instead of 254.5 the mean value of t for No. 125.[*] The sum of t for Nos. 129 and 130 is 240.1, while t for No. 128 is 241.3, and for No. 131, with a slightly different head, 238.6. Apportioning these errors in a probable manner, we have ;

| No. | H | | — t | q | Q | c |
	H_1	H_2				
125	3.157	.559	253.8	$2.598 \times A = .5214$	.002 054	.636
126	3.157	1.758	113.1	$1.399 \times A = .2808$	.002 483	.626
127	1.758	.559	140.7	$1.199 \times A = .2406$	.001 710	.645
128	3.159	.661	240.1	$2.498 \times A = .5013$	.002 088	.630
129	3.159	1.760	113.6	$1.399 \times A = .2808$	.002 472	.623
130	1.760	.661	126.5	$1.099 \times A = .2205$	.001 743	.637

The value given to c, is its mean value with H ranging from H_1 to H_2, and is deduced by $Q = c\,\alpha\,(2\,g)^{\frac{1}{2}}\dfrac{H_1^{\frac{1}{2}} + H_2^{\frac{1}{2}}}{2}$, α being the area of the orifice ($\alpha = .000\,3189$).

COMPARATIVE EFFECTS OF CHANGES IN TEMPERATURE.

The iron tank for these experiments was first filled with either hot or cold water, the plug was then withdrawn, and the time noted of the drop in the glass column from 3.157 to .559 above centre of orifice. A uniform correction of $-.009$ being made for capillarity in the glass tube. The experiments were made in the consecutive order as given in the following table. Where two values of T are given, the first is the temperature of the water at the beginning of the experiment, and the other the temperature near the close of the experiment. Only a short interval of time elapsed between the several experiments.

The jets for Experiments Nos. 132 to 141 inclusive were perfectly true, touching at no time the divergent side of the orifice. For No. 142 the jet did touch slightly the lower divergent side of the orifice, when the head was dropping from .76 to .559 ; this resulted in making the jet a little ragged for a few seconds at the close of the experiment.

---

[*] The value of t for the same lowering in the tank (3.157 − .559) was 252.1, by one determination, at end of hot water experiments, No. 142.

The temperature of the cool feeding water, before it was put in the tank, varied from 46° to 48° ; the temperature of the air was about 50°.

These experiments were made on March 5th, 1885.

TABLE CVI.

Comparative Effects of Changes in Temperature of Water upon the Flow through a Circular Orifice, with D = .020 15. The Time for all of these Experiments being taken when the Water in the Glass Tube was dropping from 3.157 to .559 above Centre of Orifice.

No.	t	T	Remarks.
125	253.8	About 48°	This being the most *probable* value of t for this temperature, as before stated ; the mean *observed* value for this temperature was 254.5.
132	257.9	132°-128°	Hot water put in tank for No. 132. Outer temperature of tank 85° (?).
133	255.6	67°-67°	Cold water put in tank for No. 133. Outer temperature of tank 80° (?).
134	257.3	114°-112°	Hot water put in tank for No. 134. Outer temperature of tank 95° (?).
135	257.9	128°-125°	Hot water again put in tank for No. 135. Outer temperature of tank 100° (?). After No. 135 the cool template, showing mean diameter of iron tube, when placed in tube, was perceptibly loose, showing expansion of tube.
136	257.4	132°. $\begin{cases} 128° \\ 125° \end{cases}$	Hot water again put in tank for No. 136. Outer temperature of tank 105° (?). In the preceding experiments the temperature of the outer side of the iron tube seemed to be slightly higher than that of the outer face of the plate in which the orifice was pierced.
137	255.2	82°-81°	For No. 137, and for the succeeding experiments, cold water was put in tank. Outer side of iron tube still warm after No. 137.
138	254.6	65°-64°	
139	253.6	56°-56°	
140	253.1	53°	After No. 140 standard cool template fits nicely in iron tube.
141	252.4	50°	
142	252.1	48°	Tank has very nearly same temperature as water.

Professor Rogers found that the diameter of this orifice increased with T, the increase being not very far from the usual co-efficient of expansion for cast-iron. We can hence assume that the area of the orifice increased in two dimensions with T. The iron tube, which was the standard of capacity, also increased in two dimensions with T, or its area. The altitude of the column of water in the iron tube was not affected by changes in T, as it was measured by the outer wooden rod, which had the temperature of air, nearly constant at 50°. The area of the glass tube was so small compared with the area of the iron tube, that the different rate of expansion for glass is immaterial. We hence can safely assume that the effect of T upon quantity, and upon the area of the orifice was practically identical, and that the differences in t in the foregoing table truly indicate the effect of changes in T upon the contracted vein.[*]

---

[*] The most careful determination of Professor Rogers showed that the diameter of the orifice increased $\frac{1}{15000}$th part for 1° of increased temperature. The co-efficient of expansion for cast iron is about $\frac{1}{16000}$. Upon these assumptions, with increasing temperatures, the orifice increased in area more rapidly than the iron tank ; hence the times should have been shorter for high values of T, provided T had no direct effect upon the flow.

The table shows a variation in t, from 257.9 with $T = 126°$, to 252.1 with $T = 48°$, a difference of about 2 per cent. .

We did not experiment as to the effect of T upon lower heads, such as a drop in the tank from 1.7 to .56, because the danger of comparative error would have been much greater owing to the shorter times.

It may be remarked that the tank was always cooler than the hot water ; this hot water hence contracted after it was poured in the tank. On the other hand, the cold water was always cooler than the tank (except Nos. 125 and 142), and hence expanded after being placed in the tank. The resulting changes in the density of the water probably chiefly took place before the commencement of an experiment, as several seconds elapsed from the instant of filling, to the beginning of t.

The indicated variation in t is so great, that we feel satisfied an increase in T considerably and directly diminished the flow from this orifice, with H ranging from 3.2 to .6. We believe that this diminishing effect was much the greatest for the least head. Unfortunately we could not with this orifice satisfactorily demonstrate this last proposition, as we could not obtain a constant flow of heated water, and hence could not make experiments with a constant head, with widely varying values of T and H.

Effect of Oiling the Orifice.

These experiments were made in the same manner as the preceding ones. The times were noted of the dropping of the water from 3.163 to .665 above centre of orifice. A constant correction of .003 is given for capillarity in the glass tube. The experiments were made in the consecutive order as given in the table. The jets were in all cases perfect.

TABLE CVII.

Comparative Effect upon the Flow of Water through a Circular Orifice, $D = .020\,15$, by Oiling Inner Edge of Orifice.

No.	T	t		Remarks.
143	46°	239.2 238.3 238.3	238.6	Orifice perfectly free from adhering grease or oil.
144	49°	243.1		Orifice before No. 144 was wet with a mineral oil of fair body ("anti-rust" oil).
145	48°	242.6		
146	47	242.1		After No. 146, water was allowed to flow from the orifice, with a head of 3.2, for 5 minutes.
147	46°	242.0		After No. 147, the orifice was well wiped from the outside by a clean cloth on a tapered wooden plug.
148	46	241.6		After No. 148, the plate was taken off, and the inner face, adjoining the orifice, found to be still wet with oil.

It is apparent, from the foregoing results, that the film of oil adhering to the inner edge of the orifice diminished the discharge. The most natural explanation is, that this film diminished the normal area of the orifice.

Flow of Quicksilver.

The experiments with quicksilver were made by noting the time of the dropping of the column in the glass tube. The height of this column was identical with the height in the iron tube, so no correction is required for capillarity. The quicksilver used was perfectly pure, never having been used before. Particles of iron rust and all traces of oxidation were carefully removed before the experiments began. The jets of quicksilver were all beautifully perfect and steady. The temperature of the air was from 42° to 38°, and of the pool of water into which the jet fell 45°; the temperature of the quicksilver was hence about 44°. A is the sum of the areas of the iron and glass tubes, being .200 69.

TABLE CVIII.

Flow of Quicksilver through a Vertical Circular Orifice. $D = .020\,15$. Free Discharge into Air, and Full Contraction. Supply from Vertical Tubes of nearly Uniform Section.

$$Q = c\, a\, (2\,g)^{\frac{1}{2}} \frac{H_1^{\frac{3}{2}} + H_2^{\frac{3}{2}}}{2}$$

| No. | H | | t | | q | Q | c |
	H_1	H_2	Experiment.	Mean.			
149	3.160	.562	267.6 268.3 268.0	268.0	$2.598 \times A = .5214$	.001 946	.602
150	3.160	1.761	118.3 118.7	118.5	$1.399 \times A = .2808$	.002 370	.597
151	1.761	.562	149.8 149.2	149.5	$1.199 \times A = .2406$	.001 609	.606

In the above experiments the sum of the mean times for Nos. 150 and 151, is exactly equal to the mean time for No. 149.

Flow of Oil.

The oil used for these experiments was a mineral lubricating oil, which became rather thick with a temperature of 52°.

The experiments were made in the same manner as with the quicksilver. The correction for the glass tube with this liquid was + .020, the elevation in the glass tube being lower than in the iron tube. T is the temperature of the oil.

TABLE CIX.

Flow of Lubricating Oil through a Vertical Circular Orifice. D = .020 15. Full Contraction. Supply Drawn from Vertical Tubes, with Head from Centre of Orifice Dropping from 3.181 to .583. T of Air 48°.

No.	T	t	Remarks.
152	68°	220.9	Jet twists somewhat.
153	65°	221.1	Jet adheres more or less to divergent sides.
154	53°	237.1	Jet twisting and irregular ; some oil dropping down on outer face of plate.
155	52°	217.3	Jet not twisting nearly as much as for No. 154. The oil more viscid than for No. 152.

Calling roughly $t = 220$, we have

H_1	H_2	t	q	Q	c
3.181	.583	220	$2.598 \times A = .5214$	.002 370	.728

The discharge of oil may possibly have been somewhat increased by its adherence to the sides of the orifice, which to some extent formed a divergent adjutage. Any such possible increase, in our judgment, could not account for the foregoing high value of c.

CORRECTIONS FOR CAPILLARITY.

The capillarity in the glass tube was determined for the three liquids employed, by filling the iron tank to the brim, the orifice being closed by the screw-plug. The height of the liquid in the two tubes was then compared, and the noted difference in elevation assumed as a correction for the observed height in the glass tube at any elevation.

Water.—The upper portion of the glass tube, where the comparison was made, was carefully wiped with a clean cloth before each determination. The differences were as follows, the water in the glass tube being always the higher ; .009, .008, .010, .008, .005, .003. This variation was irregular, and hence involved the danger of appreciable error in the observed heads. When comparisons were made, a piece of blotting paper was inserted in the top of the glass tube, with the view of removing any greasy matter which might be floating on the surface. The meniscus was generally quite perfect.

Quicksilver.—For the first determination, it was thought that the column in the glass tube was nearly .0005 higher than in the iron tube. For this test the surface of the mercury was not quite free from oxidation. Subsequent trials failed to show any appreciable difference between the height of the two columns. When the mercury had been well cleaned, these heights could be directly compared, with an error not exceeding say .0003. The meniscus in the smaller tube was very perfect.

Oil.—Before comparisons, the upper part of the glass tube was rubbed carefully with a cloth wet with the oil. The oil was from .020 to .023 *lower* in the glass tube than in the iron tube. The meniscus in the glass tube was rather irregular, but still fairly defined.

Chances of Error and Variation.

Area of Orifice.—The area of the orifice was so small, that the assumed size may perhaps be $\frac{1}{40}$ in error. Any such error would be constant, and would result in no comparative error for the Greenpoint series.

Area of Tubes.—The sum of the areas of the glass and iron tubes is given without appreciable error. This is proved by the exact measurement of the iron pail, deduced from the sum of these areas.

Irregularity in Capillary Action.—For No. 124, where the head was measured in an open tank, danger of error from this source was of course nothing. For the other experiments with water, danger of error was least with Nos. 122 and 123, with constant and considerable heads ; error from this source would be most appreciable when the lower head in the tubes, or H_a, was the least.

The experiments with quicksilver are free from this error. Those with oil are probably more in danger of error from this cause, than the ones with water.

Condition of Orifice.—The experiments with the orifice well wet with an oil of good body, show that the discharge was diminished about 2 per cent. at first, and about 1 per cent. after the escaping jet had washed away some of the oil. This indicates that when reasonable care is observed, even with so small an orifice, there is no danger of an abnormal flow, caused by greasy matter adhering to the inner edges of the orifice.

Times.—The measure of accuracy, for experiments when the supply was obtained with a steadily diminishing head from the two tubes, is the observed value of t. Errors in observation were the least at the beginning of the experiments, when the liquid column was descending pretty rapidly, and greatest with the least values of H_a, when the descent was much slower. The variations in t for the 26 experiments given in Table CV., seem to us to be beyond the limit of experimental error—certainly beyond the limit of *probable* error.* By reference to Table CVI., it will be seen that t for the final experiment, No. 142, has the comparatively low value of 252.1, with $T = 48°$; being 1.7 lower than its most *probable* value for this temperature, and 2.4 lower than its mean *observed* value given in Table CV., No. 125. When No. 142 was made, the iron tank and plate had a temperature very nearly

---

* In this series of experiments, practically the only dangers of error, were in incorrect readings in the glass tube at the beginning and ending of t, and in irregular capillarity.

identical with that of the water.   We cannot suggest a plausible reason for this marked difference.

The experiments with quicksilver show a much more satisfactory agreement of the times ; with the quicksilver, however, there was no variation between the heights in the two tubes, and hence capillary errors were not present.   The variations in t, with the quicksilver, can be reasonably attributed to errors of observation.

The marked variation of t for No. 154, with the thick lubricating oil, was perhaps due to unusually viscid particles clogging the entrance to the orifice, or possibly to the obstruction caused by some extraneous substance.

Velocity of Approach.—Theoretically in such experiments, with a descending head, some correction should be made for the head imparted by the velocity of the dropping liquid.   Such a correction for so large a value of A (area of tubes, or a_e) compared with a (area of orifice), is too slight to be worthy of consideration.

Determinations of c.

Water.—Nos. 122 to 124 afford much the most reliable data for the construction of the most probable curve for c for water, with T about $45°$; their results are plotted on Plate IV., with values of H as abscissæ, and values of c as ordinates, where the three experimental points are united by the most probable curve, which has been taken from Plate III.

On the same sheet are shown the mean values of c as deduced from the two series of experiments with descending heads.   It will be noticed that the average value of c for Nos. 127 and 130, agrees quite fairly with this curve, as does also the average value of c for Nos. 125 and 128 ; both Nos. 126 and 129 indicate values of c slightly lower than the curve.

Quicksilver.—The quicksilver experiments are plotted on the same sheet ; from the given mean values of c, the most probable curve for c has been drawn, which is harmonious with the three experimental values.

The interesting fact will be noted that the form of this curve is similar to that for the water ; also that c for the quicksilver, with H from .5 to 3, is about 5 per cent. lower than for the water.

Lubricating (Thick) Oil.—As we only have one mean value for c we cannot construct its curve.   Assuming that the oil follows the same law as water and quicksilver, the dotted curve on Plate IV. approximately indicates the varying values of c for this liquid.

We have before stated that in these experiments with oil, the jets more or less adhered to the divergent sides of the orifice.   The escaping jet, however, did not at any time completely fill this divergent "adjutage," so that probably but little increase of flow resulted, and possibly none.

The apparatus employed for the Greenpoint experiments was constructed at the Continental Works of Brooklyn. Mr. Thomas F. Rowland, the owner of this establishment, would receive no compensation for the work, and also furnished without charge the necessary assistants.

This is only one of the many instances of Mr. Rowland's liberal aid in behalf of scientific research.

FLOW OVER WEIRS—NOTES AND CORRECTIONS. (CHAPTER V.)

CORRECTIONS.

Mr. F. P. Stearns calls our attention to the following corrections:

Page 94.—The measuring vessels used for obtaining q for the experiments with the 5-foot suppressed weir given in Table XXIX. were as follows; for Nos. 17 to 26, a section of the Sudbury conduit 367 feet in length; for Nos. 27 to 34 a section of the same conduit 22 feet in length.

For these experiments, Nos. 17-34, the upper mouth of the pipe leading to the gauge-pail (in which H was determined) was placed as follows; "The head was taken through "a pipe ending at an auger hole in a planed board, 6 feet above the weir and .9 foot "above the bottom. The board was placed 1.5 feet from and parallel with the side of "the channel. Care was taken to prevent the end of the pipe from projecting beyond "the board; but it should also be stated that the auger hole was about ¼ inch larger "than the pipe."—*Transactions Am. Soc. of C.E.*, p. 58, 1883.

Mr. Stearns is of the opinion that the experimental data for No. 26 were more accurately determined than for any other experiment of the series given in Table XXIX.; next in order would be Nos. 25, 24 ... 19; for Nos. 18 and 17 the velocity of approach became considerable, and the conditions —so far as u_a was concerned—were slightly abnormal. Mr. Stearns attributes the errors or discrepancies shown by the discordant values of c for Experiments Nos. 27 to 34, and to which we have made allusion on p. 26, largely to the fact that the measuring vessel for q was of too small size, considering the methods employed.

A A A

SUPPRESSED WEIRS.

On page 130 it has been stated : "For suppressed weirs the escaping vein should be confined " by prolongations of the sides of the canal, but which must not extend lower than the level of the " crest ; in case a suppressed weir has a perfectly free discharge, the following co-efficients (c, in "Table XLVIII., and c, on Plate VII.) are about $\frac{1}{4}$ of 1 per cent. too low." This statement applies rigorously with $l = 10$. and $h = 1.0$—*vide* Experiments Nos. 6, 7 and 8, Table XXVII., and remarks on p. 125—but with h much larger in proportion to l, it is quite probable that the difference between free discharge and confined expansion* will be considerably larger than $\frac{1}{4}$ of 1 per cent.† The curves of c, for lengths from 5. to 19. have been deduced chiefly from experiments where the escaping sheet was confined, while in deducing the curves of c, for lengths less than 5. we were somewhat influenced by Experiments Nos. 1-5 of Lesbros, where the discharge was perfectly free. Hence we think it will be safer to use the values of c, given in Table XLVIII. and on Plate VII for lengths of less than 5 feet, only for weirs with perfectly free discharge.

MEASUREMENT OF H.

On p. 163 we have hardly laid stress enough on the proper precautions which should be observed in the measurement of H in order to prevent abnormal elevation or depression caused by the current of the stream flowing by the gauge-box. The reader in this connection is referred to pp. 255-256, where the results obtained by Mr. Mills with an experimental canal have been stated.

It may be remarked that with weirs having full contraction the velocity of approach will rarely in practice be large, and hence danger of error from improper openings into the gauge-box will generally be much less than with suppressed weirs. The co-efficient c, for full contraction is certainly as well established as c, for suppression ; hence if required to measure the flow of water by means of a weir, we would prefer to use a weir having full contraction, rather than one having one or both end contractions suppressed.

SHORT WEIRS.

Messrs. Donkin and Salter‡ found the following results with a sharp-edged weir having full contraction and a length of .125 ;

$h = .12$	$c = .624$	$h = .21$	$c = .617$
$h = .15$	$c = .621$	$h = .23$	$c = .619$
$h = .17$	$c = .622$	$h = .25$	$c = .618$
$h = .19$	$c = .618$		

These results agree very fairly with the Castel experiments given in Table L., Nos. 156-158 and 167-170.

---

* Lesbros states that with free discharge in Experiments Nos. 1 to 4, Table XXVI., the sheet expanded in length after escaping from the weir ; for No. 5 in the same series with a small value of H, on the contrary the sheet contracted after its escape.

† Messrs. Fteley and Stearns have proposed $l' = l + \frac{h}{25}$ when the discharge is perfectly free, l' being the length to be considered as effective, vide p. 78 Trans. Am. Soc. C.E., 1883. They hence regard the comparison of their general formula with Experiment No. 1' of Lesbros, made on p. 137, as not being a fair one.

‡ Excerpt Minutes of Proceedings of the Institution of Civil Engineers, 1885.

LONDON :

PRINTED AT THE BEDFORD PRESS, 20 AND 21, BEDFORDBURY, W.C.

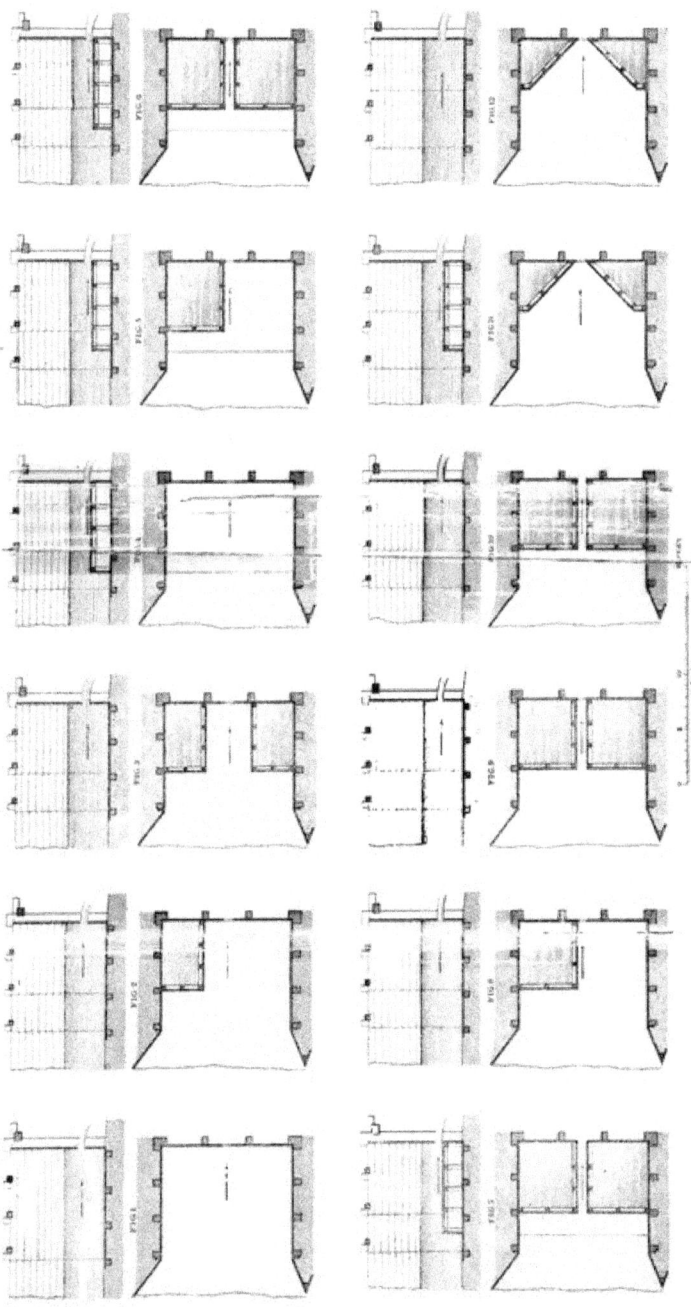

FORMS OF APPROACH USED BY LESBROS FOR ORIFICES AND WEIRS.

Plate 4

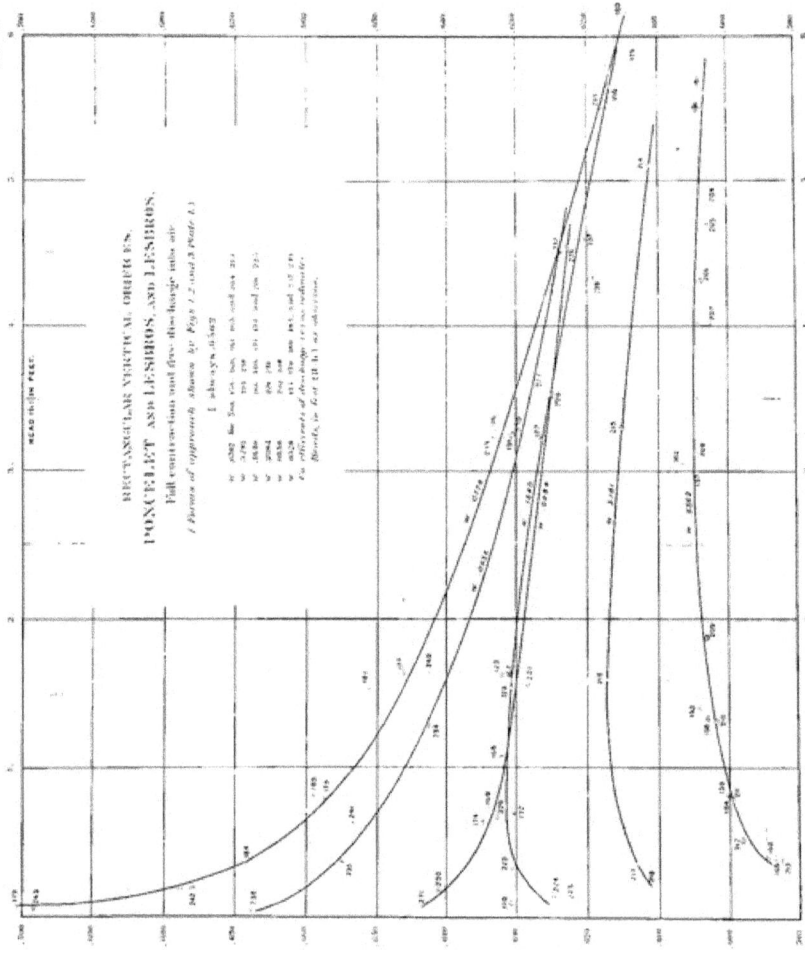

Plate II

HEAD IN SIX FEET.

RECTANGULAR VERTICAL ORIFICES.

PONCELET AND LESBROS, AND LESBROS.

Full contraction and free discharge into air.

(Forms of apparatus shown by Figs. 1, 2, and 3, Plate A.)

[alway 0.6562]

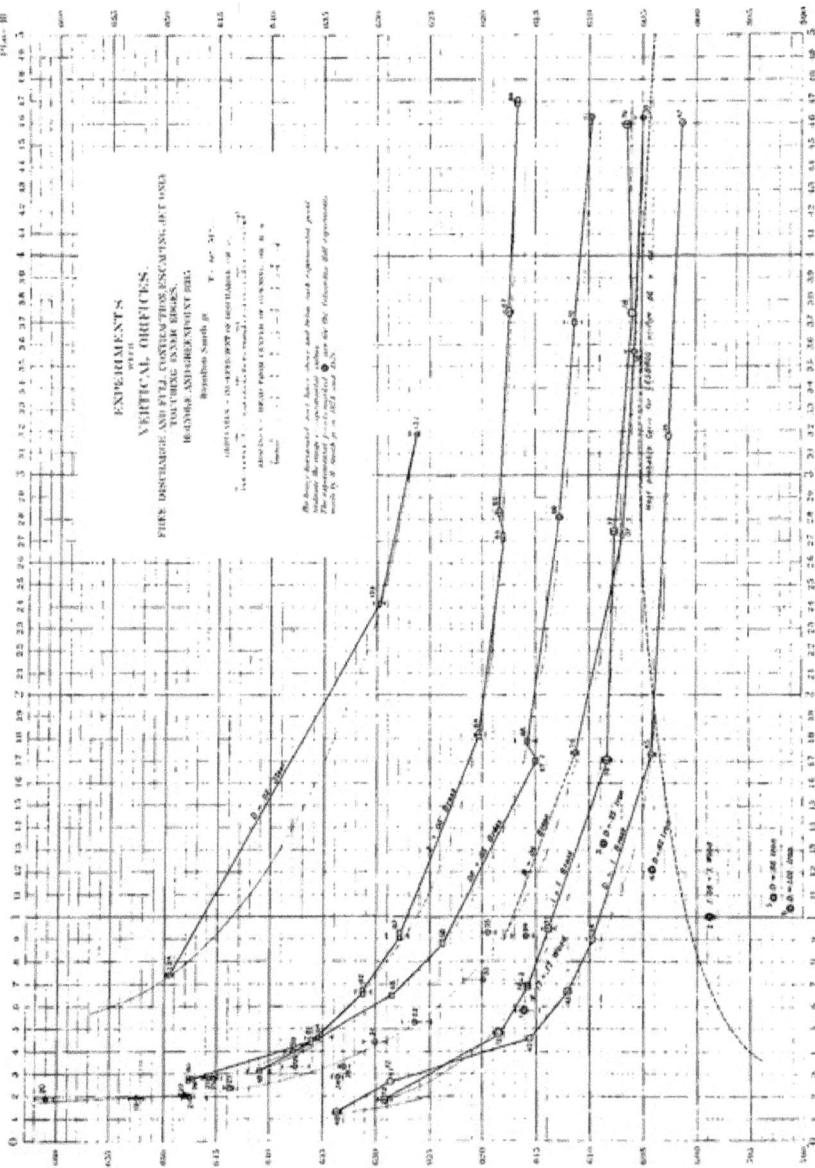

Plate III

EXPERIMENTS
WITH
VERTICAL ORIFICES.

FREE DISCHARGE AND FULL CONTRACTION ESCAPING, BUT OVER TOUCHING BANK EDGES.

READINGS ARRANGEMENT ETC.

Hamilton Smith Jr.

Plate VI.

DEPTH OF WATER ON WEIR h IN FEET

DIAGRAM

SHOWING VALUE OF CO-EFFICIENT c IN FORMULA q=c√h³h
FOR SHARP CRESTED HORIZONTAL RECTANGULAR WEIRS, THE WATER
FOR SPANS FREELY, OR BLANK THRU THIS THE AIR

Reduced from
WEIR EXPERIMENTS
by
Thomson & Lambton, Lewiston Francis,
Fteley & Stearns, and Smith.

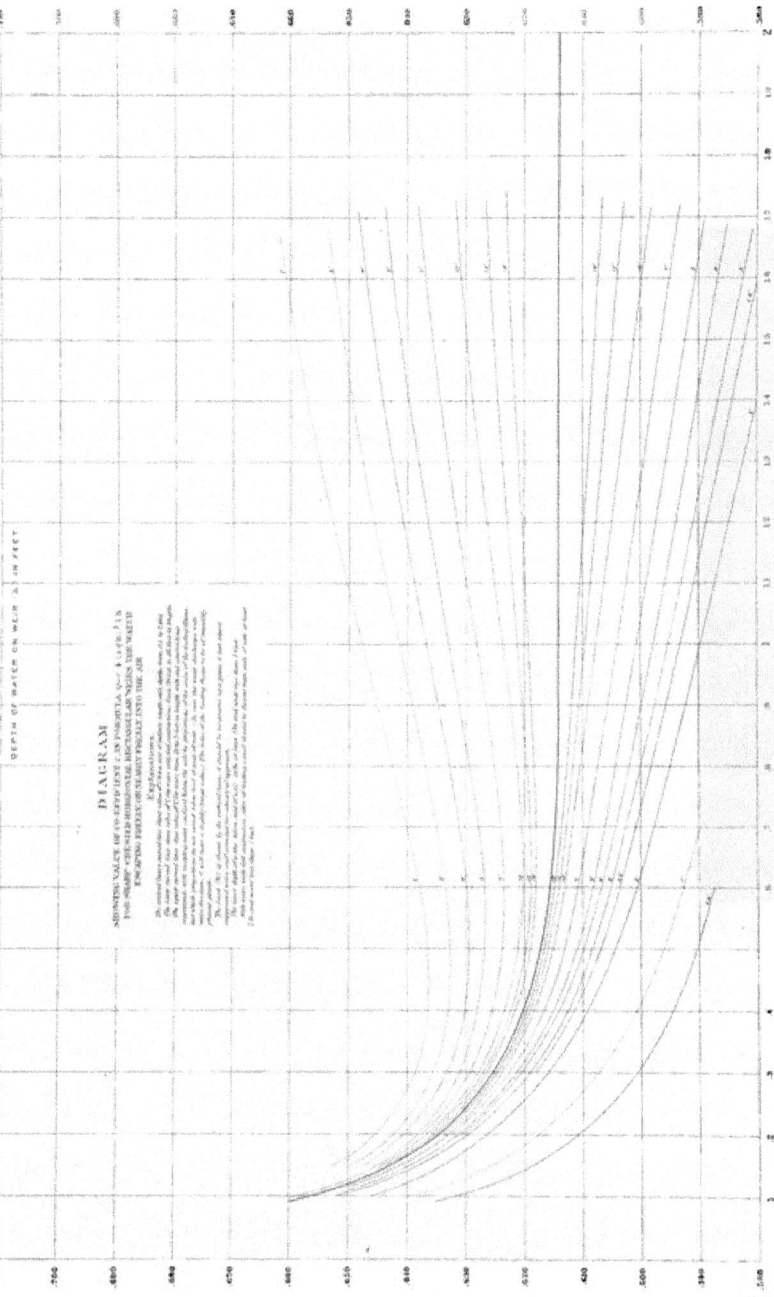

DIAGRAM

SHOWING VALUE OF CO-EFFICIENT c IN FORMULA $q = c \cdot l \cdot h \sqrt{h}$ FOR SHARP EDGED HORIZONTAL RECTANGLE WEIRS THE WATER ENTERING FREELY OR NEARLY FREELY INTO THE AIR

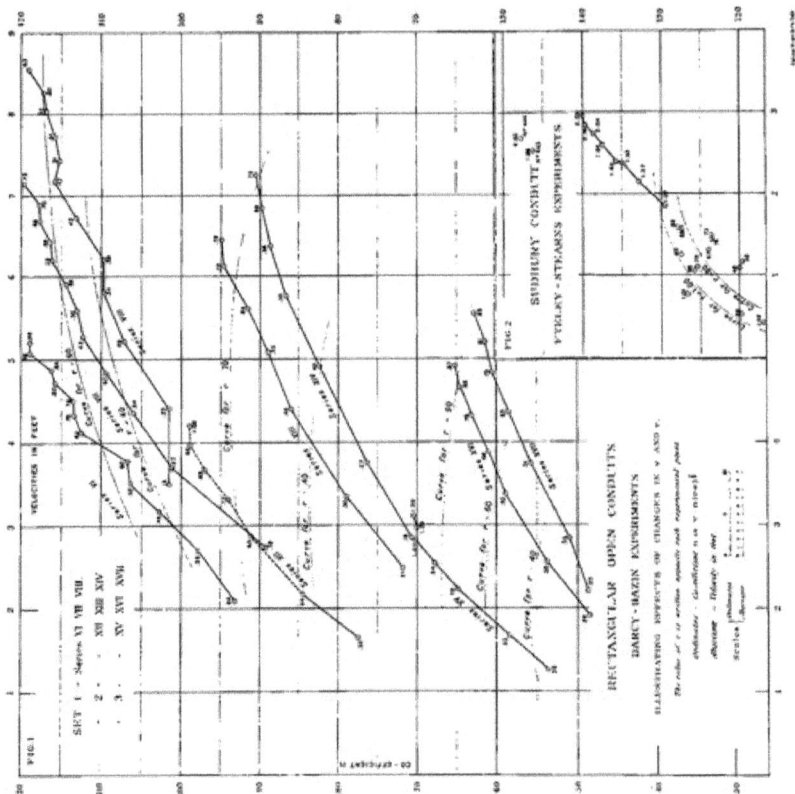

Plate XI

PLATE XII

EXPERIMENTS WITH PIPES

BY

HOSSFT, DUBUAT AND HAMILTON SMITH J[*]

VELOCITIES IN FEET

CO - EFFICIENT M

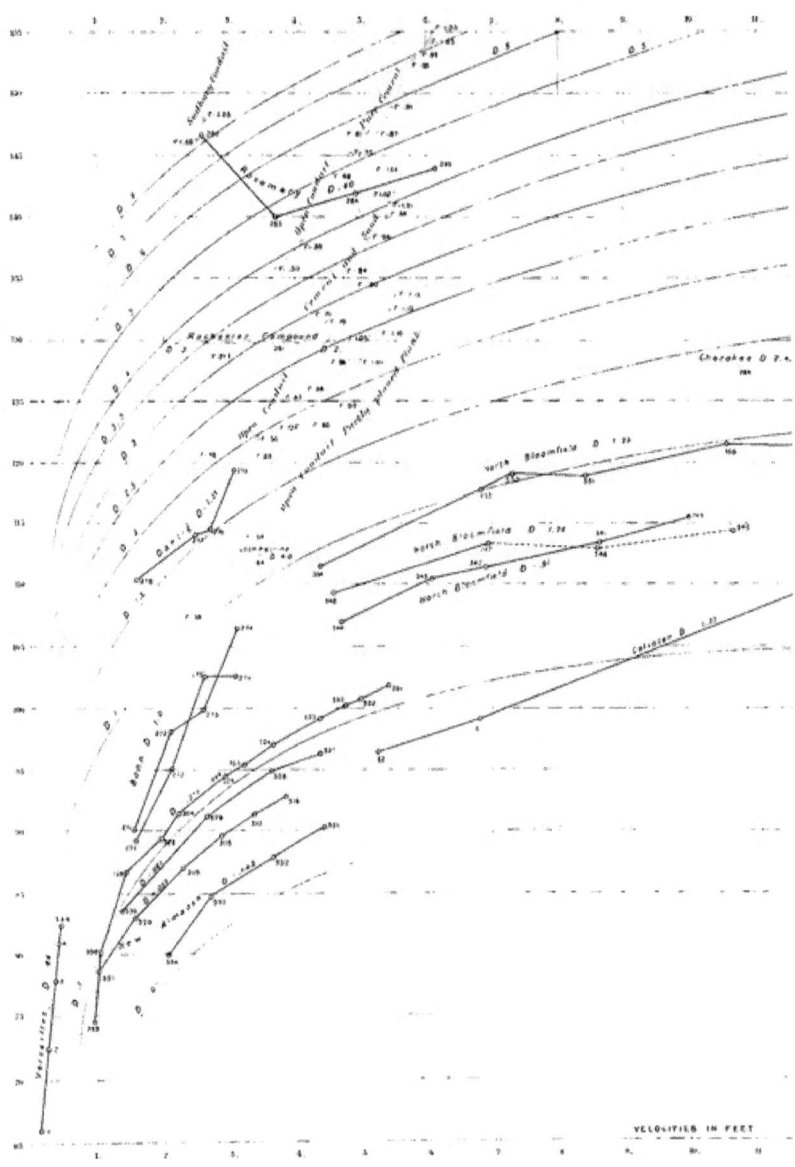

VELOCITIES IN FEET

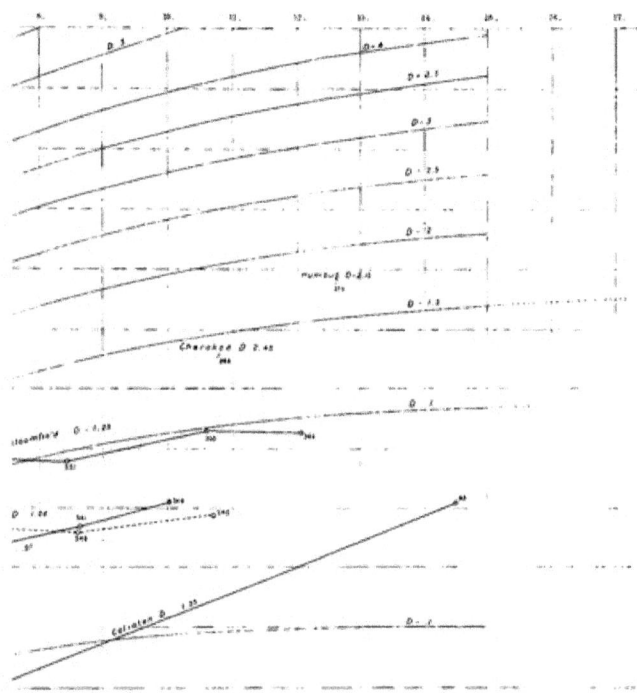

DIAGRAM

Showing values of m in $v = c \sqrt{r s}$, or $\cdot D$.
For circular Pipes, with diameters from 0.5 to 6
having quite smooth interior surfaces.
and no sharp bends.
3rd for $D = 0.5$ and 6.
2nd for $D = 1$ to 6.
Based upon experiments of Couplet et cet.
shown by heavy lines. The dotted irregular lines
represent conduit experiments of Darcy and others.
and Fteley and Stearns.

Abscissae = velocity in feet or v

Ordinates = co-efficient m.

Plate XVI.

NEW ALMADEN PIPE EXPERIMENTS, FIGS. 1-3.

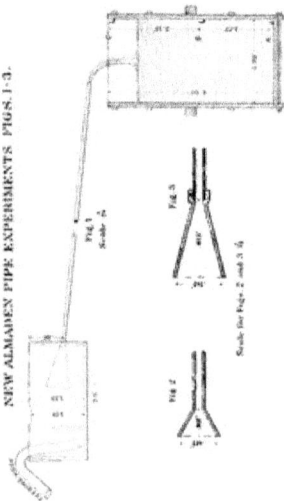

NORTH BLOOMFIELD PIPE EXPERIMENTS, FIGS. 4-12.

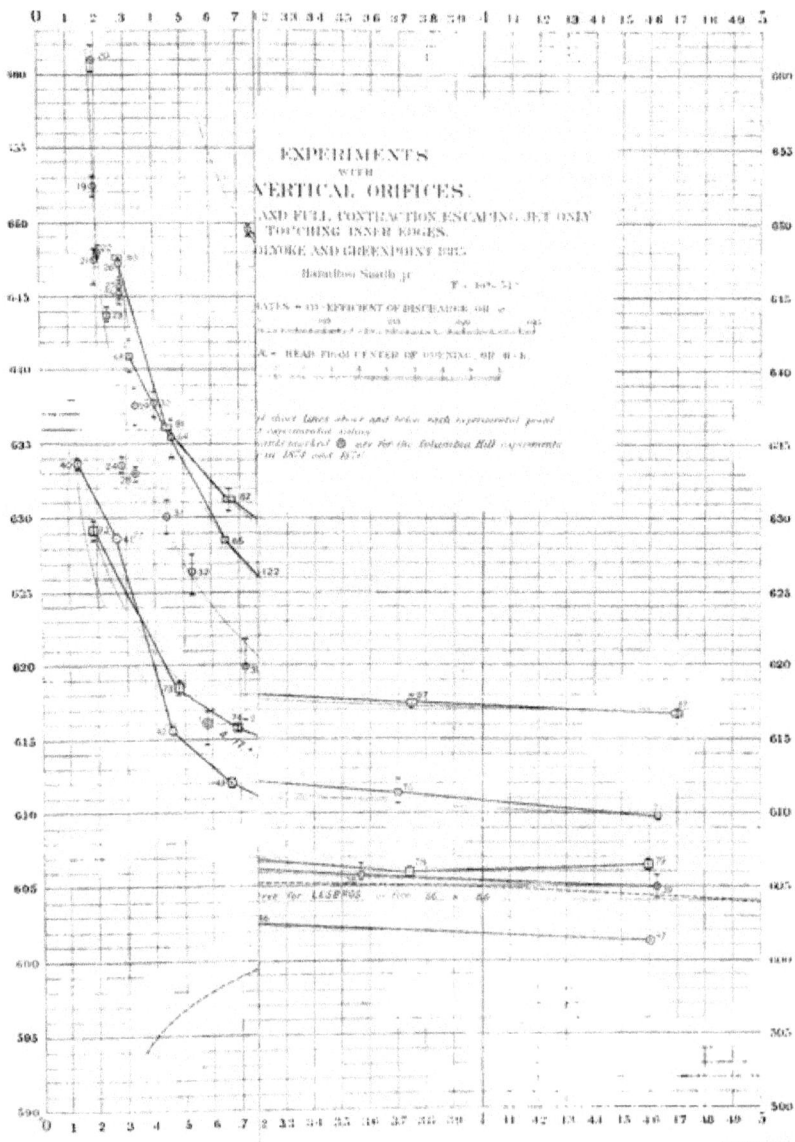

EXPERIMENTS

WITH

VERTICAL ORIFICES.

AND FULL CONTRACTION ESCAPING JET ONLY
TOUCHING INNER EDGES.

HOLYOKE AND GREENPOINT ETC.

Plate IV

Flow through a circular orifice,
with D = .02085,
of
WATER, QUICKSILVER AND THICK OIL.
All with free discharge into air, and full contraction.
Geomport Experiments 1865
Hamilton Smith, jr.

Abscissa = heads from center of Orifice, in feet
Ordinates = values of c, in $Q = c \, a \, (2 \, g \, H)^{1/2}$

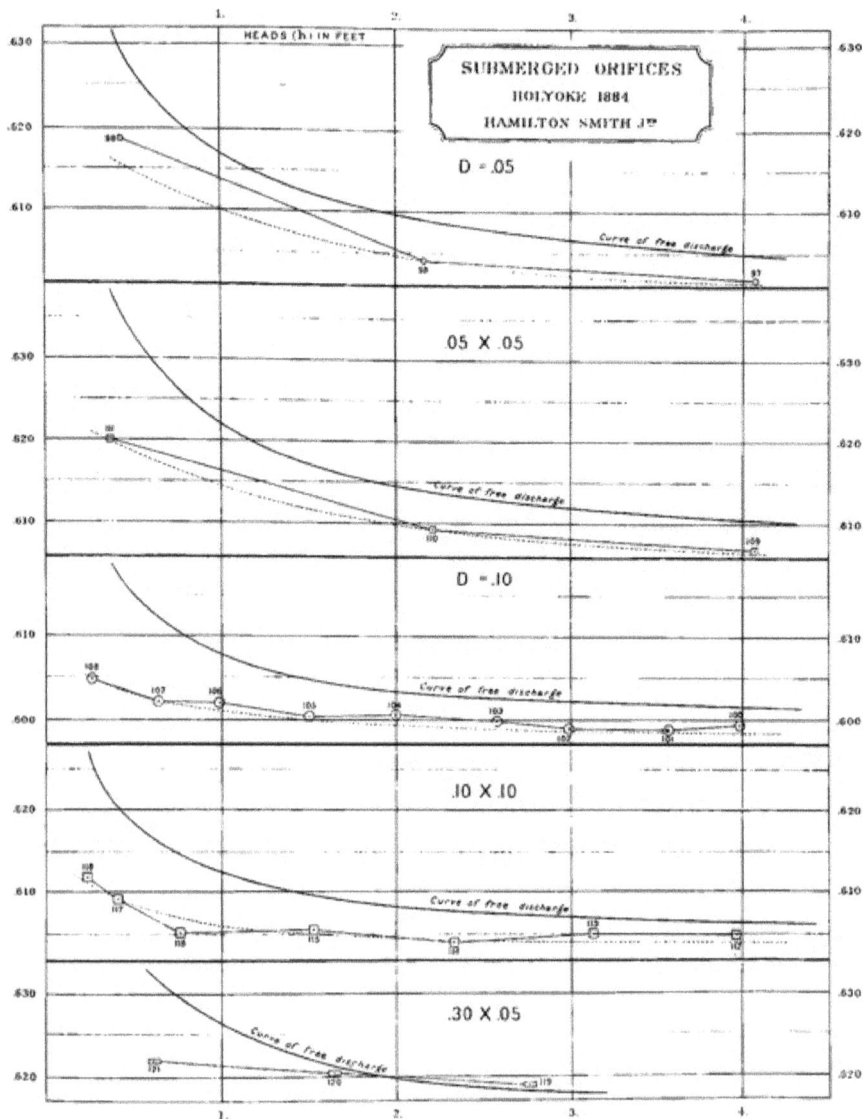

Plate V

SUBMERGED ORIFICES

HOLYOKE 1884

HAMILTON SMITH Jr

HEADS (h) IN FEET

D = .05

.05 X .05

Curve of free discharge

D = .10

Curve of free discharge

.10 X .10

Curve of free discharge

.30 X .05

Curve of free discharge

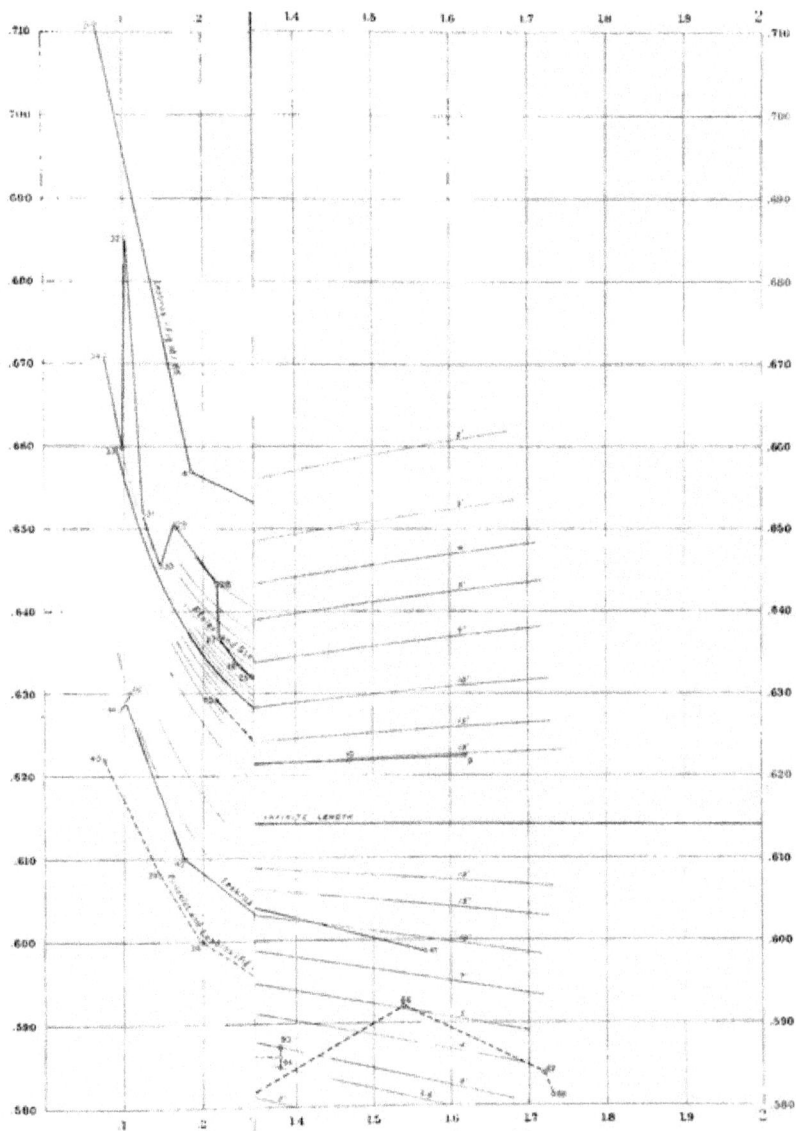

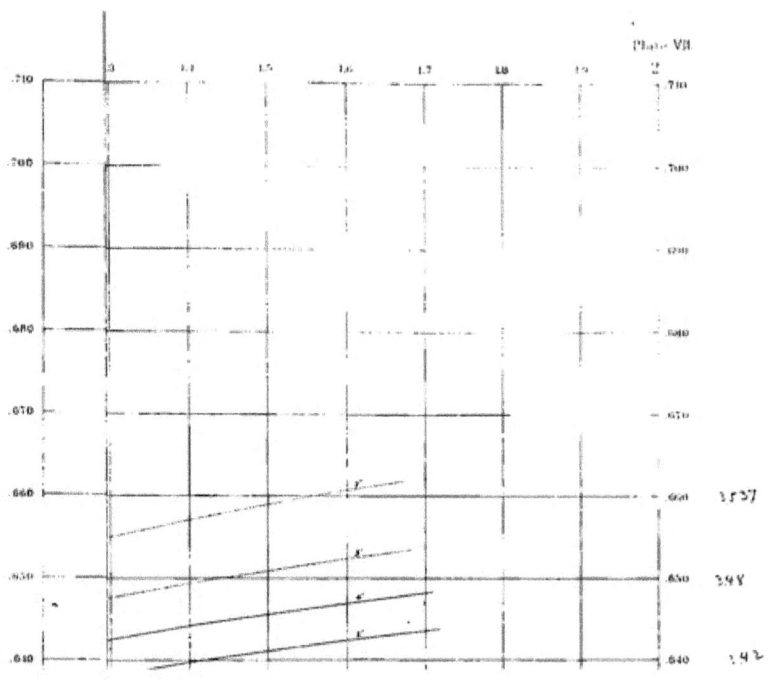

Plate VII

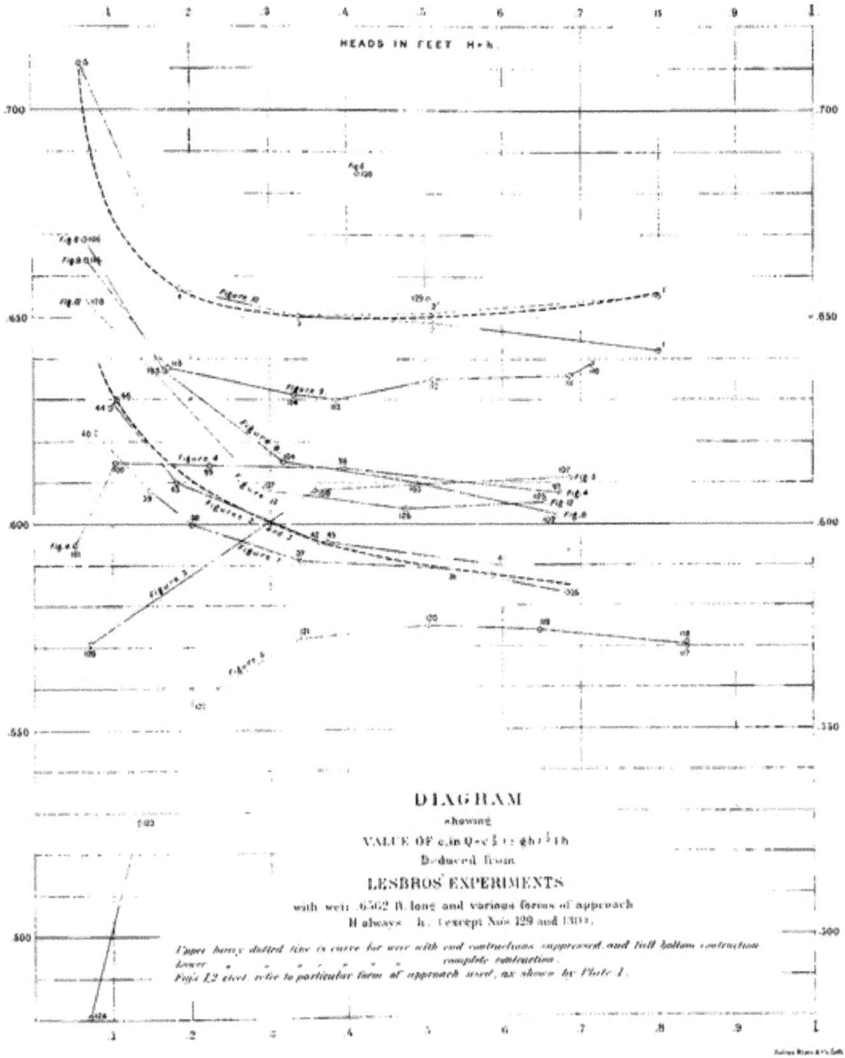

Plate VIII.

DIAGRAM
showing
VALUE OF c, in $Q = c \frac{2}{3} (2gh)^{\frac{1}{2}} lh$
Deduced from

LESBROS' EXPERIMENTS

with weir .6562 ft. long and various forms of approach
H always = h, except Nos. 129 and 130.

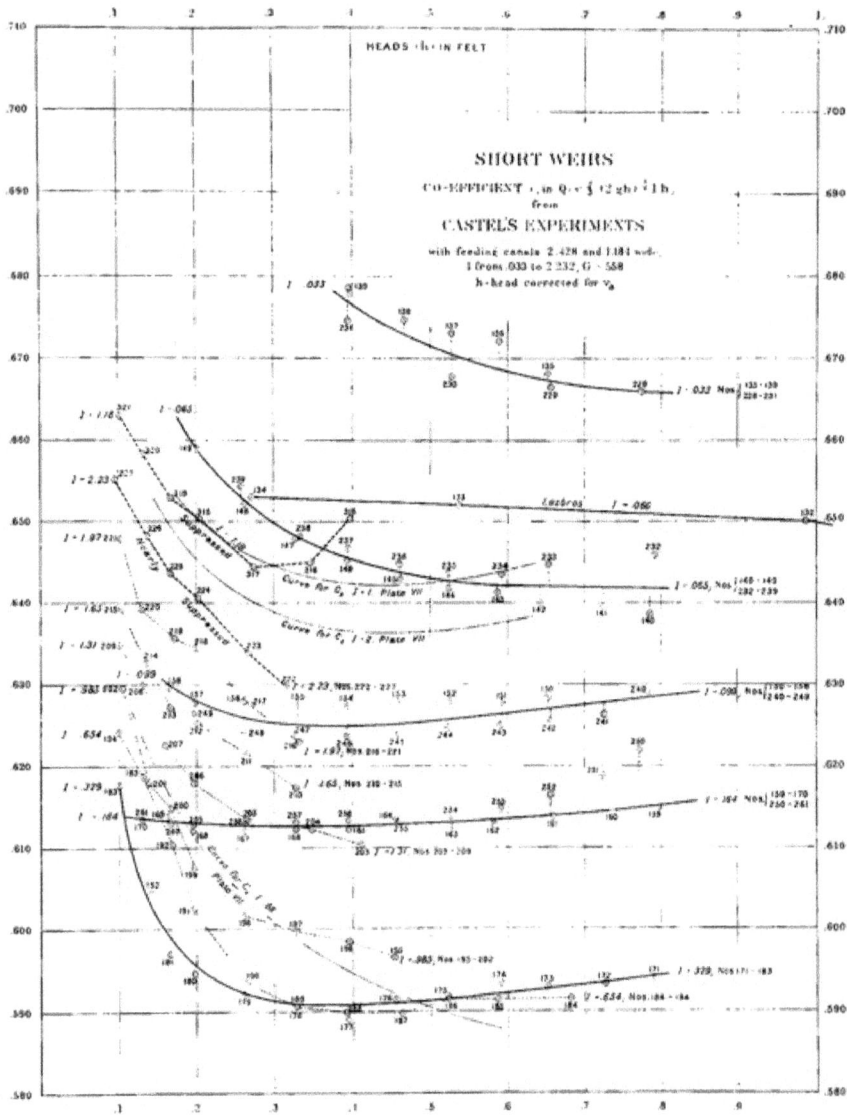

SHORT WEIRS

CO-EFFICIENT c, in $Q = c \cdot \frac{2}{3} (2gh)^{\frac{1}{2}} l h$,
from

CASTEL'S EXPERIMENTS

with feeding canals 2.428 and 1.184 wide,
l from .033 to 2.232, G = .558
h - head corrected for v_a

Plate IX

Plate X

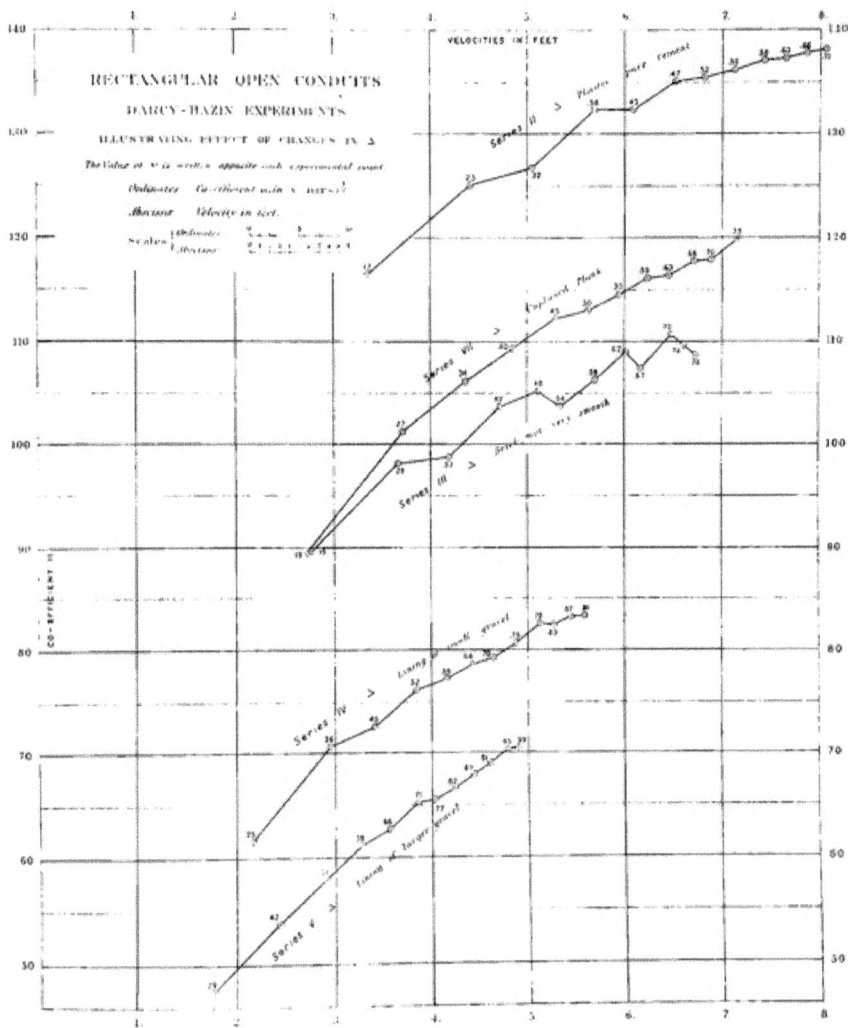

RECTANGULAR OPEN CONDUITS

DARCY-BAZIN EXPERIMENTS

ILLUSTRATING EFFECT OF CHANGES IN n

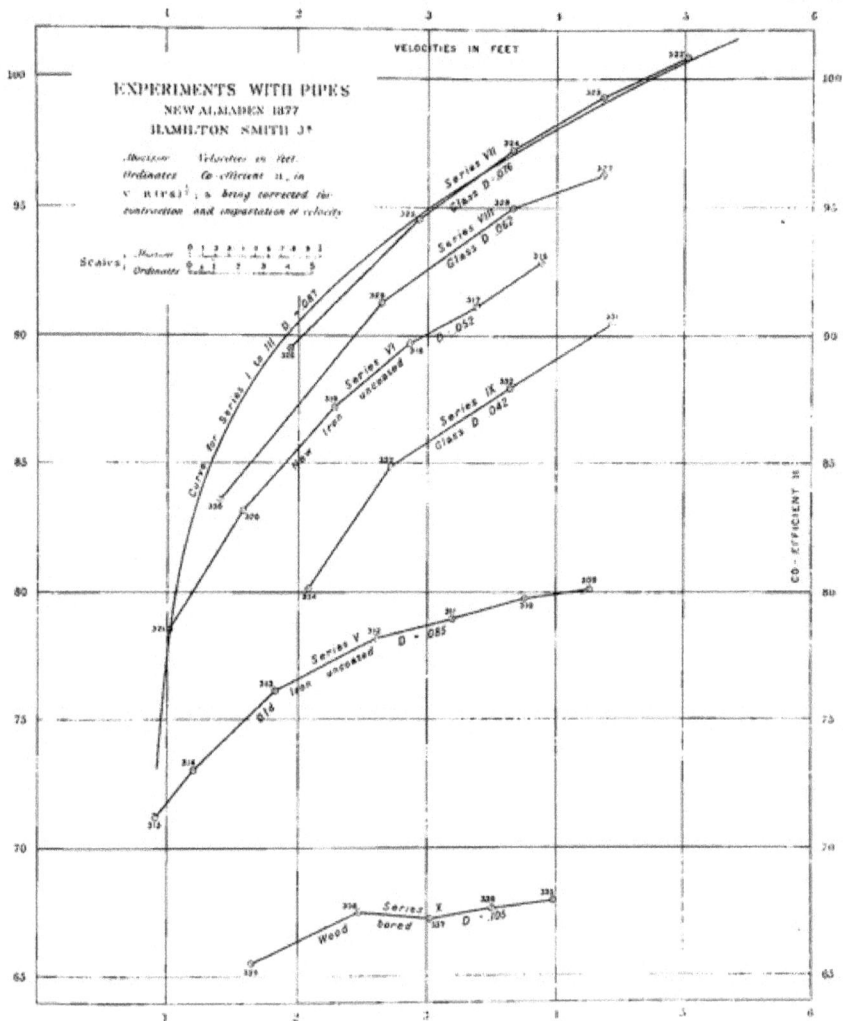

Plate XIII

VELOCITIES IN FEET

EXPERIMENTS WITH PIPES
NEW ALMADEN 1877
HAMILTON SMITH J[r]

Absciss *Velocities in feet.*
Ordinates *Co-efficient n, in*
$v = n\sqrt{H \cdot D \cdot s}$; *s being corrected for*
contraction and impartation of velocity

Scales: Absciss 0 1 2 3 4 5 6 7 8 9 1
Ordinates 0 1 1 2 3 4 5

Curve for Series I. to III. D = .087

Series VII
Glass D = .076

Series VIII
Glass D = .062

New Iron Series VI
uncoated D = .052

Series IX
Glass D = .042

Series V
Iron uncoated D = .085

Series X
Wood bored D = .105

CO - EFFICIENT n

Plate XV.

MEASURING APPARATUS AT COLUMBIA HILL. EXPERIMENTS WITH WEIR AND ORIFICES.

EXPERIMENTS WITH WEIR AND ORIFICES

HAMILTON SMITH Jr.

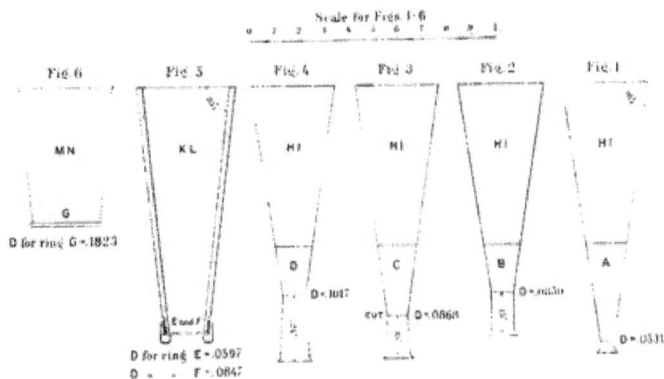

Fig. 7

MOOR GAUGE

BOTTOM OF WEIR

STOP GATE

B

OVER-FALL

GRATINGS

A

C

RESERVOIR

MEASURING TANK

GAUGE MARKS

LONGITUDINAL SECTION

Scale for Figs. 7-8

Fig. 8

GAUGE MARKS

Splash Board

Over-fall

GATE

B

A

Splash Board

C

MOOR GAUGE

WEIR

D

FALSE GATE

RESERVOIR

GAUGE MARKS

PLAN.

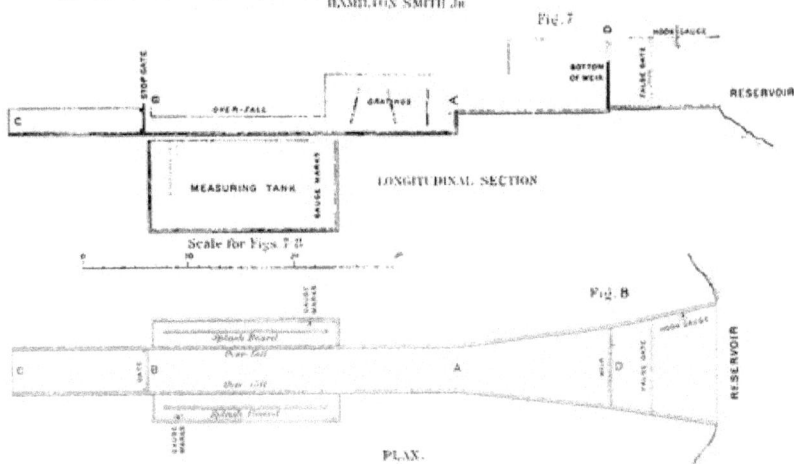

FORMS OF NOZZLES FOR EXPERIMENTS WITH ORIFICES WITH GREAT HEADS.

HAMILTON SMITH Jr.

Scale for Figs. 1-6

Fig. 6   Fig. 5   Fig. 4   Fig. 3   Fig. 2   Fig. 1

MN   KL   HI   HI   HI   HI

G

D for ring G = .1823

D = .1017

D   C

cut   D = .0866

B

D = .0630

A

D = .0511

E and F

D for ring E = .0597
D = F = .0847

Plate XVII

APPARATUS FOR EXPERIMENTS WITH ORIFICES

HOLYOKE, 1884 - 1885.

HAMILTON SMITH Jr.

LONGITUDINAL SECTION

Scale ⅛

Fig 1

Fig 2

TANKS A AND B 4 WIDE IN CLEAR

Plates set in center

HALF SIZE SECTION THROUGH PLATE